Air Disaster
&
The Fall of Boeing

How Bold Leadership Created a Global Icon and How Failed Leadership Brought It Down

Rom Solene

Published by Iron Compass Publishing

First Edition, 2026

ISBN: 979-8-9975388-1-1

The events, individuals, organizations, and historical circumstances described in this book are based on sources believed by the author to be reliable. Every reasonable effort has been made to ensure accuracy.

Photographs and illustrations are credited individually where applicable.

Printed in the United States of America

For the men and women of the Boeing that once was — and for all those who helped bring the wonder of flight to millions.

My own journey began at the age of three, aboard a Pan American Boeing 707 crossing the Atlantic. I was too young to understand what aviation would come to mean to me, but that first flight awakened a fascination that has never left.

This book is dedicated to those who made such journeys possible.

Preface

Many books have been written about The Boeing Company. Some chronicle its legendary airplanes and engineering achievements; others focus on the two 737 MAX crashes and the corporate failures exposed in their aftermath. This book takes a broader and more detailed view.

Boeing's rise was not the product of engineering brilliance alone, nor can its decline be explained by a handful of decisions surrounding a single airplane. Both resulted from forces that converged over decades — wars, government policy, technological change, competitive pressures, corporate mergers, financial priorities, and, above all, the personalities and leadership philosophies of the people who shaped the company.

Beginning with Boeing's founding in 1916, this book examines not only what happened, but why — how events influenced one another and how decisions made years earlier created consequences that were not always apparent at the time. It also explores the executives, engineers, government officials, and competitors whose ambitions, judgments, strengths, and shortcomings helped determine Boeing's direction.

Written for a broad audience, the book avoids excessive technical detail while connecting events, personalities, and outside forces too often examined in isolation. Together, they reveal how engineering excellence, technological ambition, and bold leadership transformed Boeing into the world's preeminent aerospace company — and how changing priorities, leadership, and corporate culture ultimately undermined many of the qualities that had made it exceptional.

— The Author

Prologue

Prologue

At the end of 1996, the Boeing Company stood astride the global aerospace industry, much as it had throughout the post-World War II era, when it brought commercial jet travel to the flying public on an unprecedented scale. That year, its assembly lines in the Seattle area produced more than 60 percent of the world's commercial jet aircraft, continuing a dominance Boeing had maintained for much of the postwar period.

The company offered an airplane for virtually every segment of the commercial market: the venerable 737, the iconic 747, the 757 and 767, and the new 777, which had entered commercial service only the year before, in 1995. Whatever the size of the market or the distance an airline needed to fly, Boeing appeared to have the right aircraft for the job.

From January through December 1996, Boeing recorded an extraordinary 719 commercial-aircraft orders — nearly twice the combined total secured by its two principal rivals, McDonnell Douglas and Airbus. By almost every meaningful measure, Boeing appeared to be at the height of its power.

Yet only three years later, by the end of 1999, Airbus had overtaken Boeing in annual commercial-aircraft orders, claiming the crown as the world's leading commercial-aircraft manufacturer — even though Boeing had by then absorbed McDonnell Douglas. In the years that followed, Airbus would repeatedly outperform Boeing, not only by securing more orders but eventually by delivering more airplanes as well. For much of the next quarter-century, Boeing found itself in the unfamiliar position of trying to catch its European rival.

After surrendering its position at the top of the industry, Boeing sought to leapfrog Airbus with a revolutionary new airplane, the 787 Dreamliner. Through the use of composite materials, advanced systems, and highly efficient engines, Boeing promised airlines a dramatic improvement in fuel economy, operating efficiency, range, and passenger comfort. The finished airplane largely fulfilled those promises. The 787 was technologically groundbreaking and met — or exceeded — many of the performance commitments Boeing had made to its customers.

The program that produced it, however, was mismanaged from the outset, with disastrous consequences. Boeing's radical production strategy, which transferred unprecedented responsibility for design, engineering, and manufacturing to a sprawling global network of suppliers, resulted in defective components, incomplete assemblies, repeated redesigns, enormous cost overruns, and years of delay.

Those delays gave Airbus the time it needed to develop the A350, a competing aircraft that matched the 787 in many respects and surpassed it in others. By the time the Dreamliner entered service, Boeing had forfeited much of the technological initiative and market advantage it had expected the airplane to deliver. Airbus had not only blunted Boeing's breakthrough but had also captured a substantial share of the orders and revenue Boeing had once assumed would be its own.

More consequentially, Boeing failed to confront the deeper organizational and cultural failures that had crippled the 787 program. The fixation on cost reduction, accelerated schedules, financial performance, and the outsourcing of engineering responsibility remained largely intact. The symptoms of failure were treated, but the underlying disease was not.

Less than a decade after the 787 debacle, those same forces would surface again in Boeing's next major commercial-aircraft program. This time, the airplane was not a revolutionary clean-sheet design but an upgrade of the company's oldest and most successful product line — the 737 MAX. The consequences would be far more devastating. The program — and two of the airplanes it produced — would come crashing down, figuratively and literally.

To place these two fiascos in context, one must first understand what Boeing had represented for nearly a century. The company had stood at the forefront of aviation innovation, repeatedly setting the standard that other aircraft manufacturers sought to match. No company had done more to develop, advance, and popularize commercial aviation, and Boeing had built that legacy through a succession of remarkably successful airplanes.

Yet at the very height of its power, Boeing failed to respond effectively to the rising threat posed by Airbus. Worse still, it then failed to execute the major commercial-aircraft program intended to propel it into the future. The 787 Dreamliner was supposed to reestablish Boeing's technological leadership and restore its competitive advantage. Instead, the program became synonymous with mismanagement, delays, cost overruns, and organizational dysfunction.

Even more humiliating — and far more egregious — was what Boeing's leadership did to the 737. The company took the most successful commercial jetliner program in aviation history, an airplane it had been producing since 1965, and drove it — flew it — into the ground. In doing so, Boeing damaged not only the sterling reputation the 737 had accumulated over more than half a century, but also the reputation of the company itself, a reputation built across a century of flight. It was as though Boeing's legendary Midas touch had become a kiss of death.

But commercial aviation was not the only arena in which Boeing had once excelled, only to lose its footing and tarnish its reputation. By 1996, even as the company stood at the zenith of the commercial-aircraft industry, it also occupied a commanding position in space. Boeing built some of the world's largest and most sophisticated rockets, spacecraft, and satellites for government and commercial customers, a position it had attained through decades of involvement in the American space program and its extensive partnerships with NASA and the U.S. Department of Defense.

Yet the same pattern that unfolded in commercial aviation would eventually repeat itself in space. Boeing had lost its leadership in commercial aircraft to Airbus, a European enterprise whose origins dated only to 1969, when Boeing was already more than half a century old. It would then lose its dominance in space to an even newer challenger — Space Exploration Technologies Corporation, better known as SpaceX.

Airbus required roughly three decades to dislodge Boeing from its commercial-aviation perch. SpaceX accomplished much the same in space in less than fifteen years. Through vertically integrated manufacturing, rapid development cycles, reusable rockets, and an organizational culture willing to challenge nearly every convention of the traditional aerospace industry, SpaceX came to outperform Boeing in a field Boeing had once helped define.

Thus, within the span of little more than two decades, Boeing had fallen from its position as the world's preeminent aerospace engineering and manufacturing company — an institution admired and envied around the globe — to one struggling to compete with younger, faster, and more innovative rivals in both of its principal markets.

What happened?

A logical place to begin is in Boeing's executive suite. During the eighty years between the company's founding in 1916 and 1996, only six men occupied its highest executive office. Over the next twenty-eight years, from 1996 through 2024, five different chief executives would lead the company. Four of those five would be forced out amid a major corporate crisis.

To understand how Boeing became a global industrial juggernaut — and how it was ultimately brought to its knees — one must first understand its history. From its earliest beginnings in 1916, the company endured ten decades marked by world wars, economic calamities, technological revolutions, and profound shifts in the global political order. Through them all, Boeing not only survived but repeatedly adapted, reinvented itself, and emerged stronger.

Yet the company began to unravel during one of the greatest periods of economic expansion in modern history. The world's airlines were carrying record numbers of passengers, aircraft were flying with scarcely an empty seat, and carriers could not order new airplanes quickly enough to satisfy the seemingly insatiable demand for air travel. Boeing should have been reaping the rewards of a century spent at the forefront of aviation. Instead, two crashes involving its newest commercial jetliner sent the company itself into a tailspin.

The 737 MAX disasters exposed far more than defects in a single airplane. They revealed the consequences of leadership and a corporate culture that had gradually subordinated engineering judgment, product integrity, and institutional responsibility to cost reduction, production schedules, financial targets, and shareholder returns. The global pandemic that followed only compounded Boeing's difficulties, magnifying its financial and operational problems until the company's future appeared, to some, genuinely uncertain.

Boeing had once been the world's preeminent aerospace icon, a company whose name was virtually synonymous with American engineering excellence. It became instead a cautionary tale — a stark example of how failed leadership and a corrupted corporate culture could squander a century of accumulated trust and besmirch the legacy of one of the greatest industrial enterprises in history.

Wood for Floating and Wood for Flying

The Boeing Company was born in 1916 near Seattle in the upper Northwest corner of Washington state. The company's original name was the Pacific Aero Products Company, and its creation personified the brash daring and risk-taking nature of its founder, William Edward Boeing. It was this daring and risk-taking nature that would personify the Boeing Company and its leadership throughout most of its history until the advent of the 21st century.

William "Bill" Boeing was born in Detroit, Michigan, in 1881 as a first-generation American. His father hailed from Germany and his mother was from Austria. Although the senior Boeing immigrated to the United States with very little financial means, he managed to make a name for himself, and a small fortune, through timber and mining rights that he accumulated in Minnesota when there was a rapid expansion in the timber and mining industries in that region in the second half of the 19th century.

The senior Boeing passed away when young Bill was just eight years old, and his mother remarried shortly afterwards to a wealthy businessman. Young Boeing wanted for nothing growing up after his father died, and he was even sent abroad to Europe for the remainder of his primary education. Upon his return, after a short stint in a preparatory school, he was accepted into Yale University where he studied engineering with the Class of '04. However, Bill Boeing left Yale early, in 1903, at the age of twenty-two for an opportunity in the American Northwest timber industry where, with the inheritance he had received from his late father's estate, he was able to parlay that opportunity into a lumber empire along the Duwamish River where he acquired land and built lumber mills. The blend of water and wood made Boeing take an active interest in yachting and motorboats, which he owned several of and some of which he built himself.

In 1909 Boeing chanced to glimpse his first airplane, which soon took him in a new direction. Although he had studied about the Wright Brothers when he was at Yale, he didn't exhibit a lot of interest in aviation while studying there. It took an actual sighting of a flying machine to fire up his interest and imagination in powered flight. Although aviation was spreading around the country, it was slow to make a regular appearance in the American Northwest, where the region was heavily forested and sparsely populated with vast and perilous distances between inland towns. The early airplanes were rudimentary flying machines and rarely ventured far beyond population centers.

But chance would appear again on a muggy July 4 in 1914. On that day, a flier had appeared at the local Independence Day celebrations where large crowds had gathered to put on an exhibition with an airplane. The pilot was flying a Curtiss-built amphibian biplane. Glenn Curtiss, the designer and builder of this particular model, was a pioneer in American aviation; at the time probably second only in renown to the Wright Brothers.

The pilot flying this Curtiss amphibian put on an aerial exhibition for the crowds and, after landing and mooring his plane to the dock, announced that he would take a single passenger up for a short ride. Out from the crowd stepped Bill Boeing, which surprised many because he was by then a renowned and successful businessman throughout the area, with a quiet and reserved demeanor. Boeing was a perfectionist in many ways, a man who adhered to customs and traditions, and one who naturally commanded respect. Airplanes were considered risky at the time, far from being reliable and in no way perfect, so it was out of the ordinary for someone like Bill Boeing, with his reputation, to step forward and take up the offer for a ride. But his 1910 airplane sighting had lit an intense interest in airplanes in him, and this was his opportunity to experience a flight in one.

The pilot took Boeing up for a ten-minute flight, and by the time they had landed back on the water Boeing had become captivated with flight and had decided right then and there to pursue flying as his newfound passion. His initial interest and passion about flight was focused on how an amphibian plane, such as the Curtiss he just flew in, could take him to remote lakes and rivers where he could indulge in his other passions for fishing and hunting.

The pilot also took another gentleman up for a flight that day, Conrad Westervelt, a close acquaintance of Boeing. After Westervelt returned, the two men spent the next several days talking endlessly about airplanes and how they might pursue their newfound mutual passion for flying.

Within several weeks, Boeing enrolled in flight training with another famous aviator of the era, Glenn Martin, who operated a flight school in Los Angeles while also designing and selling his own airplanes. One of those designs was the Martin Pusher, so named because its engine and propeller were mounted behind the pilot and pushed the airplane through the air rather than pulling it from the front, as would become far more common. Once Boeing had learned to fly, he and Westervelt purchased one of Martin's airplanes for about $10,000, the equivalent of roughly $334,000 in 2026 inflation-adjusted dollars.

It took a few months for the airplane they had purchased to arrive back in Seattle, transported via ship in pieces in multiple crates, and then it took several weeks more for it to be assembled. Although Boeing knew how to fly and he was considered an engineer given his education at Yale, putting a machine together would be left to someone else. That someone else was a gentleman that Boeing and Westervelt hired as their test pilot. Unfortunately, the test pilot severely damaged the airplane during an early flight shortly after it was assembled, which was not a difficult thing to do considering airplanes in those days were built with wood, and held together with bale wire, glue, and nails and with the wings covered in fabric.

Not to be discouraged by this mishap, Boeing contacted Glenn Martin to inquire about getting replacement parts for the damaged airplane. The news was not good as Boeing was told it would take more than six months to make the parts and then get them crated for shipping.

And it is here where the desire to do the difficult, or even the impossible, was born and became a part of the DNA of what would become The Boeing Company. In an era when only rudimentary knowledge was held by a few about the science of powered flight, William Edward Boeing decided he would design and build his own airplane in less time than it would take to wait for replacement parts for his damaged one.

Boeing discussed the idea with Westervelt, who quickly agreed to the plan. They set it in motion, with Westervelt getting aeronautical engineering research papers from his friends in the Navy Department — he was a Naval Reserve officer — and at MIT, and with Boeing taking care of the logistics. From the start, and despite his desire to use an airplane to reach remote areas for hunting and fishing, Boeing harbored the idea that this could be a doorway for him into a new business venture, the airplane manufacturing business.

It was an idea that he was not shy about sharing with his partner, Westervelt, who agreed to go in that direction if they could build their own plane. The design they came up with was called the "B&W," for Boeing and Westervelt, and which was also the name of their partnership, and it made its first flight on June 29, 1916, with Boeing himself at the controls. It was a historic day, and considered the birth of the Boeing Airplane Company. In the span of less than two years, Bill Boeing had gone from taking his first airplane ride to taking an airplane that he helped design and build on its first test flight.

Shortly after the B&W's first flight, Westervelt was called to Washington, D.C., as part of his reserve officer commitments. The drums of war in Europe were beating louder and the United States was not oblivious to what was happening across the Atlantic, and preparations were being made for what many feared would be the inevitable, the United States' involvement in the European war. With Westervelt leaving the area for an extended period of time, the B&W partnership was dissolved, and Boeing set up the Pacific Aero Products Company with his own investment and himself as CEO, and then set about designing a training plane that he hoped to sell to the U.S. Navy.

With help from Westervelt's connections (highly unethical in today's government contracting environment, if not illegal, but par for the course back then), the new company won a contract from the Navy for 50 of Boeing's newly designed trainers. Unfortunately, once built and delivered, the trainers were not considered very good because of inherent design shortcomings, and the Navy never placed a follow-on order. However, the Navy did have a need for a new airplane designed by Glenn Curtiss, which was a very large (for its day) flying boat.

Curtiss' plane, called an HS-2L, was highly sought-after by the Navy, but Curtiss couldn't build them fast enough. So, the Navy gave Boeing a contract to build 50 of these large planes under license from Curtiss. While these were not Boeing's design, the opportunity to build one of the largest planes flying at the time gave Boeing a lot of valuable insight and know-how in the design, engineering, and building of large aircraft. This newly gained know-how in the building of large airplanes would become a hallmark of the Boeing Company throughout most of its history.

Unfortunately for Boeing, when the Armistice was signed in November of 1918 closing the books on World War I, the Navy reduced this order for 50 of Curtiss' design down to just 25. Once the 25 were built and delivered to the Navy there were no more planes for Boeing to build. In a desperate attempt to stay afloat and keep the company going in hopes of better days, Boeing, for a brief period, turned to making office furniture. The supply of wood was cheap and endless, coming from Bill Boeing's own timbered land and mills, and the skilled labor was readily available and adaptable for making something far easier than a flying machine. But Boeing's foray into the furniture-making business wouldn't last long. Just when he thought he was going to have to shutter his factory and his dreams of making money from building airplanes, the government came through with an opportunity that was a temporary lifeline to the company.

***

This marked the close of Boeing's formative chapter. From the outset, the company reflected the temperament of its founder — bold, opportunistic, and willing to attempt what others would not. That mindset, carried forward by successive leaders and embedded in the company's culture, became the engine of its ascent. It is what propelled Boeing from a speculative venture into the dominant force in global aviation. Although Bill Boeing himself epitomized the legendary American spirit, he was not like a lot of other American legends who started from inauspicious beginnings.

William Edward Boeing was a man of means, a man of power, which is to say he was a man of money. He was a millionaire several times over by the time he was 33 when he took that first airplane ride in 1914 and he didn't become a millionaire after starting out with nothing. He started with a sizable inheritance from his father that he parlayed into his own personal fortune. And then, on a whim, he decided to learn to fly and then, again on another whim, decided to purchase an airplane for an exorbitant sum of money.

His original idea for flying his own airplane was to get him into remote lakes and rivers in and around the Northwest where he could pursue his passion for hunting and fishing. It was only when he was forced to consider building his own airplane in order to circumvent the time to wait for replacement parts that he, in a classic moment of business acumen, considered parlaying his own designed-and-built airplane into a profit-making machine. And he began to profit handsomely from his government connections. No one else could have possibly taken a home-designed-and-built airplane and sold 50 copies to the government without some means, and some means of connections.

At the close of its formative years, Boeing's identity had taken shape around five traits: boldness, inventiveness, perseverance, business acumen, and a steady stream of government contracts. That final element would prove both sustaining and corrosive over the next century.

Uncle Sam's Needs and the Need for Uncle Sam

Wilbur and Orville Wright are universally credited with achieving the first sustained, controlled, powered flight of a heavier-than-air machine. They were not the first to imagine or experiment with the airplane, but they succeeded where others had failed by solving the fundamental problems of lift, propulsion, and — most critically — control. Their brief flight at Kitty Hawk, North Carolina, in 1903 marked the beginning of practical powered aviation.

Although the United States produced that historic breakthrough, the center of gravity of early aviation remained largely in Europe. Many of the most prominent pioneers of the period were European, and in the years immediately following the Wright brothers' achievement, European designers, engineers, and aviators rapidly advanced the airplane from a fragile experimental contraption into an increasingly practical machine.

The outbreak of World War I accelerated that development dramatically. Once the airplane demonstrated its military value, the European powers poured enormous resources into improving its speed, range, reliability, maneuverability, and payload capacity. What had begun as a crude reconnaissance platform quickly evolved into a weapon of war, capable of directing artillery, attacking ground forces, intercepting enemy aircraft, and bombing targets far behind the front lines.

War, after all, has a cruel but undeniable way of accelerating technological progress. The urgent need to gain an advantage over an adversary compressed years of peacetime experimentation into months, transforming the airplane from a mechanical curiosity into an indispensable instrument of modern warfare.

When the United States formally entered World War I in April 1917, the conflict in Europe had already been raging for nearly three years. During that time, the European powers had made enormous advances in the design, production, and combat employment of military aircraft. By comparison, the American aircraft industry remained small, fragmented, and poorly equipped to manufacture modern combat airplanes in the numbers the war would require.

The United States was not entirely without aircraft manufacturers or aeronautical expertise. Companies such as Curtiss and Wright-Martin were already producing airplanes and engines. But American industry lacked both the experience and the production capacity needed to equip a rapidly expanding air service with advanced fighters, bombers, and reconnaissance aircraft. By the time the United States entered the war, European factories were producing military airplanes by the thousands, while many American designs were already technologically obsolete.

As a result, American combat pilots sent to Europe initially relied heavily upon aircraft designed and manufactured by the Allies, particularly France and Great Britain. Rather than wait for entirely new American airplanes to be designed, tested, and placed into mass production, the federal government pursued a more expedient course, which was to identify the best combat-proven Allied aircraft and arrange for them to be produced under license in the United States.

In 1917, the War Department dispatched a delegation to Europe headed by Colonel Raynal C. Bolling. The Bolling Commission studied Allied aircraft, engines, manufacturing methods, and production requirements, then recommended which designs should be adopted for American manufacture. Its best-known selection was the British-designed de Havilland DH-4, which became the only American-built airplane to see significant combat service with the American Expeditionary Forces before the war ended.

The broader objective extended beyond meeting the immediate demands of the conflict. The government hoped to use the emergency as a catalyst for creating a large and technologically capable American aviation industry — one that could eventually design and manufacture its own military aircraft rather than depend upon foreign suppliers.

Several Allied aircraft were selected for American production, and examples, drawings, specifications, and manufacturing information were brought back from Europe. American manufacturers then adapted the designs for domestic mass production. Of the European aircraft selected, only one American-built model — the British-designed de Havilland DH-4 — reached combat units in significant numbers before the war ended. Even then, the pace of aeronautical development was so rapid that aircraft only a few years old would be approaching obsolescence by the time they entered large-scale production.

The American-built DH-4 was not a simple, bolt-for-bolt copy of the British airplane. It was modified extensively to accommodate American manufacturing methods, equipment, and the powerful American-designed Liberty engine. Thousands were ordered from several manufacturers, and nearly 1,900 reached France before the Armistice. The DH-4 became the only American-built airplane to see substantial combat service with the American Expeditionary Forces during World War I.

The DH-4 was already the stuff of legend. Designed as a two-seat reconnaissance aircraft and light bomber, it had distinguished itself during the war through its speed, range, and versatility. Alongside it was the Curtiss JN-4 "Jenny," a slower and more docile airplane that had become the principal trainer for American military pilots.

When the war ended in November 1918, the federal government found itself possessing enormous numbers of aircraft for which it no longer had an immediate military need. Production contracts were canceled, military appropriations were slashed, and many surplus airplanes — particularly JN-4s — were sold to civilians for a fraction of what they had cost the government. The sudden availability of inexpensive airplanes and thousands of demobilized military pilots helped ignite the barnstorming era and expand public interest in aviation. The Jenny, in particular, became the characteristic airplane of the barnstormers who crisscrossed the country giving passenger rides, performing aerobatics, walking on wings, and staging aerial exhibitions at county fairs and public gatherings.

The DH-4 also found important uses after the war. Its greater speed, range, and payload made it suitable for military observation, aerial photography, experimentation, and the rapidly expanding airmail service. Rather than discard all the DH-4s remaining in its inventory, the Army began modifying some of them to improve their safety, utility, and performance.

The resulting DH-4B incorporated several important changes. Most notably, the fuel tank — which in the original airplane had been positioned between the pilot and observer — was moved forward, allowing the two cockpits to be placed closer together and reducing the danger posed by the centrally located tank. The landing gear was also repositioned, and other changes were made to improve the airplane's balance and operational usefulness.

In 1919, the Army awarded Boeing a contract to modernize fifty DH-4s to the new DH-4B standard. The order was modest, but its importance to the struggling company was immense. With wartime contracts having evaporated, Boeing had been forced to manufacture furniture and other products merely to keep its workers employed and its factory operating. The DH-4 contract placed Boeing firmly back in the airplane business. The company could finally stop building office furniture and return to building airplanes.

After these initial 50 were built, some were sent back to Boeing to incorporate even more changes and other improvements because of newer advancements in aviation engineering know-how. Again, another lifeline had been thrown to Boeing by the U.S. government. And when that contract was fulfilled, once again another lifeline was tossed to Boeing by the U.S. government.

After World War I, the Army studied the battles it had fought in Europe to assess the impact the airplane had exerted on the conflict and the role it might play in future wars. One area in which military planners believed the airplane could provide a significant advantage was close air support, a still-nascent concept that would prove pivotal in later conflicts. Just as artillery could suppress or destroy enemy defenses before ground troops advanced, aircraft could be used for much the same purpose.

Because of Boeing's experience constructing the large Curtiss-designed flying boats a few years earlier — and because it submitted the lowest bid — the company was awarded a contract to build twenty copies of an experimental ground-attack aircraft designed by the U.S. Army. The order was later reduced to ten. Designated the GAX, for Ground Attack Experimental, the airplane was heavily armored and powered by two large engines. Weighing more than 10,000 pounds, it pushed the limits of aircraft size for its time.

The GAX was an ungainly machine—slow, cumbersome, and burdened by the weight of its armor and armament. Yet, like the Curtiss flying boats Boeing had previously constructed, it gave the company invaluable experience in aircraft manufacturing, particularly in the construction of large and complex airplanes. Just as importantly, the contract provided Boeing with a steady stream of revenue. Once again, that revenue came from the U.S. government.

Ground-attack aviation, however, remained largely experimental, and the GAX demanded more than the engine and aircraft technology of the period could deliver. Its performance fell well short of the Army's operational requirements, and further development of the airplane was eventually abandoned.

The Boeing Company soon found itself in need of yet another lifeline, and this time its situation was becoming increasingly desperate. The government contracts that had sustained it were growing shorter and less substantial, providing enough work to keep the factory alive but not enough to ensure its future. Bill Boeing needed something larger — an order capable not merely of keeping his airplane business operating, but of allowing it to grow.

Once again, the lifeline came from the U.S. government. This time, however, it was substantial enough to transform Boeing's fortunes and help make it the largest aircraft manufacturer in the United States.

The contract called for Boeing to build 200 examples of a small fighter aircraft, known at the time as a pursuit plane. The sleek, single-engine biplane had been designed by the Thomas-Morse Aircraft Corporation of Ithaca, New York, and was designated the MB-3. Impressed by the performance of the prototype, the Army initially ordered fifty airplanes from Thomas-Morse. Shortly afterward, however, the service determined that it needed an additional 200 pursuit aircraft and opened the larger production order to competitive bidding.

Thomas-Morse was stunned when the contract to manufacture its own airplane was awarded not to itself, but to Boeing. By then, however, Boeing had accumulated considerable experience building aircraft in quantity and had learned how to translate an existing design into an efficient industrial production process. Bill Boeing had also developed a keen understanding of manufacturing costs. His company enjoyed ready access to inexpensive lumber, much of it supplied through his own timber interests, as well as the milling capacity needed to turn that raw material into finished aircraft components.

Those advantages allowed Boeing to submit a lower bid than Thomas-Morse while convincing the Army that it possessed the facilities and production expertise necessary to complete the order. The contract was awarded in 1920, and Boeing immediately set about manufacturing the MB-3s as quickly and efficiently as possible.

The order represented far more than another temporary reprieve. It established Boeing as a major producer of military aircraft and elevated the company into the front rank of American airplane manufacturers. Thomas-Morse, despite having designed the MB-3, would gradually recede into aviation history. Boeing, by mastering its manufacture, was beginning its ascent.

As significant as the MB-3 contract was, it also exposed an important weakness within the Boeing Company. By 1920, Boeing had manufactured hundreds of airplanes, but only about fifty of them had been based on a Boeing design, the Model C trainers produced for the Navy during World War I. Most of the company's work had involved building or modifying aircraft conceived by others. Unless that pattern changed, Boeing risked becoming little more than a contract manufacturer of other companies' designs.

Bill Boeing wanted something more. Although the company had survived largely through military contracts, he believed the long-term future of aviation lay not solely in government work but in commercial flight. That outlook would be dramatically reversed at the beginning of the 21st century, when Boeing became increasingly dependent upon its defense and government businesses. In the early 1920s, however, Boeing saw civil aviation as the field offering the greatest potential for growth.

He remained convinced that the fascination he had experienced during his first airplane ride in 1914 would eventually be shared by the broader public. Flight possessed an inherent ability to captivate people, but its true commercial promise lay in something even more consequential, which was the power to reduce journeys that required days by train or ship to a matter of hours.

Ironically, the opportunity that would allow Boeing to move toward commercial aviation would once again originate with the federal government. This time, however, government work would provide more than a temporary production contract. It would give Boeing an opening that could be transformed into a genuine commercial enterprise — one involving civil aircraft designed and built entirely by Boeing.

***

From its establishment in the late 18th century until the dawn of the 21st century, the United States postal system played a central role in the nation's economic development and in the daily lives of its citizens. Letters carried through the mail allowed Americans separated by vast distances to remain connected, while the reliable movement of documents, newspapers, goods, and packages helped commerce expand across an increasingly large and interconnected country.

Because time has always carried an economic cost, faster delivery meant lower expenses, quicker transactions, and greater commercial efficiency. Businesses could receive orders sooner, settle accounts more rapidly, and move information across the country with less delay. As a result, inventors, entrepreneurs, transportation companies, and government officials continually searched for a better postal "mousetrap" — a faster and more efficient means of moving the mail.

By the early 20th century, the airplane appeared to offer the most dramatic improvement yet. It promised to compress journeys that had taken days by rail into mere hours and, in doing so, to transform not only the delivery of mail but the speed at which American commerce itself could operate.

Because the carriage of mail was principally the responsibility of the federal postal system, it fell largely to the government to explore new methods of improving—and, above all, accelerating—its delivery. Just as the steam locomotive had revolutionized the movement of mail during the nineteenth century, the airplane appeared poised to carry that revolution to an entirely new level. By transporting mail through the air rather than over the ground, it promised literally to shave days from journeys across the country.

The Post Office Department inaugurated the nation's first regularly scheduled airmail service on May 15, 1918, during the final year of World War I. The initial route connected Washington, D.C., Philadelphia, and New York, linking the seat of the federal government with two of the nation's most important commercial and financial centers. The volume of correspondence moving between those cities made the corridor an ideal proving ground for the new service.

Because no commercial airline industry yet existed to operate the route, the Post Office Department turned to the Army for aircraft, pilots, mechanics, and operational expertise. Army Air Service pilots flew specially modified Curtiss JN-4H "Jennies," carrying the mail between the three cities according to a published daily schedule. The Army regarded the assignment as useful practical training for its pilots, but the primary purpose was to demonstrate that mail could be carried reliably and routinely by airplane.

The service got off to an appropriately rocky start. On the inaugural day, one pilot became disoriented shortly after departing Washington and landed in a Maryland field after flying in the wrong direction. Despite that highly publicized mishap and the inherent unreliability of early aircraft, the operation soon settled into a workable routine and demonstrated that scheduled airmail service was practical.

After three months, on August 12, 1918, the Post Office Department assumed complete control of the operation, replacing the Army pilots with its own civilian aviators and beginning the gradual expansion of the route system. It initially acquired purpose-built JR-1B aircraft, but increasingly turned to surplus de Havilland DH-4s made available by the Army after the war.

The DH-4 required substantial modification for its new civilian mission. Its structure, landing gear, cockpit arrangement, and fuel-tank placement had been designed for military operations rather than the relentless demands of carrying heavy loads of mail in all weather. Once strengthened and reconfigured as the improved

DH-4B, however, the airplane became the dependable workhorse of the expanding U.S. Airmail Service, carrying hundreds of millions of letters and helping establish the operational foundation upon which American commercial aviation would be built.

Bill Boeing followed the Post Office Department's early airmail operations along the northeastern corridor of the United States with great interest. He spent considerable time contemplating how his company might profit from this emerging form of transportation. The carriage of mail between destinations within the United States remained the exclusive responsibility of the federal postal system, which also controlled the contracts awarded to private transportation companies carrying the mail by rail, road, and sea. Because scheduled airmail service was still in its infancy, the government continued to operate it directly.

International mail, however, presented a different opportunity. It was through that opening that an ingenious idea emerged — one that would set in motion a chain of events that dramatically expanded Boeing's size, reach, and influence within the aviation industry.

Edward "Eddie" Hubbard joined Boeing during the company's earliest years as a mechanic but soon advanced to become one of its test pilots. He and Bill Boeing came to know one another well, developing a close personal and professional relationship that endured after Hubbard left the company in 1919 to pursue opportunities of his own.

Like Boeing, Hubbard was convinced that aviation possessed enormous commercial potential. He earned a living flying demonstration flights for the public, carrying passengers, and pursuing aviation "firsts" that would showcase what airplanes could accomplish. In searching for another such opportunity, Hubbard turned his attention to the passenger liners traveling between Seattle and ports across the Pacific.

Ships arriving from Asia frequently stopped at Victoria, on Vancouver Island in British Columbia, before continuing to Seattle. During that stop, passengers and cargo bound for Canada were unloaded, while additional passengers and freight destined for the United States were taken aboard. The ships would then cross the Strait of Juan de Fuca and proceed through Puget Sound to their final destination in Seattle. Vessels traveling in the opposite direction followed much the same pattern, stopping at Victoria before continuing across the Pacific to Asia.

Those stopovers would typically last an entire day. Mail carried aboard the ships — whether arriving from Asia and bound for the United States or traveling in the opposite direction — remained in the vessels' holds throughout the stop, making no progress toward its destination.

Hubbard recognized that an airplane could eliminate much of that lost time. Mail could be removed from a ship during its stop at Victoria and flown directly to Seattle, reaching the United States many hours before the vessel arrived. On westbound journeys, outbound mail could similarly be flown from Seattle to Victoria and placed aboard a departing ship before it began its transpacific voyage.

Hubbard's idea was to rendezvous with inbound passenger liners from Asia when they stopped at Victoria, remove the mail bound for the United States, and fly it directly to Seattle. Once there, it could be transferred into the conventional postal network and routed to its final destination. The flight from Victoria to Seattle would take roughly an hour and could save nearly an entire day compared with leaving the mail aboard the ship until it completed the final leg of its voyage.

The same advantage could be realized in the opposite direction. Mail originating in the United States and destined for Asia could be collected in Seattle, flown to Victoria, and placed aboard an outbound liner shortly before it departed across the Pacific. In both directions, the airplane would eliminate the long delay created by the ship's stopover and shaving an entire day off of the normal transit time.

Bill Boeing and Edward Hubbard agreed that Hubbard would present the proposal to the U.S. Post Office Department and attempt to secure a contract to carry the mail between Seattle and Victoria. There was, however, one significant obstacle and that was they needed an airplane capable of performing the work.

The transpacific liners carried several thousand pounds of mail. Even if only a small portion of that total were assigned to the expedited air service — presumably at a premium rate — it could still amount to a few hundred pounds on each flight. That represented a substantial payload for Boeing's existing aircraft, whether the original B&W seaplane or the Model C trainers the company had built for the Navy during World War I.

Using an airplane designed and manufactured by another company, such as Curtiss, would also undermine the broader purpose of the venture. Boeing and Hubbard were not merely seeking a postal contract. They wanted to generate publicity, earn a profit, demonstrate the commercial potential of aviation, and create a market for Boeing-built airplanes. To make the plan work, Boeing would need a new aircraft — one designed specifically to carry a meaningful load of mail reliably between Seattle and Victoria, and one bearing the Boeing name.

The B-1, known within the company as the Model 6, was an entirely Boeing design. From outward appearances, however, it looked much like a scaled-down version of the enormous Curtiss flying boat that Boeing had built under license during the closing months of World War I. Like the Curtiss design, it was a wood-hulled biplane flying boat with its engine mounted above the fuselage in a pusher configuration.

Boeing originally conceived the B-1 as a commercial aircraft for private owners. It could carry a pilot and two passengers, along with a modest quantity of cargo, and the company hoped to sell it to affluent individuals seeking an airplane for transportation or recreation. It was Boeing's first aircraft designed expressly for the civil market.

Unfortunately, the B-1 could not have appeared at a worse time. The end of World War I flooded the market with thousands of inexpensive surplus military airplanes, including DH-4s and Curtiss JN-4 Jennies. Prospective buyers had little incentive to pay thousands of dollars for a new Boeing flying boat when a serviceable surplus airplane could be purchased for only a few hundred dollars.

Only one B-1 was built. With no customers on the horizon, Boeing had no reason to invest scarce funds in constructing additional examples. The airplane appeared destined to become an unsuccessful experiment — a sound design for which no viable market yet existed.

Hubbard's proposal for an international airmail service between Victoria and Seattle suddenly offered the solitary B-1 a new purpose. An airplane that had failed to find a market as a private recreational craft might instead prove its value carrying mail and cargo across the waters of the Pacific Northwest.

Less than a year after the B-1's first flight, the U.S. Post Office Department awarded Edward Hubbard a contract to carry international mail between Seattle and Victoria. Although Boeing built only one B-1, Hubbard's exploits in the airplane would help burnish the young company's reputation and demonstrate the practical commercial value of a Boeing-designed aircraft.

The first flights attracted extensive coverage in newspapers on both sides of the border. Here was a Boeing airplane carrying international mail across the waters of the Pacific Northwest and delivering it nearly a full day sooner than the steamships on which it had arrived. By January 1923, a little more than two years after the contracted service began, Hubbard had transported approximately fifty tons of mail between Victoria and Seattle.

The flights were not without danger. Hubbard regularly confronted strong winds, fog, darkness, mechanical difficulties, and the treacherous waters of Puget Sound and the Strait of Juan de Fuca. In March 1923, the B-1 was badly damaged in an accident shortly after departing Victoria Harbor. Hubbard escaped serious injury, the mail was recovered, and the airplane was returned to Seattle, where Boeing rebuilt it and restored it to service.

Through these exploits, Hubbard made a name not only for himself but also for Boeing. The B-1 had originally been conceived as the company's gateway into the commercial-aircraft market. Had the postwar market not been flooded with inexpensive military-surplus airplanes, it might have succeeded in that role through conventional sales to private owners. Instead, the sole B-1 fulfilled a different and ultimately more consequential purpose. The airplane became Boeing's first meaningful step toward commercial aviation. It demonstrated that a Boeing-designed aircraft could perform useful transportation work, generate revenue, and operate reliably in demanding conditions. More importantly, it showed that government-sponsored mail service could be transformed into a genuine commercial opportunity.

That modest mail route between Seattle and Victoria marked the beginning of a transformative journey. Over time, Boeing would use government airmail contracts as a bridge from its dependence on military aircraft production toward the design and manufacture of commercial transports, and eventually toward the operation of an airline itself.

Ironically, nearly a century later, Boeing would attempt something resembling that journey in reverse. As its commercial-aircraft business faltered, the company increasingly looked to its defense, space, and government operations to provide the stability and revenue that commercial aviation had once supplied.

***

As the Post Office's domestic airmail network expanded during the early 1920s, so did its accident rate, leaving a trail of dead Army pilots and generating mounting public criticism. Much of the service still depended on surplus DH.4s, aircraft whose design dated to the early years of World War I and whose age was beginning to show. The planes had been lightly built from the outset, and years of hard use only worsened their condition. Nor did military discipline always help because Army pilots sometimes flew in dangerous weather or with mechanical defects when less regimented non-military judgment might otherwise have kept them on the ground.

The Post Office wanted to make airmail a regular and reliable premium service, but Congress tied further funding to the eventual transfer of the routes to private contractors. As part of that transition, the Post Office invited aircraft manufacturers in 1925 to submit designs for a modern, purpose-built mail plane to replace the aging DH.4s. To contain costs, however, it required the new aircraft to use the same liquid-cooled Liberty engines that powered the DH.4. Thousands were already available, and engines were then, as they remain today, among the most expensive components of an aircraft.

Boeing responded with an entirely new design, the Model 40, incorporating lessons the company had learned since the war while meeting the requirement to use the Liberty engine. The Post Office purchased Boeing's sole prototype, but awarded the production contract to a relative newcomer, the Douglas Aircraft Company. Douglas' victory marked an early milestone in what would become a decades long rivalry between the two manufacturers.

For Boeing, however, the loss was not entirely a setback. The mail plane competition was only one part of the government's broader plan to privatize airmail. The Post Office also intended to award routes across the country to private operators, generally on the basis of the lowest bid. Boeing now had an aircraft that could be adapted to compete for them.

The 1925 competition had been for a new mail plane. The following year, the Post Office solicited bids for airmail routes across the country, effectively opening the service to commercial operators under government contract. Bill Boeing wanted a piece of the action. He saw airmail not merely as a source of revenue, but as a stepping stone toward a broader commercial aviation industry — and a means of securing his company's future. First, however, he had to determine which route to pursue and how to submit a bid low enough to win without making the venture unprofitable.

Once again, Edward Hubbard played an influential role. He urged Boeing to bid on the San Francisco–Chicago route, one of the longest and most difficult in the system. It crossed vast stretches of the sparsely populated West and required aircraft to traverse the Rocky Mountains. Because the route promised high operating costs and uncertain profits, Hubbard believed it would attract relatively little competition.

The idea was intriguing, but it raised an obvious question: If the route was undesirable because of its poor profit potential, how could Boeing operate it successfully? Into that discussion stepped Clairmont Egtvedt, known to everyone as "Claire."

***

Clairmont "Claire" Egtvedt was born in 1892 to Norwegian immigrant parents. His family moved to Seattle in 1911, and he later attended the University of Washington's College of Engineering. Bill Boeing hired him immediately after his graduation in 1917 as a draftsman, making him one of the company's earliest employees.

Egtvedt quickly established himself as a forward thinker who was constantly searching for Boeing's next big opportunity. He had championed the sole B-1 flying boat, which later enhanced the company's reputation by inaugurating an international airmail service. He had also persuaded Boeing to pursue the Army's GAX ground-attack prototype after World War I and to bid on production of the Thomas-Morse MB-3 fighter, the contract that became a financial boon for the young company.

His ability both to conceive ideas and execute them steadily increased his stature, and by 1926 he had risen to vice president. At the time, he was overseeing Boeing's efforts to design new fighters for the Army Air Corps when Bill Boeing asked him to determine how the company could win the Chicago–San Francisco airmail route and still operate it profitably.

Egtvedt's answer was to resurrect the Model 40, which had lost the Post Office mail plane competition to Douglas the previous year, and redesign it around a new engine, called the Wasp, that was in development by engine manufacturer, Pratt & Whitney. The Wasp was more powerful than the Liberty engine originally installed in the aircraft, but it was also air-cooled, which meant it was smaller and lighter than the bulky liquid-cooled Liberty engine. Its reduced size and weight created additional space behind the firewall and ahead of the pilot's cockpit.

Egtvedt proposed using that extra space for a small enclosed cabin — dubbed the "limousine compartment" — with side-by-side seating for two passengers. Their fares would supplement the revenue from carrying mail, allowing Boeing to submit a competitive bid for the otherwise marginal route while still earning a profit. In effect, Egtvedt had devised an airplane in which passenger service would subsidize airmail service — and, in doing so, helped lay the foundation for Boeing's entry into commercial aviation.

It was an ingenious plan, and Bill Boeing was sufficiently confident of its success that he decided not to burden his aircraft company with operating the route. He wanted Boeing to remain focused on designing and building airplanes rather than be distracted by the demands of running an air transport service. Instead, he formed a separate company, Boeing Air Transport, Inc., which would purchase airplanes from Boeing and operate the San Francisco–Chicago route. In one stroke, Boeing had created not only a new airline, but also what would become one of his aircraft company's most prominent customers.

The plan worked. When the sealed bids were opened in Washington, D.C., in January 1927, Boeing's was the lowest submitted for the route. At $3 per pound of mail, it was nearly half the amount sought by the next-lowest bidder. The bid was so aggressive that the government required Boeing to post a $500,000 performance bond — more than $9 million in 2026 dollars — to guarantee that the company could fulfill the contract.

The celebration in Seattle was muted because an enormous amount of work remained. Under the terms of the contract, Boeing Air Transport had to begin carrying mail within six months using a fleet of 24 airplanes. Yet the redesigned aircraft, now designated the Model 40A, had not even flown. Boeing nevertheless met the deadline. The first Model 40A took to the air on May 20, 1927, and by midnight on June 30, all 24 airplanes were completed and ready for service.

***

On the evening of July 1, 1927, two Boeing Air Transport flights departed almost simultaneously. One from Chicago heading west and the other from San Francisco heading east. The eastbound aircraft carried only mail, while the westbound flight also carried a single passenger, Chicago Herald-Examiner reporter Jane Eads. Carrying a reporter on the flight, a female one at that, proved highly valuable from a publicity standpoint.

Passenger demand would soon prove sufficient to justify another modification to the Model 40A. Pratt & Whitney had developed a more powerful engine, the Hornet, which allowed Boeing to enlarge the passenger cabin from two seats to four, arranged in two rows of two, while increasing the aircraft's mail capacity from 1,200 to 1,500 pounds. The resulting Model 40B-4 became an important step in Boeing's transformation from one of many small aircraft manufacturers into a recognized leader in commercial aviation. It also marked the beginning of a long and profitable relationship between Boeing and Pratt & Whitney.

Boeing's access to Pratt & Whitney's engines was itself the product of a shrewd bargain. The Wasp engine used in the original Model 40A had been developed primarily for the Navy, and Pratt & Whitney was obligated to satisfy the Navy's requirements before offering it commercially. When Egtvedt first proposed redesigning the Model 40 around the Wasp, he doubted that Boeing could obtain enough engines in time to fulfill the airmail contract.

Bill Boeing approached Pratt & Whitney's president, Frederick Rentschler, with a proposition. If Rentschler could secure the release of the necessary Wasp engines, Boeing would provide Pratt & Whitney with a Model 40A at no cost for use as a flying engine testbed. Such an aircraft would be invaluable to Pratt & Whitney as it developed future engines, including those intended for the Navy. Rentschler used his Navy connections to obtain approval for the release of 24 production Wasp engines to Boeing. In return, Pratt & Whitney received the flying testbed it needed. Once again, Bill Boeing had converted a seemingly insurmountable obstacle into an arrangement that benefited everyone involved.

The Model 40, first in its A configuration and later as the enlarged B-4, proved an unqualified success and laid the foundation for Boeing's future in commercial aviation. Boeing produced 77 aircraft across the two versions, making the Model 40 one of the most successful — and innovative — commercial airplane programs of its era.

Commercial Flight Takes Flight

As the 1920s progressed, Boeing's Seattle factory hummed with activity while Boeing Air Transport carried mail and passengers between San Francisco and Chicago. Its Model 40As and 40B-4s regularly flew at capacity, and the publicity surrounding the service helped fuel public enthusiasm for commercial aviation.

Boeing was hardly alone in capturing the public's imagination. Former Army pilots crisscrossed the country in surplus DH.4s, JN-4s, and almost anything else capable of leaving the ground. Their exploits demonstrated that airplanes were more than military machines or objects of spectacle. They offered the possibility of transforming long-distance travel.

At the time, trains remained the fastest and most economical form of dependable transportation, but airplanes could reduce journeys that took days to a matter of hours. The aircraft of the era were still limited in range, capacity, and reliability. Mechanical failures were common, and forced landings in farm fields or on dirt roads were an accepted hazard of flight. Even so, the potential was obvious. If engineers could extend an airplane's range, increase its capacity, and make it dependable enough to operate on a regular schedule, commercial aviation could become a practical form of mass transportation. It did not require a visionary to imagine that future — but it would require considerable engineering skill to make it a reality.

To encourage the engineering advances needed to make that future possible, newspapers, philanthropists, and other organizations offered cash prizes for aviation "firsts." The most famous was the Orteig Prize, established in 1919 by New York hotelier Raymond Orteig for the first nonstop flight between New York and Paris in either direction.

The year is worth noting. Orteig offered the prize only sixteen years after the Wright brothers' first flight, yet people were already contemplating nonstop air travel between two of the world's greatest cities on opposite sides of the Atlantic. Eight years later, in May 1927, Charles Lindbergh accomplished the feat. His flight electrified the world, particularly the United States.

The 1920s — the "Roaring Twenties" — were heady years in America. The United States had emerged from World War I without the physical devastation suffered by its European allies, and its economy was booming. Prosperity helped create a culture eager to embrace each new novelty, although public enthusiasm for the latest sensation could disappear almost as quickly as it arose.

Aviation may still have seemed novel, but it was no fad. It occupied much the same position that commercial spaceflight would at the end of the century: at the frontier of human technology, yet beginning to appear practical for broader use. In the 1920s, a steamship voyage from New York to Cherbourg, a principal European port for travelers bound for the Continent, typically took four to seven days. Lindbergh crossed the Atlantic in just over 33 hours.

The potential savings in travel time were easy to grasp. Draw a straight line on a map between St. Louis and Chicago, for example, and an airplane could cover the distance in less than three hours, compared with seven or more by train. Automobiles were becoming increasingly common during the 1920s, but poor roads and limited infrastructure outside major cities still made long-distance travel by car a slow, often multi-day undertaking.

The public's enthusiasm for air travel was therefore more than a passing fascination. Airplanes promised to transform transportation, provided they could be made safe, reliable, and capable of carrying passengers on regular schedules. The desire was there. What remained was to develop the aircraft that could satisfy it.

By then, Boeing had accumulated a wealth of design, engineering, and manufacturing experience since building its first airplane, the B&W. Particularly valuable were the lessons gained from working on larger aircraft, including the Curtiss flying boats it built under contract for the Navy and the massive GAX ground-attack aircraft produced for the Army. Boeing had also strengthened its industrial capabilities through the large-scale production of the Thomas-Morse MB-3 pursuit plane. The company was no longer merely an ambitious newcomer. It had emerged as one of the country's leading aeronautical manufacturers.

Under Claire Egtvedt's leadership, Boeing had designed the PW-9, a highly successful fighter purchased by the U.S. Army in quantities approaching 160 aircraft and regarded as one of the finest military airplanes of its era. That success was followed by the innovative Model 40A and 40B-4 mail-and-passenger planes. Boeing had grown increasingly confident in pushing the boundaries of aeronautical design, engineering, and manufacturing. Over the next several years, a succession of groundbreaking aircraft would further solidify its position, beginning with the Model 80 and its larger successor, the Model 80A.

The Model 80 was born of both necessity and urgency. By 1927, Boeing Air Transport's San Francisco–Chicago service was operating at capacity, and passenger fares were becoming nearly as important as mail revenue. The airline needed a larger aircraft capable of carrying more people. Whereas the Model 40 had been designed primarily to haul mail, with passengers providing supplemental income, the Model 80 would be designed first and foremost as a passenger airplane, with space remaining for mail.

The Model 80 incorporated several features intended to make scheduled air travel more practical and appealing. It had three engines — one in the nose and one mounted on each wing — providing enough power to cross the high mountain ranges of the West while offering an added measure of safety if one engine failed. Its enclosed cabin accommodated twelve passengers in relative comfort, while the enlarged Model 80A eventually carried eighteen. Two pilots sat in an enclosed cockpit, protected from the wind and weather that earlier aviators had endured.

The aircraft also made greater use of metal than its predecessors. Its wings retained the fabric-covered wooden construction that had been standard since the earliest days of flight, but portions of the fuselage were covered in sheet metal over the structural framework. The Model 80 remained a biplane because that configuration provided the lift needed to operate from the short, high-elevation airfields along Boeing Air Transport's western routes. Large and imposing for its day, it represented a substantial step toward the modern airliner.

Yet one of the Model 80's most enduring innovations involved not the airplane itself, but the service provided aboard it. As flights grew longer and aircraft began carrying a dozen or more passengers at a time, Boeing Air Transport placed registered nurses aboard its airplanes. Their presence reassured travelers still wary of flying and provided trained assistance to anyone suffering from motion sickness or another ailment. They also served beverages and snacks during the journey.

The idea was a sensation. What began as an effort to make nervous passengers feel safer created an entirely new profession — one that would come to symbolize commercial air travel for decades.

By 1929, Bill Boeing was watching the aviation industry expand at a remarkable pace. A businessman above all else, he was determined to capture as much of that growth as possible. Boeing was prospering, but so was its most important supplier, Pratt & Whitney. Their collaboration on Boeing's newest aircraft had helped drive the revenues of both companies sharply upward.

To Boeing, the next step seemed obvious: merge them. Pratt & Whitney's engines would power Boeing airplanes, while Boeing would gain assured access to the company's latest engine technology. He then expanded the idea, bringing Boeing Air Transport, Hamilton Standard Propeller Company — the era's leading propeller manufacturer — and several other aviation businesses into the new enterprise. The resulting conglomerate was named United Aircraft & Transport Corporation, or UATC.

UATC gave Boeing control over nearly every stage of the aviation business. Its companies designed and manufactured airplanes, built the engines and propellers that powered them, and operated the airlines that carried passengers and mail. Each subsidiary became a natural customer or supplier of the others, creating a vertically integrated aviation empire.

In 1929, Boeing viewed the arrangement as a masterstroke — an enterprise that few others possessed the vision or influence to assemble. Yet the very concentration of power that made UATC so formidable would eventually draw the attention of the federal government and help drive Bill Boeing from the industry he had helped create.

***

Despite its innovations, Boeing's Model 80 was essentially a catch-up design, created in response to a formidable new competitor. Transcontinental Air Transport, backed by prominent investors from the railroad, oil, and other industries, had begun offering service between New York and Los Angeles. Its transcontinental system combined daytime flights with overnight rail travel, allowing passengers to cross the country far faster than by train alone. The company's emergence was also one of the developments that encouraged Bill Boeing to assemble UATC.

Transcontinental Air Transport operated the Ford Trimotor, a three-engined transport built by the Ford Motor Company. Henry Ford believed airplanes could bring rapid transportation to the public much as his automobiles had democratized travel on the ground. The Trimotor drew heavily from the all-metal aircraft pioneered by the German manufacturer Junkers, including its three-engine arrangement and corrugated metal construction.

Compared with Boeing's Model 80 and 80A, the Ford appeared distinctly modern. It was an all-metal monoplane rather than a fabric-covered biplane, making it sleeker and faster than Boeing's transport. Ford also achieved far greater commercial success, producing approximately 200 Trimotors between 1926 and 1933. Boeing built fewer than 20 Model 80s and 80As, nearly all for its own airline. Although the Model 80 served Boeing Air Transport well, it never seriously challenged the Ford in the broader market.

The Trimotor's metal skin was durable and required less maintenance than fabric, but it did not yet function as a primary load-bearing structure. The underlying framework still carried the aircraft's aerodynamic and structural loads, while the aluminum merely enclosed it. Because the skin had to remain thin and light, it was corrugated to prevent it from flexing and losing its shape. The corrugations added strength to the covering but also created aerodynamic drag, even though they ran parallel to the airflow.

For all its modern appearance, the Trimotor remained a slow and cumbersome airplane, cruising at approximately 107 miles per hour. Its control cables and pulleys were exposed to the slipstream, its wing-mounted engines hung from external frameworks reinforced by bracing wires, and its fixed landing gear protruded permanently beneath the fuselage. Each feature added drag.

Boeing's Model 80A was even less aerodynamically efficient. Its biplane configuration required two wings, numerous supporting struts, and a web of bracing wires, all of which disturbed the airflow. As commercial aviation gained momentum, the industry needed an aircraft that was faster, cleaner, and far more efficient than either the Model 80 or the Ford Trimotor.

Boeing was about to provide it. The company's next transport would not merely improve upon the airplanes already in service. It would introduce an entirely new approach to airliner design and disrupt the commercial aviation industry.

Led by Claire Egtvedt, Boeing's engineers set out to design an airplane that would leapfrog anything then in commercial service.

Their objective was straightforward but technically demanding: reduce aerodynamic drag, lighten the airframe, and produce an aircraft capable of transforming passenger travel. The result was the Model 247, introduced in 1933.

The 247 carried ten passengers, along with two pilots and a cabin attendant, and was powered by two engines mounted on the leading edges of its wings. Although many of its features had appeared separately on earlier aircraft, Boeing brought them together — and refined them — in a single design.

The engines were enclosed in streamlined nacelles rather than suspended beneath the wings on exposed frameworks. Closely fitted cowlings directed air smoothly around the radial engines, reducing drag while improving cooling. The landing gear retracted into the nacelles behind the engines, removing the wheels and supporting struts from the slipstream during flight. The airplane was also constructed almost entirely of aluminum, including both its internal framework and exterior skin.

Individually, these were evolutionary advances. Their integration into one airplane was unprecedented. Yet the Model 247's most important innovation went deeper than its streamlined appearance.

Earlier metal aircraft such as the Ford Trimotor relied primarily on an internal framework to carry aerodynamic and structural loads. Their aluminum skin served mainly as a covering. On the Model 247, Boeing employed stressed-skin construction, in which the smooth aluminum exterior became an integral part of the load-bearing structure. In effect, stresses that would otherwise have been concentrated within a heavy internal framework were distributed across much of the airplane's outer surface. Because the skin helped carry those loads, the underlying structure could be made lighter without sacrificing strength. The smooth exterior also eliminated the drag-producing corrugations found on aircraft such as the Ford Trimotor.

The result was an airplane that was lighter, stronger, more durable, and dramatically more streamlined than the transports it replaced. With a top speed of approximately 200 miles per hour, the Model 247 was roughly 50 percent faster than its principal commercial competitors and could outrun many contemporary military aircraft.

The Model 247 established the basic architecture of the modern airliner: an all-metal, stressed-skin, low-wing monoplane with enclosed engines and retractable landing gear. Those principles would shape commercial aircraft through the piston-engine era and continue well into the jet age.

Boeing had an immediate hit on its hands, and airlines lined up to order the new airplane. Through UATC, Boeing Air Transport received priority, ordering 60 aircraft for a then-unprecedented $3.5 million. Boeing began filling the order immediately. Transcontinental and Western Air, or TWA, also wanted the 247, but Boeing told the rival airline it would have to wait until Boeing Air Transport's order was completed. The decision would prove disastrous.

TWA knew that Boeing Air Transport, equipped with the 247, could gain a commanding competitive advantage. Desperate for an equivalent airplane, it turned to the Douglas Aircraft Company. Douglas was equally concerned that the 247 might allow Boeing to dominate commercial aircraft manufacturing and drive it from the market. With TWA's backing and investment, Douglas began developing a competitor.

The result was the DC-2, which incorporated many of the 247's innovations in a larger, faster, and more capable airplane. (The DC-1 was a prototype and only one was ever built for testing and evaluation.) The DC-2 carried fourteen passengers compared with the 247's nine and could fly both farther and faster. Once Douglas introduced it, orders for the 247 slowed to a trickle and then stopped. Boeing built only 75 Model 247s, while Douglas produced nearly 200 DC-2s.

The DC-2 soon evolved into one of the most celebrated airplanes in aviation history: the DC-3. Larger and more capable still, the DC-3 carried roughly three times as many passengers as the 247. Including its military derivative, the C-47, it became one of the most-produced transport aircraft in history. If the 247 revolutionized air travel, the DC-3 transformed it on an even greater scale. The DC-2 and DC-3 propelled Douglas past Boeing and established it as the world's preeminent commercial-aircraft manufacturer, a position it would hold for nearly three decades, until the jet age reshaped the industry.

Although the 247 never approached the success of the DC-2 or DC-3, its production run of 75 aircraft was respectable, and its importance extended far beyond its sales. It established Boeing as a leader in aircraft design and manufacturing and demonstrated the company's growing engineering ingenuity. But Boeing understood that remaining competitive would require another game-changing airplane. To leapfrog Douglas, it would have to address one of the most pressing challenges confronting the rapidly expanding commercial aviation industry.

Unfortunately, UATC's dominance, vertically integrated structure, and refusal to supply the 247 to competing airlines until Boeing Air Transport received its full order of 60 aircraft attracted government scrutiny. That scrutiny intensified during the Air Mail Scandal of 1934, which began with revelations of collusion in the awarding of federal airmail contracts but soon expanded into a broader investigation of the nation's rapidly growing aviation industry.

After congressional hearings and multiple investigations, President Franklin D. Roosevelt canceled all private airmail contracts in February 1934 and temporarily assigned the routes to the U.S. Army Air Corps. Congress subsequently passed the Air Mail

Act of 1934, which, among other provisions, prohibited aircraft manufacturers from controlling airlines. The legislation forced the breakup of UATC into three independent companies. Boeing Airplane Company would continue manufacturing aircraft. United Aircraft Corporation, which included Pratt & Whitney and Hamilton Standard, would concentrate on engines and aircraft components. United Air Lines would operate as an independent airline. No longer affiliated, each company would have to compete on its own merits within its respective market.

Disgusted by the investigations and the breakup of the enterprise he had built, Bill Boeing withdrew from the company and largely left the aviation business. Philip Johnson, UATC's president, was implicated in the scandal, forced to resign, and temporarily barred from participating in the U.S. airline industry. Leadership of the Boeing Airplane Company passed to Claire Egtvedt.

***

During the 1930s, commercial aviation remained largely the province of the wealthy. Few average Americans could afford to fly. A round-trip transcontinental ticket cost the equivalent of roughly $7,500 in 2026 dollars—nearly ten times the average American's annual per-capita income. Airlines nevertheless found ways to operate profitably. Their affluent passengers valued time and were willing to pay handsomely to cross the country in a fraction of the time required by train. Speed therefore became one of the principal measures by which new airliners were judged. Because passengers were paying premium fares, comfort mattered as well. Safety, however, remained paramount. Commercial aviation was still in its infancy, accidents were relatively common, and fear of flying remained widespread.

Douglas' DC-3 established new standards in all three areas. It was faster, safer, and large enough to carry as many as 30 passengers in relative comfort. Airlines even offered sleeper versions for overnight flights. Because airplanes competed directly with trains, manufacturers sought to reproduce many of the amenities available aboard the finest railcars. By the eve of World War II, the DC-3 had become the standard against which every competing airliner was measured — and it was an extraordinarily high standard.

Boeing had no intention of conceding the commercial market to Douglas. Its engineers studied the principal obstacles to wider air travel and considered how a new design might overcome them. Yet the company was also drawn to the more immediate profitability of government work. Military contracts had become almost narcotic: a ready source of revenue that could satisfy the company's financial needs while reducing the risks of the commercial market. And the Army Air Corps was already searching for an airplane to meet its next anticipated requirement.

The Winds of War Lift the Wings of Commercial Aviation

Aviation had made remarkable strides in the years following World War I. By 1933 — only 30 years after the Wright brothers' first powered flight — airplanes had evolved far beyond the flimsy, fabric-covered wooden machines of the early 20th century. The largest commercial and military aircraft were now built primarily of aluminum and could carry substantial loads over increasingly long distances with far greater reliability and safety. The U.S. Army Air Corps was eager to capitalize on these advances and, during the 15 years following the war, ordered progressively larger, faster, and more capable aircraft.

In 1933, the same year Boeing introduced the Model 247, the Army quietly asked the company to study the feasibility of an exceptionally large bomber capable of carrying 2,000 pounds of bombs over a distance of 5,000 miles. Boeing responded with the Model 294, which received the Army designation XB-15.

The XB-15 was enormous by the standards of its time. Its immense wing carried four engines and was thick enough to contain internal passageways through which crew members could crawl to perform in-flight engine maintenance. Designed beginning in 1934 and first flown in 1937, it was then the largest airplane ever built in the United States.

Only one XB-15 was produced, and its performance fell short of some of the Army's original ambitions. Nevertheless, it succeeded as a proof of concept, demonstrating that an aircraft of such unprecedented size and range could be designed and built. The project further established Boeing as more than an innovative producer of commercial transports; it was also capable of engineering the large strategic bombers the Army envisioned for the future.

The XB-15 strengthened Boeing's standing within the War Department at a moment when the government's need for military aircraft was about to rise dramatically. In Boeing's early years, the company had often approached Washington hat in hand, searching for contracts to keep its factories operating. With the XB-15, the relationship appeared to be changing: now the government was turning to Boeing to determine what was possible.

***

By 1934, the world remained mired in economic depression while political divisions threatened the international order established after World War I. The United States' geography imposed a unique requirement on its military planners. Separated from Europe and Asia by vast oceans, the country would need aircraft with exceptional range — or endurance, in aviation parlance — to fight another major overseas war. Airpower had already demonstrated its growing importance, but carrying a war to a distant enemy would require airplanes capable of traversing immense distances.

While Boeing was still developing the XB-15, the Army Air Corps issued a tender for a new bomber that could deliver a powerful long-range punch in any future conflict. The pace of aviation development was evident in the fact that the proposed aircraft was intended to replace the Martin B-10, which had only begun entering service that same year. The B-10 was itself revolutionary, but the Army wanted to build upon its advances with a larger, faster, and more capable bomber. Boeing's experience designing large aircraft — and particularly its work on the XB-15 — made the competition a natural opportunity for the company.

Unlike the experimental XB-15, which had been conceived largely to demonstrate what was technically possible, the new bomber was intended for production and frontline service. The Army envisioned deploying it from bases in Alaska, Guam, and Panama, locations from which American interests could be defended only by crossing vast stretches of ocean. Although the tender left many details unspecified, it was clear that the Army wanted an aircraft larger and more capable than anything then in service.

Claire Egtvedt and Boeing's engineers initially believed the Army expected an evolutionary improvement over the twin-engine B-10. Boeing, however, had already gained valuable experience from the XB-15 and saw no reason to confine itself to such a modest advance. Rather than merely refine an existing bomber, its engineers decided to redesign the XB-15 concept from nose to tail, producing a smaller, faster, and more practical aircraft. The result was the Model 299.

The Army required the new bomber to carry at least 2,000 pounds of bombs over what it described only as "a great distance," fly faster than 200 miles per hour, and operate above 10,000 feet. Boeing designed the Model 299 to exceed every one of those requirements.

The competition culminated in a fly-off among three contenders: Martin's improved version of the B-10, Douglas' new DB-1, and Boeing's Model 299. Martin and Douglas submitted the type of aircraft the Army appeared to expect — larger and more capable twin-engine bombers descended conceptually from the B-10. Boeing arrived with something altogether different: an imposing four-engine aircraft derived from the lessons of the XB-15.

The Model 299 surpassed the Army's requirements in every major category. It could fly at more than 250 miles per hour, travel over 2,000 miles, climb above 25,000 feet, and carry more than 4,000 pounds of bombs — more than twice the minimum specified. Boeing had not merely improved upon the existing generation of bombers. It had produced an aircraft that rendered them nearly obsolete.

The competing aircraft were flown to Wright Field near Dayton, Ohio, for evaluation. Boeing's Model 299 immediately became the center of attention and appeared to be the clear front-runner. Its most consequential feature was its enormous wing, which provided both the lift needed to carry a heavy bomb load and the internal volume required for the fuel that gave the aircraft its exceptional range. The Model 299 also bristled with defensive armament, with five .30-caliber machine guns positioned around its fuselage. Because contemporary fighters lacked the range to escort bombers deep into enemy territory, a long-range bomber had to defend itself. After seeing the aircraft, a reporter for the Seattle Times described it as a "flying fortress." The name stuck.

The Model 299 also incorporated gust locks, devices that secured its control surfaces while it was parked. The aircraft's immense wings and tail required correspondingly large ailerons, elevators, and rudder, which could be violently battered by strong winds on the ground. Locking them in place protected both the control surfaces and the surrounding aircraft structure.

Those gust locks would prove disastrous. On October 30, 1935, during a later evaluation flight at Wright Field, the pilots failed to disengage them before takeoff. The Model 299 lifted from the runway, climbed steeply, stalled, and crashed. Boeing chief test pilot Leslie Tower and Army pilot Major Ployer Hill, who was flying the aircraft, were killed; several others aboard survived with serious injuries.

Because Boeing had built only one Model 299, the crash ended its participation in the competition. Douglas won the contract with its twin-engine DB-1, which entered Army service as the B-18 Bolo. Its lower cost also weighed heavily in its favor, and the Army ultimately ordered more than 300 of them.

Yet the Army remained convinced that Boeing's four-engine design represented the future of long-range bombing. Because the crash had resulted from pilot error rather than an inherent flaw in the aircraft, the Air Corps found a way to keep the design alive. Using funds available for experimental procurement, it ordered 13 improved aircraft, designated YB-17s, for service evaluation. The "Y" identified them as test aircraft rather than fully operational production models.

A fourteenth aircraft, designated Y1B-17A, was fitted with turbo-supercharged engines. The modification dramatically improved its high-altitude performance, allowing it to exceed 300 miles per hour and reach approximately 38,000 feet. It demonstrated that the basic design possessed far greater potential than even the original Model 299 had revealed.

The small YB-17 order kept Boeing in the bomber business and preserved a design that would eventually become one of the most important military aircraft in history. It also gave the company a technological foundation it could apply to the commercial market, where it was still searching for an answer to Douglas' dominant DC-3. Boeing would adapt much of the B-17's engineering into its next commercial transport, the pressurized Model 307 Stratoliner.

***

Weather was one of the greatest obstacles to early air travel — and remains a factor even today. Modern jetliners can fly above or through most weather systems, although severe conditions still cause delays and cancellations. In the 1930s, however, even relatively mild weather could disrupt an entire route. The greatest challenge was often not the weather at the departure or destination airport, but the storms, clouds, and turbulence encountered along the way.

Airliners frequently had to make lengthy detours around weather systems. Because their cabins were not pressurized, they generally remained below 10,000 feet, above which the thinning air could cause passengers and crew to suffer from oxygen deprivation. Flying at these lower altitudes also exposed aircraft to more turbulence generated by uneven terrain, surface heating, and winds disturbed by mountains and other obstacles. One apparent solution was to fly higher, above much of the weather and turbulent air.

Aircraft were already reaching extraordinary altitudes. By 1935, the world altitude record was approaching 50,000 feet. But such flights were isolated feats performed by specially equipped pilots wearing pressure suits or using supplemental oxygen. The challenge was to make high-altitude flight routine and to carry ordinary passengers safely and comfortably at 20,000 feet or higher. Claire Egtvedt directed Boeing's engineers to find a practical solution.

Their answer was to turn the airplane's fuselage into a pressure vessel. Air compressed from the power of the engines would maintain a comfortable cabin atmosphere even as the aircraft climbed into the thin air of the upper atmosphere. But pressurization imposed entirely new structural demands. The fuselage had to withstand the outward force created by the difference between the cabin pressure and the surrounding atmosphere. It also had to endure repeated cycles of pressurization during flight and depressurization after landing without developing cracks, loose rivets, or weakened joints.

The result was the Model 307 Stratoliner, the first commercial airliner designed with a pressurized cabin. Its broad, nearly circular fuselage gave it the appearance of a giant metal cigar, a shape that helped distribute the stresses created by internal pressure. Boeing then married that new fuselage to wings, engines, landing gear, and tail surfaces derived largely from the YB-17. The result combined the structural strength and high-altitude capability of a bomber with the comfort and accommodations of a commercial transport.

The Model 307, named the Stratoliner for its high-altitude capabilities, represented another major aviation milestone produced under Claire Egtvedt's leadership. It could carry as many as 33 passengers in a pressurized cabin and operate at altitudes approaching 25,000 feet, allowing it to fly above much of the turbulence and weather that disrupted lower-flying airliners. Once again, Boeing had established a new standard in commercial aviation. The Stratoliner was among the most advanced passenger aircraft of its era and the first pressurized airliner to enter commercial service.

Its timing, however, could hardly have been worse. The Stratoliner entered service in 1940, just as World War II was consuming the aircraft industry's production capacity. Boeing built only ten before military requirements brought the program to an end.

The Model 307 also illustrated how Boeing's government work strengthened its commercial business. Military contracts financed the development of large, technologically ambitious aircraft and gave Boeing's engineers experience that could then be applied to commercial designs. The company used the knowledge gained from the XB-15 and B-17 programs to produce civilian airplanes that were in some respects more advanced than those of its competitors. The Stratoliner was not the only example.

Boeing also drew heavily upon its XB-15 experience in developing the Model 314, the enormous flying boat built for transoceanic passenger service. Its immense wing was derived from work on the XB-15 and provided both the lift and fuel capacity needed to cross vast stretches of ocean. The Model 314 could carry as many as 68 passengers, although considerably fewer were accommodated on overnight flights so that sleeping berths could be provided.

A journey such as San Francisco to Honolulu could keep passengers aboard for more than 20 hours. Because long-distance air travel remained largely the preserve of the wealthy, the 314 was designed to offer the trappings of an ocean liner. Passengers were served elaborate meals prepared by trained chefs, slept in convertible berths, and had access to separate dressing rooms for men and women. The aircraft's spacious cabin made the journey itself part of the experience rather than merely a means of reaching a destination.

World War II also curtailed the Model 314's career. Only 12 were built. Nine initially entered service with Pan American Airways, while three were later transferred to Britain's BOAC.

Despite its brief production run, the 314 again demonstrated Boeing's ability to design aircraft of exceptional size, range, and technical sophistication.

By the end of its third decade, Boeing's military and commercial businesses had become closely intertwined. Government contracts supplied the money and technical challenges that expanded the company's engineering and manufacturing capabilities. Boeing then carried much of that expertise into its commercial designs. In turn, the government became increasingly dependent upon Boeing to produce the large, long-range aircraft required by the Army Air Corps. The relationship was becoming symbiotic: government work strengthened Boeing, while Boeing's expanding capabilities made it increasingly valuable to the government.

Yet technological leadership did not translate into commercial dominance. Boeing had failed to regain the foothold it once held with the Model 40 series, and neither the Stratoliner nor the Model 314 was produced in sufficient numbers to challenge Douglas in the land-based airliner market. During the latter half of the 1930s, military contracts became increasingly important to Boeing's financial health. World War II would transform that dependence into an unprecedented expansion, making Boeing one of the largest and most strategically important suppliers to the United States government.

The Generosity of War

Philip G. Johnson was born in Seattle in 1894, the son of Swedish immigrants. Mechanically inclined from an early age, he attended the University of Washington and earned a degree in mechanical engineering. In 1917, while Johnson was completing his senior year, William Boeing recruited him as a draftsman for his fledgling airplane company. As Boeing grew, Johnson rose rapidly through its ranks, becoming company president in 1926 at only 31 years old, while Boeing moved into the chairman's role.

When Boeing and several other aviation companies were consolidated into United Aircraft and Transport Corporation in 1929, Johnson became one of its senior executives and eventually assumed leadership of its airline operations. His departure from the company came amid the airmail scandal of 1934, when the Roosevelt administration canceled existing airmail contracts and Congress prohibited executives associated with the disputed awards from participating in the reorganized airline industry. Johnson was barred without a trial or any finding that he had committed a crime.

Unable to work in the American airline industry, Johnson moved to Canada in 1937 and helped organize Trans-Canada Air Lines, serving as its vice president of operations. Two years later, after the restrictions against him were rescinded, he returned to Seattle and resumed the presidency of Boeing. His return came at a pivotal moment, both for the company and for the world.

Germany's invasion of Poland on September 1, 1939, ignited World War II in Europe. The United States was still recovering from the Great Depression, and most Americans wanted no part in another European war. Protected by two oceans and haunted by the losses of World War I, the country remained deeply isolationist.

President Franklin D. Roosevelt and members of his administration nevertheless understood that the conflict could not easily be contained within Europe. Germany threatened Britain, France, and the broader balance of power upon which American security and trade depended. Roosevelt could not yet lead the United States directly into the war, but he could begin preparing the country for the possibility that it would eventually be drawn in. For Boeing — and for Philip Johnson — the coming mobilization would transform everything.

Even before war erupted in Europe, Washington was watching events across the Pacific with growing alarm. Japan had annexed Korea in 1910, occupied Manchuria in 1931, and launched a full-scale invasion of China in 1937. Its leaders sought to establish an empire throughout East Asia, in part to secure access to the oil, minerals, and other natural resources that Japan's industries and military required but the home islands lacked.

As Japanese forces advanced deeper into China, the United States began strengthening its defenses in the Pacific. In 1939, the Army Air Corps resumed its purchases of Boeing's bomber with an order for 39 B-17Bs. Their turbo-supercharged engines substantially improved high-altitude performance, making the aircraft faster and more capable than the experimental versions that preceded them. The bombers were assigned to units on both coasts, reflecting their envisioned role in defending the continental United States and its distant Pacific possessions.

That same year, Philip Johnson returned to Boeing as president after the restrictions imposed upon him during the airmail scandal were lifted. Claire Egtvedt moved into the chairman's role and entrusted Johnson with preparing the company for the production expansion both men believed was coming. Johnson was renowned for his organizational and production-planning abilities — precisely the skills Boeing would need if the United States began ordering military aircraft on an unprecedented scale.

By 1940, Germany had overrun much of Western Europe, while Japan had begun moving into French Indochina. President Roosevelt called upon American industry to produce 50,000 military aircraft a year, and in July the Army placed its first major heavy-bomber production order: 512 of the substantially redesigned B-17E. (Between the B-17B and the large-scale B-17E order, Boeing produced small numbers of B-17C and B-17D models, progressively adding armor, self-sealing fuel tanks, improved engine cooling, and heavier defensive armament. These transitional versions provided valuable operational and combat experience that shaped the substantially redesigned B-17E.)

The irony could hardly have been lost on Boeing. Five years earlier, the crash of the Model 299 had eliminated the company from a competition whose winner initially received an order for approximately 200 bombers. Now the Army was ordering more than twice that number of Boeing's aircraft in a single procurement. The airplane that had lost the competition was rapidly becoming indispensable to the nation's defense.

Although Boeing's B-17 was among the world's most capable bombers, its range would be inadequate for sustained operations across the immense distances of the Pacific. Anticipating that the Army Air Corps would eventually require something larger, faster, and longer-ranged, Boeing began privately studying a successor in 1938 without waiting for a government contract. Using its own funds, its engineers initially explored an extensively modified B-17 equipped with more powerful engines, a pressurized cabin, and tricycle landing gear. The concept, designated the Model 334, never advanced beyond the study stage, but it laid the foundation for what followed.

In January 1940, the Army Air Corps invited manufacturers to submit proposals for a new class of "very heavy" bomber. Its extraordinarily ambitious objectives included a speed approaching 400 miles per hour, a combat radius of roughly 2,500 miles, and the capacity to carry as much as 20,000 pounds of bombs. Douglas, Lockheed, Consolidated Aircraft, and Boeing submitted proposals.

Douglas and Lockheed eventually withdrew, leaving Boeing's Model 345 and Consolidated's Model 33 as the principal contenders.

Boeing entered the competition with an important advantage. Its engineers had already spent nearly two years studying the problems involved in producing a bomber substantially more capable than the B-17. That privately financed work gave the company a head start, but the Army did not immediately declare Boeing the sole winner. In September 1940, it ordered two experimental aircraft from Boeing, designated XB-29s, while also ordering prototypes of Consolidated's competing design, which became the XB-32. Consolidated's aircraft would remain under development as insurance in case Boeing's far more ambitious design failed.

The XB-29 represented an extraordinary technological leap. It would be larger and faster than the B-17, carry a far heavier bomb load, and fly much farther. Its design incorporated pressurized crew compartments, remote-controlled gun turrets, powerful new engines, and numerous electrical and mechanical systems of unprecedented complexity. The Army understood that developing such an aircraft would entail enormous technical risk, but it also recognized that no existing bomber could meet the demands of a truly global war.

In May 1941 — more than a year before the first XB-29 flew — the Army ordered 250 production aircraft. It was a remarkable gamble: Boeing was instructed to prepare the bomber for mass production while its design was still being refined and before a prototype had even left the ground. Following the Japanese attack on Pearl Harbor on December 7, 1941, the urgency increased dramatically. In January 1942, the order was doubled to 500 aircraft.

The new bomber became known as the B-29 Superfortress, a name that proclaimed its superiority over the B-17 Flying Fortress. In speed, range, altitude, defensive firepower, and bomb-carrying capacity, it promised to surpass every American bomber that had preceded it. Although the precise bases from which it would eventually operate had not yet been determined, its strategic purpose was clear: the United States needed an aircraft capable of crossing the vast reaches of the Pacific and striking the Japanese home islands. The B-29 was being created for that mission.

The Army Air Corps had no intention of replacing the B-17 with the new bomber. The B-17 was well suited to the European theater, where it could fly from bases in England to targets in Germany and occupied Europe, or from North Africa and Italy against targets in Southern and Central Europe. The B-29, by contrast, was designed for the immense distances of the Pacific, where reaching the Japanese home islands would require far greater range.

Even the B-29 could not strike Japan directly from the continental United States. American forces therefore needed bases progressively closer to the Japanese homeland. The island-hopping campaigns conducted by the Army, Marines, and Navy served several purposes: bypassing heavily defended Japanese strongholds, securing sea and air routes, and establishing bases from which further advances could be supported. The capture of the Mariana Islands in 1944 proved especially important because airfields on Saipan, Tinian, and Guam placed Tokyo within the B-29's operational reach.

The bomber's speed, range, and service ceiling above 30,000 feet made it exceptionally well suited to the mission. At its intended operating altitude, many Japanese fighters struggled to reach it, while those that could often had little time to intercept before exhausting their fuel or performance advantage. High altitude did not make the B-29 invulnerable, but it complicated the task of defending against it.

The B-29 also incorporated technologies far beyond those found in the B-17. Its crew occupied heated and pressurized compartments, allowing them to fly long missions without remaining continuously exposed to extreme cold and thin air. Instead of placing gunners directly in most of its turrets, Boeing developed a centralized fire-control system in which gunners aimed remotely through optical sights. An analog computer calculated corrections for range, airspeed, temperature, and the movement of the attacking fighter. Through a system of switches, a gunner could direct more than one turret against the same target. The compact, unmanned turrets also produced less aerodynamic drag than conventional manned installations.

The aircraft's technological sophistication came at a tremendous cost. Its Wright R-3350 engines were powerful but immature, with chronic cooling and reliability problems that caused repeated failures and fires. The remote gun system, pressurization equipment, electrical network, and other systems were vastly more complex than anything Boeing had previously placed into production. The decision to manufacture the bomber while it was still being developed magnified every defect, forcing engineers to redesign components even as factories were assembling operational aircraft.

The dangers became starkly apparent on February 18, 1943, when an engine fire broke out aboard the second XB-29 during a test flight near Seattle. The aircraft crashed into a meatpacking plant, killing Boeing chief test pilot Edmund "Eddie" Allen, all ten other men aboard, and 19 people on the ground. The disaster exposed the severity of the engine problems but did not slow the program for long. The military considered the B-29 too important to the emerging Pacific strategy to permit a conventional development schedule.

The program continued at extraordinary speed despite its technical problems. The B-29 offered the United States something no other operational bomber could provide — the ability to carry a heavy bomb load across the vast distances separating newly captured Pacific bases from the Japanese home islands. It became the centerpiece of the Army Air Forces' strategic bombing campaign against Japan, first from bases in China and later, on a far larger scale, from the Marianas.

Development and production of the B-29 ultimately cost approximately $3 billion, making it the most expensive American weapons program of World War II. It exceeded even the roughly $2.2 billion spent on the Manhattan Project. In a final convergence of the two enormous undertakings, two specially modified B-29s carried the atomic bombs dropped on Hiroshima and Nagasaki in August 1945. The bomber built to overcome the distances of the Pacific ultimately delivered the weapons that helped bring the war to its conclusion.

The B-29's staggering cost reflected the complexity of its technology and the unprecedented manufacturing effort required to place it into combat. Philip Johnson was the man responsible for leading that effort. He worked relentlessly throughout the war, traveling among Boeing's plants and the factories of its suppliers to resolve production problems, eliminate bottlenecks, and accelerate the flow of aircraft to the front.

Johnson possessed an exceptional talent for organization and production planning. At a time when Boeing was expanding faster than ever before, he helped transform the company from a relatively modest airplane manufacturer into the center of a nationwide production network. His efforts, however, exacted a severe personal toll. On September 14, 1944, while overseeing operations at Boeing's Wichita plant, the 49-year-old Johnson suffered a cerebral hemorrhage and died.

His sudden death was felt throughout Boeing and across the wartime aviation industry and the country. Boeing workers, military generals, and even the President of the United States paid tribute to his contributions to the war effort. Johnson had guided the company through the most dramatic expansion in its history and had become one of the country's most important industrial leaders. Claire Egtvedt temporarily reassumed the duties of president while remaining chairman, serving until a new leader could be found.

Even before the United States entered the war, military orders had begun transforming Boeing's business. Wartime demand then expanded it on an almost unimaginable scale. By the end of production, 12,731 B-17s had been built by Boeing and its manufacturing partners, Douglas and Lockheed-Vega. Another 3,970 B-29s were produced by Boeing, Bell, and the Glenn L. Martin Company. Aircraft that had once been ordered by the dozen were now emerging from American factories by the thousands.

World War II did more than provide Boeing with a temporary surge in revenue. It transformed the company's workforce, factories, production methods, and relationship with the federal government. By the war's end, Boeing was no longer merely an innovative airplane manufacturer. It had become one of the indispensable pillars of America's military-industrial capacity.

The War Drums of a Fragile Peace

With the end of World War II in 1945, Boeing's torrent of government revenue began to recede almost as quickly as it had risen. Production of the B-17 ended in May, shortly after Germany's surrender rendered the bomber's principal mission in Europe unnecessary. B-29 production continued, however, because the war against Japan appeared likely to last well into the following year.

American planners were preparing for Operation Downfall, the proposed invasion of the Japanese home islands. Its first phase, the invasion of Kyushu, was scheduled for November 1945, with a larger assault against Honshu to follow in the spring of 1946. Until Japan could be forced to surrender, the United States expected to need a continuing supply of B-29s for the strategic bombing campaign and for supporting any eventual invasion.

Then came the dawn of the nuclear age. On August 6, 1945, the specially modified B-29 Enola Gay dropped an atomic bomb on Hiroshima. Three days later, another B-29, Bockscar, dropped a second bomb on Nagasaki. Japan announced its acceptance of the Allied surrender terms on August 15, bringing the war to an abrupt conclusion and forever linking the B-29 to the opening of the nuclear era.

The first atomic bombs were exceptionally large and heavy, and only specially modified B-29s could carry them in combat. Those aircraft, developed under the Silverplate program, were stripped of unnecessary equipment and extensively altered to accommodate the new weapons. The B-29 had therefore demonstrated that it could perform the nuclear mission, although its range, reliability, and payload remained limited compared with what military planners would soon require.

Even before the war ended, Boeing had begun developing a substantially improved version of the Superfortress. Initially designated the B-29D, it incorporated more powerful Pratt & Whitney R-4360 engines, a strengthened and lighter airframe, a larger vertical tail, and numerous systems improvements. Because the changes were so extensive the airplane was redesignated the B-50.

The B-50 first flew in June 1947 and entered service with the newly independent United States Air Force, which had been established that September. It became an important early component of America's postwar strategic-bomber force and was capable of carrying both conventional and nuclear weapons. Yet it remained fundamentally an evolutionary development of the B-29 rather than the wholly new intercontinental bomber the Air Force ultimately sought.

Only a few hundred B-50s were built, a fraction of the thousands of B-17s and B-29s produced during the war. The decline reflected the abrupt contraction of military demand after Japan's surrender. Contracts were canceled, production lines slowed or closed, and tens of thousands of aircraft workers across the country faced layoffs. Boeing now had to adjust from the demands of total war to the uncertainties of peace.

Unlike the devastated industrial powers of Europe and Asia, the United States emerged from the war with its factories intact, its industrial capacity greatly enlarged, and its economy poised for expansion. Boeing began turning its attention once again toward peacetime opportunities, even as it sought to preserve the military business and technical expertise it had accumulated during the war.

Claire Egtvedt had temporarily reassumed the duties of president after Philip Johnson's sudden death in 1944, but he had no intention of holding the position indefinitely. By 1945, he was searching for a permanent successor. The choice was not obvious. Many of Boeing's most capable engineers were still comparatively young and lacked the business experience required to manage a company of Boeing's new size and complexity. Most of the veterans who had helped build the company during its formative years were approaching retirement.

Egtvedt faced one of the most consequential decisions of his career. Boeing needed more than an accomplished engineer. It needed a leader capable of guiding an organization built for wartime mass production into an uncertain postwar market.

***

William McPherson "Bill" Allen, born in 1900, was an unlikely choice to lead one of America's foremost engineering companies. A graduate of Harvard Law School, he had begun his career with a prominent Seattle law firm rather than in an aircraft factory. His association with Boeing began in 1930, when he joined the company's board of directors and became its legal counsel while continuing to practice law.

Over the next 15 years, Allen became deeply involved in Boeing's affairs and gained a close-up view of the decisions that guided the company through economic depression, political upheaval, and war. He worked alongside Claire Egtvedt and Philip Johnson as they directed the development and production of some of the world's most advanced aircraft. Although Allen had never designed an airplane, he understood the complete life cycle of an aircraft program — from the enormous financial commitment required at its inception to the engineering, production, and market risks that followed.

Allen was intelligent, experienced, and intimately familiar with the business of aircraft manufacturing. He also possessed the judgment and leadership skills needed to manage a company that had grown far beyond its roots as a small collection of engineers and craftsmen. Boeing still depended upon engineering excellence, but its wartime expansion had demonstrated that technical expertise alone was no longer sufficient to run an enterprise of such size and complexity.

Allen was initially reluctant to accept the presidency, and Egtvedt reportedly had to approach him several times before he agreed. On September 1, 1945, Allen became president of the Boeing Airplane Company.

Neither Allen nor those who selected him could have known what lay ahead. Yet in choosing a lawyer rather than an engineer, Boeing had found the leader who would guide it through the uncertain transition from war to peace, take it into the jet age, and secure its place at the forefront of global aviation.

The technology Boeing had developed for the government before and during the war — aircraft capable of carrying heavy payloads over vast distances — could also be applied to commercial aviation. With peace came enormous production capacity, a greatly expanded workforce, and a wealth of engineering and manufacturing expertise. One of Bill Allen's first major challenges was to harness those resources to meet the public's pent-up demand for comfortable, long-distance air travel.

Boeing's Model 307 Stratoliner, introduced shortly before the war, represented the point at which the company's commercial-airliner development had effectively paused. It had been the first pressurized passenger airplane and one of the most advanced transports of its time. Yet five years of intensive wartime development had rapidly overtaken it. Advances in engines, aerodynamics, pressurization, structures, and large-scale manufacturing had made the Stratoliner appear antiquated almost overnight. Boeing needed a new commercial aircraft that embodied everything it had learned during the war.

A promising foundation already existed. In 1942, Boeing had begun developing a military transport derived from the B-29. Designated the C-97 Stratofreighter, it retained the bomber's wings, engines, landing gear, and tail but replaced its narrow fuselage with a much larger cargo body. The new fuselage consisted of two overlapping circular sections, producing the distinctive "double-bubble" cross section that resembled an inverted figure eight. This arrangement created a broad upper deck for cargo while preserving a narrower lower section beneath it.

The first XC-97 prototype flew in November 1944, too late to make a significant contribution to World War II. Nevertheless, the design demonstrated considerable promise as a long-range military transport, and the Army Air Forces continued its development after the war. The C-97 could carry large loads across intercontinental distances and offered the spacious internal volume needed to support America's growing worldwide military commitments.

As Boeing searched for a postwar commercial airliner, its engineers recognized that the C-97 offered an obvious starting point. Its cavernous fuselage could be fitted with a pressurized passenger cabin, sleeping berths, dining areas, and other accommodations suitable for long-distance travel. Its B-29-derived wings and systems had already benefited from years of intensive military development. By adapting this proven military technology for civilian service, Boeing could reduce development time and risk while producing an airplane substantially larger and more capable than the prewar Stratoliner.

From that combination of wartime engineering and postwar ambition emerged the Model 377 Stratocruiser. It would become Boeing's attempt to translate the range, size, and technological sophistication of the B-29 and C-97 into a luxurious commercial transport for the new era of global air travel.

The Model 377 represented the pinnacle of commercial aviation in the immediate postwar era. Its unusually broad upper cabin made it a forerunner of the modern widebody airliner, offering passengers substantially more space than most contemporary transports. Even more remarkable was its lower-deck lounge, located behind the wing and reached by a circular staircase. Passengers could leave the main cabin, descend to a cocktail lounge, and socialize in a setting unlike anything previously offered aboard a land-based airliner. The arrangement anticipated the distinctive upper-deck lounges Boeing would introduce on the 747 more than two decades later.

One of Bill Allen's first major decisions as Boeing's president was also one of his boldest. With Boeing's wartime contracts disappearing, he authorized the construction of 50 Stratocruisers even though no airline had yet placed a firm order. The decision committed Boeing's own resources to an aircraft whose market remained unproven, but Allen believed that the company needed a commercial product to keep its factories operating and preserve the capabilities it had developed during the war.

His gamble soon received a crucial endorsement from Juan Trippe, the president of Pan American World Airways. Trippe already knew Boeing from Pan Am's successful operation of the Model 314 flying boats and recognized the potential of the new aircraft for luxurious, long-distance service. In November 1945, Pan Am ordered 20 Model 377s for $24.5 million — then described as the largest commercial-aircraft order ever placed.

Orders followed from Northwest Airlines, British Overseas Airways Corporation, and United Airlines. For a moment, Allen's judgment appeared fully vindicated. He had correctly anticipated the enormous pent-up demand for air travel following the war and had positioned Boeing to serve the rapidly expanding international market. Northwest, for example, used the Stratocruiser on both trans-Pacific and domestic routes, taking advantage of its large capacity, pressurized cabin, sleeping berths, and lower-deck lounge.

Boeing was not alone, however, in recognizing the opportunity. Douglas and Lockheed had also emerged from the war with advanced aircraft, enlarged factories, and experienced engineering teams. Both were preparing commercial transports of their own, setting the stage for an intense struggle to dominate the postwar airline market.

The Model 377 faced formidable competition from Douglas and Lockheed, both of which brought new four-engine, pressurized airliners to market at approximately the same time. Like Boeing, Douglas had gained valuable experience developing military transports during the war and applied that knowledge to the commercial market. Lockheed followed a somewhat different path. It had originally designed the Constellation for Trans World Airlines before the war, but the military took over early production and operated the aircraft as the C-69 transport. When peace returned, Lockheed resumed development of the civilian version.

Douglas' answer was the DC-6, a pressurized four-engine transport derived from the wartime XC-112A. It lacked the Stratocruiser's cavernous double-deck fuselage and luxurious lower-deck lounge, but it was a capable, dependable airplane that proved adaptable to a wide range of routes. Lockheed offered the graceful L-049 Constellation, commonly known as the "Connie," and subsequently developed a succession of larger, faster, and longer-ranged variants. Neither competitor could match the Stratocruiser's interior volume or distinctive accommodations, but both offered airlines a more practical balance of performance, capacity, cost, and reliability.

The market's verdict was decisive. Douglas ultimately produced 704 aircraft in the DC-6 family between 1946 and 1958. Lockheed built 856 Constellations and their derivatives, including military versions, over approximately the same period. Boeing, by contrast, produced only 56 Stratocruisers before production ended in 1950. An airplane that appeared to represent the pinnacle of postwar luxury had been overwhelmingly outsold by its less extravagant competitors. What had gone wrong?

The Stratocruiser's advantages came at a considerable price. Its size, complexity, and elaborate passenger accommodations made it more expensive to purchase than competing aircraft. It was also costly to operate, requiring large flight crews, substantial maintenance, and enormous quantities of fuel. Its economics worked best on long international routes where airlines could charge premium fares and make full use of its range, sleeping berths, and luxurious cabin.

On shorter routes, the aircraft was far more difficult to operate profitably. Northwest Orient managed to employ its Stratocruisers on some domestic services, but revenue from carrying mail helped offset their high operating costs. United Airlines fared less successfully and disposed of its small fleet after only a few years, transferring the aircraft to British Overseas Airways Corporation.

At the heart of many of the Stratocruiser's difficulties were its four Pratt & Whitney R-4360 Wasp Major engines. Among the largest and most powerful piston engines ever placed into mass production, each contained 28 cylinders arranged in four radial rows and produced approximately 3,500 horsepower. That power was necessary because the Stratocruiser was substantially larger and heavier than the B-29 from which its basic architecture had descended.

The R-4360 engines gave the Stratocruiser impressive performance, but they were extraordinarily complex and maintenance-intensive. Its tightly packed rear cylinders were difficult to cool and service, while failures involving cylinders, propellers, and associated systems contributed to delays, costly repairs, and several serious accidents. The same sophistication that made the Stratocruiser an engineering marvel also made it an operational burden. The result was that the Model 377 experienced a high rate of in-flight emergencies. It also experienced a high accident rate — of the 56 copies that were produced, 13 of them, over 20 percent of the entire fleet, were lost to accidents over the years.

Douglas and Lockheed had pursued a more commercially successful balance. Their airplanes were less spacious and less opulent, but they were easier for airlines to integrate into existing route networks and could earn money across a broader range of missions. Boeing had built the most impressive passenger airplane of the three. Douglas and Lockheed had built the airplanes airlines could more readily afford to operate.

Ultimately, Bill Allen's gamble on the Model 377 did not pay off — at least not in the commercial market. Yet the aircraft's military predecessor, the C-97 Stratofreighter, continued to serve with the newly established United States Air Force. Although relatively few were initially ordered as transports, the advent of long-range jet bombers soon gave the basic design a far more important second life.

The Air Force's new jet bombers consumed fuel at prodigious rates, and their range could be extended only through aerial refueling. Boeing adapted the C-97 into the KC-97, fitting it with additional fuel tanks and a retractable flying boom through which fuel could be transferred to another aircraft in flight. The Air Force ultimately ordered 811 KC-97 tankers — far more than the number of C-97 cargo transports or civilian Stratocruisers produced.

The irony was striking. The commercial airplane that had failed to challenge Douglas and Lockheed became the foundation for one of the most important military support aircraft of the early Cold War. The KC-97 enabled the Air Force's growing fleet of gas-guzzling jet bombers to remain aloft longer and strike targets far beyond their normal operating range. And many of those bombers would also come from Boeing.

The KC-97 was not an ideal match for the jet aircraft it supported. Its piston engines made it considerably slower than the bombers it refueled, sometimes forcing the receiving aircraft to descend and reduce speed to remain connected. Nevertheless, until a purpose-built jet tanker became available, the KC-97 provided the Air Force with the global reach demanded by its emerging Cold War strategy. In that sense, the underlying design proved far more consequential in military service than it ever had in the commercial marketplace.

***

To preserve a credible means of delivering its growing arsenal of nuclear weapons, the Air Force continued developing Boeing's B-50. Yet even that substantially improved Superfortress remained an interim solution. The first atomic bombs were immense: the weapons dropped on Hiroshima and Nagasaki each weighed roughly 10,000 pounds. Carrying such a load consumed much of the B-29's — and even the B-50's — available payload and sharply reduced the distance it could fly. The Air Force needed a bomber with true intercontinental range.

It had already been working with Consolidated Aircraft, later renamed Convair, on an airplane designed for precisely that purpose. The resulting B-36 Peacemaker became the ultimate expression of the piston-powered strategic bomber. The B-36 dwarfed nearly every aircraft that had preceded it. Its wingspan measured 230 feet — greater than that of the later Boeing 747 — and its six Pratt & Whitney R-4360 engines were mounted in an unusual pusher configuration, with the propellers behind the wings. The enormous aircraft could carry a maximum bomb load approaching 86,000 pounds and possessed sufficient range to strike targets on another continent without relying on overseas bases. Under favorable conditions, its maximum range approached 10,000 miles, although its operational radius with a significant bomb load was shorter.

The B-36 became the principal nuclear-delivery aircraft of the Air Force's newly established Strategic Air Command. It was the first operational American bomber capable of carrying the earliest nuclear weapons from bases in North America to targets deep inside the Soviet Union. Its immense size and capabilities placed it in a class of its own. With the introduction of the B-36, the Air Force reclassified the smaller B-29 and B-50 as medium bombers, even though both had been regarded as heavy bombers during and immediately after World War II.

Piston engines, however, were entering their twilight as powerplants for strategic aircraft. World War II had demonstrated the potential of jet propulsion, and the Air Force was determined to exploit the new technology. Early jet engines offered far greater speed than piston engines, but they consumed fuel at prodigious rates and initially lacked the efficiency required for intercontinental missions. The B-36 therefore represented a bridge between two eras. It possessed the range of a giant propeller-driven aircraft while gradually incorporating jet power to improve its performance. Beginning with later versions, four General Electric jet engines were added in twin pods near the wingtips, supplementing the six piston engines and giving rise to the expression "six turning and four burning." The jets provided additional thrust during takeoff and allowed the bomber to climb faster or make a high-speed dash through defended airspace, or to climb rapidly over its target before dropping its bomb load. With all ten engines operating, later B-36 variants could exceed 430 miles per hour.

The additional jet engines improved the B-36's performance, but they also underscored the aircraft's transitional nature. The future belonged to an all-jet strategic bomber — one capable of combining the B-36's intercontinental reach and enormous payload with far greater speed. The Air Force would turn to Boeing to produce it.

The Air Force's pursuit of an all-jet bomber could be traced to 1943, when the Army Air Forces asked the nation's leading aircraft manufacturers to study designs powered by the new turbojet engines then under development. Boeing initially considered fitting jet engines to a modified B-29, but the concept offered too little improvement to justify pursuing it. A successful jet bomber would require an entirely new aircraft.

Boeing's early designs retained the straight wings common to existing bombers, but the company's thinking changed dramatically after the war. As Allied investigators examined captured German aircraft, wind-tunnel data, and engineering research, they discovered how much progress German aerodynamicists had made in understanding swept wings. By angling a wing rearward, designers could delay the sharp rise in aerodynamic drag that occurred as an aircraft approached the speed of sound. Swept wings would therefore allow a jet-powered airplane to exploit its greatest advantage: speed.

A Boeing engineer who reviewed the captured German research while serving on a technical mission in Europe quickly communicated its significance to Seattle. Boeing's engineers redesigned their proposed bomber around a thin wing swept sharply rearward. Six turbojet engines were suspended beneath it in streamlined pods: four mounted in paired nacelles closer to the fuselage and one mounted farther out beneath each wing. Suspending the engines below and ahead of the wing reduced structural weight, simplified maintenance, and helped prevent aerodynamic interference between the engines and the wing.

The resulting aircraft, Boeing's Model 450, was so unconventional that it bore little resemblance to any bomber that had preceded it. In April 1946, the Army Air Forces ordered two prototypes under the designation XB-47. The first aircraft emerged from Boeing's factory in September 1947 and made its maiden flight on December 17.

The XB-47 was revolutionary. Its slender fuselage, swept wings, bicycle-style landing gear, and pod-mounted engines gave it an appearance unlike that of any existing bomber. Its basic configuration would eventually influence not only later military aircraft but also the first generation of successful commercial jetliners. Boeing had created the architectural blueprint for the modern large jet aircraft.

The B-47 was designed primarily for speed and altitude rather than the enormous payload and intercontinental range of the B-36. Those priorities imposed compromises. Its fuselage was relatively narrow, its minimum basic crew numbered only three, and its unrefueled combat radius was insufficient for many missions against the Soviet Union from bases in North America. Nevertheless, its ability to approach 600 miles per hour and operate at high altitude made it considerably more difficult to intercept than the lumbering piston-powered bombers it was intended to supplement.

The B-47 entered service as a medium bomber and operated alongside the much larger Convair B-36. The two aircraft fulfilled complementary roles. The B-36 provided the range and payload needed for intercontinental missions, while the B-47 provided the speed required to penetrate increasingly sophisticated Soviet defenses. Aerial refueling eventually extended the B-47's reach, although the speed mismatch between the jet bomber and Boeing's piston-powered KC-97 tanker made the process difficult.

The Air Force ultimately acquired more than 2,000 B-47s, with Boeing, Douglas, and Lockheed sharing production. During the 1950s, the Stratojet became one of the principal nuclear-delivery aircraft of Strategic Air Command and a visible symbol of American airpower. Production ended in 1957, after 2,042 aircraft of all versions had been built. Yet the B-47 remained a medium bomber. Its range depended heavily on overseas bases and aerial refueling, and its payload was modest compared with that of the B-36. Meanwhile, advances in Soviet radar, jet interceptors, and antiaircraft weapons were rapidly eroding the B-36's usefulness. For all its size and range, the Peacemaker was slow and increasingly vulnerable.

The Air Force therefore still needed an all-jet heavy bomber — one capable of combining the B-36's intercontinental reach and payload with the B-47's speed and operating altitude. Boeing's experience with the revolutionary Stratojet positioned the company to produce it.

***

The Air Force's search for an all-new intercontinental bomber had begun before the B-36 even entered service. Boeing won an initial development contract in 1946, but the design changed repeatedly as the Air Force revised its requirements and propulsion technology advanced. By 1948, Boeing was proposing a large swept-wing bomber powered by turboprop engines, which appeared to offer a workable compromise between the speed of jets and the fuel economy needed for intercontinental range.

Then came one of the most celebrated weekends in aviation history. In October 1948, a small group of Boeing's leading engineers traveled to Wright-Patterson Air Force Base near Dayton, Ohio, to present the turboprop design. Air Force Colonel Pete Warden was unimpressed. The jet-powered B-47 had demonstrated the potential of swept wings and turbojet propulsion, and Warden wanted Boeing to determine whether the same principles could be applied to a much larger intercontinental bomber.

It was Friday, and Boeing's engineers were given until Monday to produce a new proposal. Rather than return to Seattle, the team gathered in a room at Dayton's Van Cleve Hotel and began redesigning the airplane from the ground up. The engineering team worked through the weekend with slide rules, drafting equipment, and the performance data Boeing had accumulated from the B-47 program. The design that emerged incorporated a thin wing swept rearward 35 degrees, similar in principle to the B-47's but considerably larger. Eight turbojet engines were suspended beneath the wings in four twin-engine pods. The airplane was projected to carry a 10,000-pound bomb load, fly approximately 6,750 miles, reach speeds of around 560 miles per hour over its target, and operate at altitudes approaching 50,000 feet.

While the others completed the engineering calculations and drawings, one engineer visited a local hobby shop and purchased balsa wood, glue, paint, and carving tools. He fashioned a small model of the proposed bomber, approximately 14 inches long, so Air Force officials could visualize the radical new configuration. By Monday morning, the team had produced a 33-page proposal, supporting drawings, performance calculations, and the handmade model.

The Air Force was sufficiently impressed to authorize Boeing to proceed with the new design, designated the Model 464-49. The proposal did not represent the final B-52, and several years of additional engineering and redesign would follow. Nevertheless, the essential configuration had been established — swept wings, eight jet engines mounted in paired pods, and an enormous airframe capable of combining intercontinental range with high-subsonic speed. The episode demonstrated not only Boeing's engineering agility, but also the Air Force's confidence that the company could convert a weekend concept into an operational weapon system.

The resulting airplane became the Boeing B-52 Stratofortress, one of the most enduring military aircraft ever built. The first prototype flew on April 15, 1952, and the bomber entered operational service in 1955. Boeing produced progressively more capable versions until the last of 744 B-52s rolled off the assembly line in 1962.

The final production version, the B-52H, could carry approximately 70,000 pounds of nuclear or conventional weapons, fly roughly 8,800 miles without aerial refueling under typical mission conditions, and operate at altitudes approaching 50,000 feet. Its enormous range, adaptable weapons capacity, and rugged airframe allowed it to evolve from a high-altitude nuclear bomber into a versatile platform for conventional bombing, maritime strike, close-air support, and the launch of long-range cruise missiles.

Roughly 75 B-52Hs remain in the Air Force inventory more than six decades after they were built. They are now undergoing the most extensive modernization in the bomber's history. Their eight 1960s-era Pratt & Whitney TF33 engines are being replaced by modern Rolls-Royce F130 turbofans, while new radar, communications, avionics, electrical systems, and weapons capabilities are also being installed. The modernized aircraft will be redesignated the B-52J.

The new engines are expected to improve fuel efficiency substantially, increase range, reduce maintenance requirements, and provide greater electrical power for future systems. The Air Force intends to operate the modernized bomber into the 2050s. Should that plan be fulfilled, airplanes manufactured in the early 1960s will still be flying nearly a century after they rolled off of their assembly line. And that would make the basic design of those aircraft more than a century old.

While the B-47 and B-52 would provide much of the "meat" of Boeing's military business, their missions created an equally important requirement for the "gravy" of support aircraft. Both bombers depended heavily upon aerial refueling to fly intercontinental missions deep into Soviet territory, remain airborne for extended periods, or approach their targets by indirect routes intended to evade enemy defenses.

The Air Force had already ordered hundreds of Boeing KC-97 tankers to provide that capability, but the piston-powered aircraft was poorly matched to the new generation of jet bombers. During refueling, the heavily laden KC-97 had to fly near the upper limit of its speed range while the receiving B-47 or B-52 slowed toward the lower limit of its own. The narrow margin between the tanker's maximum speed and the bomber's minimum safe speed made the operation difficult and potentially hazardous.

Refueling a B-52 was especially demanding. The bomber had to descend to the KC-97's lower operating altitude and slow considerably, sometimes extending its flaps and landing gear to remain safely behind the tanker. Under some conditions, the two aircraft performed the maneuver in a shallow descent, allowing the KC-97 to gain speed while the B-52 remained above its stall margin. Even then, two or more KC-97s could be required to provide a single B-52 with the fuel needed for a long-range mission.

The KC-97 also complicated fuel logistics. Its four piston engines burned aviation gasoline, while the jet bombers it supported used kerosene-based military jet fuel. The tanker therefore carried separate supplies and distribution systems: gasoline for its own engines and jet fuel for transfer to the receiving aircraft. The arrangement added weight, complexity, and maintenance requirements that an all-jet tanker using the same fuel as the aircraft it supported could avoid.

The speed disparity created additional operational problems. Because the KC-97 cruised far more slowly than the bombers, tankers had to depart well in advance and be carefully positioned along the planned route. A B-47 or B-52 could rapidly overtake a tanker that arrived late or was displaced from its assigned rendezvous point. Coordinating a large strike therefore required tankers to be airborne, properly spaced, and waiting at precise locations before the bombers arrived.

Aerial refueling allowed the bombers to depart with combinations of weapons and fuel that would otherwise have exceeded their maximum takeoff weights. After climbing and burning some of their initial fuel, they could rendezvous with tankers and replenish their tanks before beginning the long journey toward their targets. The system greatly extended Strategic Air Command's reach, but it also made a successful mission dependent upon an elaborate chain of tankers, crews, bases, and precisely timed airborne rendezvous.

What the Air Force needed was an all-new jet-powered tanker capable of flying at the same speeds and altitudes as the aircraft it supported. Such an airplane would simplify refueling, reduce the need for elaborate speed-management maneuvers, and eliminate the burden of carrying two different types of fuel. It would also need enough capacity to refuel the enormous B-52 without requiring a procession of piston-powered tankers.

As the 1950s progressed, Boeing's business continued to benefit from the government's willingness to finance aircraft that pushed the boundaries of engineering and manufacturing. The relationship remained symbiotic, much as it had during the preceding decades. The Air Force relied upon Boeing to provide the aircraft needed to execute its evolving strategic doctrine, while Boeing relied upon military contracts to finance the technological advances that would expand its capabilities. The need for a jet tanker would soon provide the company with another opportunity — one that would ultimately transform both its military and commercial businesses.

Turnabout

As Boeing entered the 1950s, it found itself almost entirely dependent on government contracts and effectively shut out of the commercial aircraft market — a position reminiscent of its predicament in the 1930s before World War II. Bill Boeing's dream, which was part of the DNA of the company, of building airplanes for the traveling public appeared to be slipping away. The Model 377 Stratocruiser had failed to gain broad acceptance among the airlines, leaving Douglas, with its DC-6, and Lockheed, with its Constellation, as the dominant suppliers of large passenger aircraft. Boeing had not produced a commercially successful airplane in more than two decades.

The B-47 and B-52 bombers had demonstrated the enormous advantages of jet propulsion in speed, power, and performance. Bill Allen became convinced that jet power represented the future of commercial aviation. The airlines, however, were a bit wary. Although they continually sought larger and faster aircraft, turbine propulsion was still new, and adopting it would require costly changes to maintenance, training, airport facilities, and fueling infrastructure.

By 1950, Lockheed was developing more advanced versions of the Constellation, while Douglas, enjoying the success of the DC-6, was preparing an even larger four-engine piston airliner that would become the DC-7. If Boeing was going to remain a contender in the commercial aircraft market, Allen believed Boeing needed something far more radical — an aircraft that could restore the company's commercial relevance before Douglas and Lockheed locked it out of the market permanently.

After flying aboard a B-47, Allen was struck by the aircraft's speed and by the relative quietness and smoothness of its flight. The experience reinforced his conviction that the future belonged to jet-powered passenger aircraft. He envisioned a Boeing jet transport that would return the company to the forefront of commercial aviation.

So it was natural that Allen would watch with growing concern as Britain's de Havilland developed the Comet, the world's first jet-powered commercial airliner, which made its initial flight in 1949. Boeing had recognized the future, but another manufacturer appeared poised to reach it first.

De Havilland could assume the enormous risks of developing a jet airliner partly because the Comet was intended for British Overseas Airways Corporation, or BOAC, a government-owned airline. Many of the world's flag carriers were similarly state-owned, and the British government bore much of the Comet's financial risk. Indeed, the government had called for the development of such an aircraft after World War II as part of a broader effort to preserve Britain's leadership in aviation.

Boeing faced a very different proposition. Although the company had earned enormous sums of money from government-financed military programs, any commercial aircraft it developed would be financed largely with its own money and undertaken at its own risk. Allen nevertheless believed that a sufficiently advanced airplane could overcome those risks.

When the Comet entered regular passenger service in 1952, Allen feared that Boeing's opportunity to establish itself in the emerging jet-airliner market was slipping away. He moved quickly. Despite having no airline customers committed to such an aircraft, he directed Boeing's engineers to begin developing a jet-powered transport that could be marketed to airlines around the world. Once again, Boeing would bet its own money on the belief that an engineering breakthrough would bring customers to its door.

Allen had reason to be cautious. His earlier gamble on the Model 377 Stratocruiser had failed to pay off, leaving Boeing with a loss on the commercial side of the program. This time, he devised a way to reduce the financial exposure: the new airplane would serve two markets. One version could become a commercial airliner;

another could be offered to the U.S. Air Force as an aerial-refueling tanker.

The aircraft would therefore serve as a proof of concept for both civilian and military use. Boeing's engineers initially considered fitting the existing KC-97 with jet engines but quickly abandoned the idea. Instead, they developed an entirely new aircraft — the Model 367-80, commonly referred to as the Dash 80. It was a radical design that would become the foundation for Boeing's future in both commercial and military aviation.

Like the British Comet, the Dash 80 was powered by four jet engines, but the resemblance largely ended there. The Dash 80's engines were suspended in pods beneath its wings, similar to the B-47 and B-52, while the Comet's were embedded in the wing roots near the fuselage. Its wings were sharply swept, drawing upon Boeing's experience with the B-47 and B-52, whereas the Comet's were comparatively straight. The Dash 80's tail surfaces were also swept back, giving the aircraft a distinctly modern appearance compared with the Comet's more conventional configuration.

In little more than eighteen months, Boeing's engineers designed and built the new aircraft. Because production tooling and assembly jigs had not yet been developed, much of the prototype was fabricated and assembled by hand. The Dash 80 was rolled out at Boeing's Renton, Washington, plant in May 1954 and made its first flight two months later.

Boeing would not be the first manufacturer to place a jet airliner into commercial service, but Allen remained convinced that jets represented the future of air transportation. He was determined that Boeing would enter that market before the opportunity passed it by.

Then Boeing received an extraordinary stroke of good fortune. The same month the Dash 80 was unveiled, the Air Force issued a request for proposals for a new jet-powered aerial-refueling tanker, with an anticipated requirement for more than 800 aircraft. Six major manufacturers were invited to compete — Boeing, Convair, Douglas, Fairchild, Lockheed, and Martin — but only Boeing already had a suitable jet transport designed, built, and preparing to fly. The timing could scarcely have been better.

Boeing's competitors were still sharpening their engineering pencils when the Air Force, unwilling to wait several years for a new tanker, moved ahead with an interim order for 29 Boeing aircraft. The military version of the Dash 80 would be designated the KC-135 Stratotanker.

Lockheed was farther along than the other competitors because its proposal drew upon a commercial jet transport it had been studying since 1949. When the Air Force completed its formal evaluation, it actually selected Lockheed's design as the winner. The plan was for Boeing's KC-135 to meet the immediate need until the Lockheed tanker could enter production and eventually become the Air Force's principal aerial-refueling aircraft.

That transition never occurred. Even after selecting Lockheed's proposal, the Air Force continued ordering KC-135s, soon raising Boeing's total to approximately 250 aircraft. The Boeing tanker proved highly capable, and its enormous head start meant it could reach operational service far sooner than Lockheed's still-unbuilt design. Faced with the expense and complexity of operating two tanker types, the Air Force abandoned the Lockheed aircraft and standardized its fleet around the KC-135.

What had begun as an interim purchase became one of the largest and most enduring military-jet aircraft programs in Boeing's history.

***

The Douglas Aircraft Company had dominated commercial aviation for more than two decades. Beginning with the DC-2 in the early 1930s, it had introduced a succession of increasingly capable piston-powered airliners — the DC-3, DC-4, and DC-6 — and was now marketing the DC-7. Each had helped set the standard for its era, and Douglas saw little reason to rush headlong into commercial jet transportation. (The DC-5 was a niche aircraft that only saw 12 built in total before World War II interrupted its production.)

Douglas had proposed a jet-powered tanker to the Air Force and, like Boeing, envisioned adapting its design into a commercial transport. Yet company leaders believed the civilian jet market would develop gradually. They were also reluctant to undermine the highly successful DC-6 and the new DC-7, which was attracting orders from major airlines. Piston-powered aircraft were still profitable, familiar, and supported by the existing airline infrastructure.

Boeing's head start changed that calculation. The Air Force's interim orders for the KC-135 gave Boeing the means to place a jet transport into production while its competitors were still developing their designs. Once Douglas recognized that Boeing intended to transform that military advantage into a commercial airliner, it accelerated work on its own jet transport, the DC-8. What had initially seemed like a distant successor to the DC-7 was becoming an urgent competitive necessity.

Lockheed chose a different course. Although it had studied several jet-transport concepts, it concluded that competing directly against Boeing and Douglas with another large pure-jet airliner offered little prospect of success. Instead, Lockheed sought a distinct place in the emerging market with the L-188 Electra, a four-engine turboprop designed primarily for medium-range routes.

A turboprop used a turbine engine to drive a propeller rather than relying principally upon exhaust thrust. It could not match the cruising speed of a pure jet, but it offered two important advantages. First, early turboprops consumed considerably less fuel than the turbojet engines then entering service. Second, their propellers produced substantial thrust at low speeds, allowing them to operate from shorter runways than the first generation of large jet airliners.

Those early jets were optimized for high-speed flight. Their swept wings and relatively thin airfoils reduced drag near the speed of sound but produced less lift at lower speeds, requiring faster takeoffs and landings. Faster speeds translated into longer runway requirements, limiting the new jets largely to major airports serving long-distance routes. The Electra could serve a much broader network while still offering airlines much of the smoothness and reliability associated with turbine power.

The Electra initially attracted considerable interest, but its promise was shattered by two catastrophic in-flight breakups. Investigators traced the failures to a phenomenon known as whirl-mode flutter, in which oscillations originating around an engine and propeller installation could interact destructively with the wing. Lockheed redesigned and strengthened the aircraft, but public and airline confidence never fully recovered. Production ended in 1961 after only 170 Electras had been completed.

The underlying design nevertheless achieved far greater success in military service. Lockheed adapted the Electra's basic airframe into the P-3 Orion, a long-range maritime-patrol and antisubmarine aircraft. More than 750 were ultimately produced by Lockheed and under license in Japan, and the type would remain in service around the world for decades.

Boeing, meanwhile, had financed the Dash 80 entirely with its own money, anticipating that the Air Force would eventually require jet tankers to support its growing fleets of B-47 and B-52 bombers. Allen also sought to hedge that enormous wager by creating the foundation for a commercial airliner. But when Boeing's engineers began designing the Dash 80, neither the Air Force nor the airlines had defined precisely what they wanted. The aircraft therefore had to be built around a series of informed assumptions — assumptions that would soon be tested as Boeing attempted to turn one experimental airplane into two very different production aircraft.

One of the assumptions Boeing's engineers had to make concerned the diameter of the fuselage. The Dash 80 was built with a relatively narrow fuselage approximately 132 inches wide. When the Air Force ordered the production tanker that would become the KC-135, however, it required a larger fuselage capable of carrying standardized military cargo pallets when the aircraft was not performing aerial-refueling missions. Boeing therefore widened the outside diameter of the fuselage to 144 inches.

Because the Air Force was Boeing's first customer — and was expected to order the aircraft in far greater numbers than any airline would — the KC-135's dimensions initially shaped Boeing's plans for the commercial version. Using the same 144-inch fuselage would allow the two aircraft to share much of the same tooling, manufacturing equipment, and production techniques, substantially reducing development and production costs.

A 144-inch fuselage could comfortably accommodate five-abreast seating in a three-plus-two cross-section configuration, but a three-plus-three cross-section seating configuration would be very cramped. Douglas, meanwhile, was marketing its proposed DC-8 with a 147-inch-wide fuselage and a more comfortable three-plus-three seating arrangement. Airline executives quickly recognized that Douglas could offer the same number of seats in each row while giving passengers more shoulder room and providing a wider center aisle.

Pan American and other prospective customers pressed Boeing to match the DC-8's cabin capacity and comfort. Boeing could no longer expect airlines to accept a narrower cabin merely because it would be cheaper for Boeing to manufacture. The company abandoned its plan to give the KC-135 and the commercial aircraft a common fuselage and authorized another expensive redesign.

Boeing widened the commercial aircraft's fuselage to 148 inches — one inch more than the proposed DC-8. The change required separate tooling and assembly equipment, sacrificing many of the economies Boeing had hoped to achieve by building the tanker and airliner around the same cross-section. But the decision gave Boeing a commercially competitive six-abreast cabin and removed one of the strongest advantages Douglas held over the emerging 707.

The commercial airliner developed from the Dash 80 would be designated the Boeing Model 707. Boeing wanted the new aircraft's designation to distinguish it from the company's earlier commercial airplanes, which had carried model numbers in the 300 series. Because the 400, 500, and 600 series had already been assigned to missiles, engines, and other products, Boeing reserved the 700 series for its new generation of jet transports.

The first aircraft could logically have been called the Model 700, but Boeing's marketing department thought "Seven-Oh-Seven" sounded more distinctive and appealing. The designation was adopted, establishing the naming convention that would identify Boeing's commercial jetliners for generations to come.

The entire Dash 80 project — from its design and construction to its subsequent modifications — proceeded with remarkable efficiency. Its timing was equally fortuitous, for the aircraft could scarcely have been conceived at a more opportune moment. As Boeing refined the design and prepared it for commercial service, events unfolding across the Atlantic would help validate Allen's gamble and accelerate the airlines' acceptance of Boeing's new jet.

***

The de Havilland Comet entered commercial service in 1952. Over the next two years, the revolutionary airliner suffered five major accidents, four of them fatal to everyone aboard.

The first two occurred during takeoff. In October 1952, a BOAC Comet failed to become airborne at Rome and ran beyond the end of the runway. Five months later, a Canadian Pacific Comet crashed during takeoff from Karachi, killing all eleven aboard. Investigators initially emphasized pilot over-rotation, but the accidents also exposed a dangerous aerodynamic characteristic — at an excessive nose-up attitude, airflow over the Comet's wing deteriorated sharply, reducing lift at precisely the moment the aircraft needed it most. De Havilland subsequently modified the wing's leading edge and added wing fences, while pilots were cautioned against over-rotating during takeoff.

The third fatal accident occurred in May 1953, when a BOAC Comet broke apart shortly after departing Calcutta in a severe thunderstorm. Investigators concluded that violent turbulence, compounded by excessive aerodynamic loads imposed as the crew attempted to control the aircraft, had caused its structural failure. Because the weather appeared to explain the accident, it did not yet suggest a fundamental weakness in the Comet's design.

The next two crashes would prove far more consequential. On January 10, 1954, BOAC Flight 781 disintegrated in clear weather over the Mediterranean near the island of Elba, killing all thirty-five people aboard. With no obvious explanation, British authorities grounded the Comet while de Havilland and government investigators examined the aircraft and introduced a series of modifications.

Yet political and commercial pressure to return the prestigious airliner to service was intense. Britain had staked both national pride and the reputation of its aircraft industry on the Comet, and a prolonged grounding threatened to destroy confidence in the program. After approximately ten weeks, the aircraft was cleared to fly again.

The decision proved premature. On April 8, less than three weeks after service resumed, a South African Airways Comet operated by BOAC broke apart over the Mediterranean near Naples, killing all twenty-one aboard. The Comet was immediately grounded again — this time indefinitely, until investigators could identify a definitive cause.

Evidence recovered from the Elba crash provided the first crucial clues. Examination of the wreckage and the victims indicated that the aircraft had undergone a sudden and violent loss of cabin pressure. Investigators considered the possibility of a bomb but found no evidence of an explosive device. Instead, the pattern of damage showed that the pressurized fuselage itself had ruptured at altitude, releasing its stored energy with almost explosive force.

The question was no longer whether the aircraft had broken apart in flight, but why a pressure cabin designed to withstand repeated flights at high altitude had failed so catastrophically. Investigators turned their attention to the Comet's design, construction, and the effects of repeatedly pressurizing and depressurizing its fuselage.

By the 1930s, aluminum alloys had become the principal structural material in large aircraft. They offered exceptional strength for their weight and were well suited to stressed-skin construction, in which the aircraft's outer skin carried part of the structural load. By the end of World War II, aluminum formed most of the fuselage, wings, and tail surfaces of modern transport aircraft.

Pressurized piston airliners such as the Boeing 377, Douglas DC-6, and Lockheed Constellation routinely cruised above the altitudes at which passengers could breathe comfortably without supplemental oxygen, typically in the 18,000 - 22,000 foot range. Their pressure cabins generally maintained an interior environment equivalent to an altitude of approximately 8,000 feet. The difference between the pressure inside the cabin and the thinner atmosphere outside placed the fuselage under stress every time the aircraft climbed and was pressurized, then relieved that stress when it descended.

The Comet carried this challenge into largely unexplored territory. It routinely cruised at approximately 35,000 feet, where the outside air pressure was far lower than at the operating altitudes of most piston airliners. Maintaining a comfortable cabin environment therefore required a pressure differential nearly twice that of many contemporary aircraft. Each flight caused the fuselage to expand slightly under pressure and contract again during descent. Repeated hundreds or thousands of times, that cyclic loading could initiate microscopic cracks and gradually enlarge them through metal fatigue.

De Havilland's engineers understood that the Comet's pressure cabin would be highly stressed, but the behavior of aluminum structures under repeated pressurization was not yet fully understood. The analytical tools available to engineers were rudimentary by modern standards, and there was little operational experience upon which to base predictions of the fatigue life of a high-altitude jet transport.

The problem was not simply that engineers reinforced some areas and thereby transferred the loads elsewhere. Every opening in a pressure vessel — including doors, windows, and antenna apertures — interrupts the smooth flow of stress through its skin. Abrupt changes in shape create stress concentrations, particularly around corners, fastener holes, and manufacturing imperfections. Reinforcement can strengthen such an opening, but it must be designed so that the loads are carried smoothly into the surrounding structure.

The Comet's large rectangular passenger windows became the most visible symbol of the failure. Although their corners were rounded rather than perfectly square, their relatively small corner radii produced greater stress concentrations than the oval windows later adopted by the industry. De Havilland reinforced the window surrounds, but the design, fastening methods, and manufacturing process still left areas vulnerable to fatigue cracking.

The investigation went far beyond visual examination of the recovered wreckage. British engineers placed an entire Comet fuselage in a water tank and repeatedly pressurized it while hydraulically loading the wings to simulate successive flights. After several thousand pressure cycles, the test fuselage ruptured at the corner of a passenger window, confirming that fatigue cracks could develop around fuselage openings and propagate until the pressure cabin ruptured.

The wreckage of the Comet lost near Elba revealed a closely related but important distinction. Investigators traced the initiating fatigue crack not to a passenger window but to a corner of an automatic direction-finding antenna aperture in the roof of the fuselage. Repeated pressurization had caused a crack to grow from this highly stressed opening until the cabin failed catastrophically. The official inquiry concluded that the Elba accident resulted from structural failure of the pressure cabin caused by fatigue. Although too little wreckage was recovered from the Naples aircraft to establish its cause conclusively, investigators considered a similar failure possible.

The inquiry transformed the understanding of aircraft structural fatigue. Designers began paying far greater attention to stress concentrations, crack propagation, pressure-cycle testing, fastener-hole preparation, and the shape of every opening in the fuselage. Rounded and oval passenger windows became standard because their contours distributed pressure loads more evenly and reduced concentrated stresses.

The original Comet 1 was permanently withdrawn from passenger service, but de Havilland did not abandon the aircraft. The company extensively redesigned and strengthened the pressure cabin, altered its windows and other fuselage openings, and incorporated the lessons of the investigation into later versions. The Comet 2 had already been under development, but it never entered service in its intended form; modified examples were instead supplied primarily to the Royal Air Force.

The substantially improved Comet 4 finally entered airline service with BOAC in 1958, more than four years after the grounding. It was safe and capable, and it inaugurated the world's first scheduled transatlantic jet service. But the delay had cost de Havilland its commanding lead. Boeing's larger 707 entered transatlantic service only weeks later, followed by the Douglas DC-8, and the center of the commercial jet-aircraft industry shifted decisively to the United States.

Boeing designed and built the Dash 80 during the same period in which the Comet suffered its disastrous accidents. Bill Allen, convinced that jet aircraft represented the future of aviation, watched events across the Atlantic with both interest and concern. Boeing's engineers followed the investigations closely, as did their counterparts at Douglas.

When British investigators revealed that repeated pressurization had produced fatigue cracks around openings in the Comet's fuselage, both manufacturers subjected their emerging jetliners to renewed scrutiny. Because the 707 and DC-8 were still under development, Boeing and Douglas could incorporate the lessons of the Comet directly into their structural designs and testing programs. They placed greater emphasis on fatigue life, stress concentrations, crack propagation, fuselage openings, and repeated full-scale pressure testing.

The Comet's grounding gave the American manufacturers an extraordinary, if tragic, advantage. Boeing and Douglas were able to correct potential weaknesses before their aircraft entered service while continuing to develop and market them aggressively. De Havilland, by contrast, had to reconstruct both its airplane and the airlines' confidence in it.

The substantially redesigned Comet 4 did not enter commercial service until 1958, more than four years after the original model had been grounded. By then, many prospective customers had canceled their orders or turned to the forthcoming American jets. Although BOAC and several other airlines would operate the improved Comet successfully, de Havilland had lost the commanding lead it once held.

Britain had been the first nation to place a jet airliner into commercial service, but the United States would reap much of the commercial benefit from the lessons learned through the Comet's failure. Boeing and Douglas were poised to dominate the emerging jet age. For Boeing, practically excluded from the commercial-airliner market for more than two decades, the transformation was especially dramatic. Under Bill Allen's leadership, the company was about to move from the commercial industry's margins to its very center.

Taking the Lead

In 1951, the Douglas Aircraft Company stood near the pinnacle of the commercial aviation industry. Its DC-6 had been ordered by dozens of airlines around the world and represented the latest in a succession of successful Douglas transports stretching back to the DC-2 of the early 1930s. Its only serious rival was Lockheed's Constellation, which had evolved into increasingly capable versions and was gaining favor among the world's airlines.

Determined not to lose its momentum, Douglas developed a successor to the DC-6 — the DC-7 — which promised exceptional range, performance, and comfort at the peak of the piston-powered era. Douglas initially believed that airlines would continue buying advanced piston aircraft for years before embracing the expense and uncertainty of jet propulsion. With the DC-6 selling well and orders accumulating for the DC-7, the company had little incentive to hasten the demise of its own profitable product line.

The arrival of the British Comet and Boeing's work on a commercial jet transport forced Douglas to reconsider. If Boeing succeeded in establishing the 707 before Douglas responded, the company that had dominated commercial aviation for two decades risked surrendering the industry's future to its old rival. Douglas accelerated development of its own jetliner, the DC-8, and marketed it aggressively as a more capable and spacious alternative to the Boeing aircraft.

The DC-8 would reach the market after the 707, but Douglas possessed something Boeing lacked: deep relationships with airlines that had operated generations of Douglas transports. It also initially offered a wider cabin capable of comfortable six-abreast seating. Pressure from airlines considering the DC-8 prompted Boeing to widen the 707's fuselage, eliminating one of Douglas' most important advantages. With that change, orders for the Boeing aircraft began to accumulate.

The competition intensified dramatically in October 1955, when Pan American World Airways ordered twenty 707s — and, in a startling move, twenty-five DC-8s at the same time. Purchasing two expensive and competing jetliners was unprecedented. Pan Am president Juan Trippe was managing the technological and commercial risks of entering the jet age while ensuring that his airline would have access to whichever aircraft proved superior or became available first. His decision also placed Boeing and Douglas in direct competition for the prestige of inaugurating Pan Am's jet service.

Other airlines quickly followed. American Airlines, Air France, BOAC, Lufthansa, Qantas, and TWA were among those ordering the 707, while United, Delta, Eastern, KLM, Japan Air Lines, and Scandinavian Airlines selected the DC-8. By the beginning of 1958, Boeing held orders for approximately 150 707s, while Douglas had sold 133 DC-8s. Boeing was ahead, but Douglas remained close enough that the outcome was far from certain.

A third American manufacturer also entered the contest. Convair developed the Model 880 and its larger derivative, the 990, as faster alternatives to the Boeing and Douglas jets. The Convair aircraft were smaller, with five-abreast seating and less capacity and range, but they were among the fastest subsonic airliners ever built. Their speed was not enough to overcome their limited capacity, high fuel consumption, and the competitive advantages Boeing and Douglas had already established. Production ended in 1963 after Convair had built only sixty-five 880s and thirty-seven 990s.

De Havilland also returned to the market with the substantially redesigned Comet 4. On October 4, 1958, BOAC used the aircraft to inaugurate the world's first scheduled transatlantic jet service, flying simultaneously eastbound and westbound between London and New York. The Comet had reclaimed one final aviation first, but its commercial lead was already gone. Although seventy-four aircraft of the Comet 4 series were eventually built, the type was smaller than its American competitors and could not match their sales.

Boeing's defining moment came later that same month. On October 16, First Lady Mamie Eisenhower christened Pan Am's first 707, Jet Clipper America, during a highly publicized ceremony at Washington National Airport. Ten days later, on October 26, Pan Am inaugurated scheduled American jet service when a 707 departed New York's Idlewild Airport for Paris. The flight carried 111 passengers and required a fuel stop at Gander, Newfoundland, but it demonstrated that mass jet travel across the Atlantic had arrived.

The DC-8 did not enter scheduled service until September 1959, when Delta Air Lines and United Airlines began operating the type. By then, the 707 had enjoyed nearly a year in which to capture headlines, establish routes, and become identified in the public imagination with speed, modernity, and the beginning of a new age in commercial aviation. The delay did not doom the DC-8, which would become a highly successful and remarkably durable aircraft, but it allowed Boeing to seize the initiative.

The Comet had demonstrated that jet-powered passenger travel was possible, but it entered service in small numbers and disappeared just as the market was beginning to develop. The 707 arrived after airlines and passengers had spent several years anticipating the transition. Rather than introducing the idea of jet travel, Boeing released the pent-up demand for it. Faster flights, higher altitudes, smoother rides, and a glamorous new image transformed public expectations almost overnight.

Boeing also brought enormous credibility to the undertaking. Its wartime bombers and the jet-powered B-47 and B-52 had established the company as one of the world's foremost builders of large, advanced aircraft. To the public, a company capable of producing giant bombers that could cross continents at extraordinary speeds appeared equally capable of building a world-class passenger jet.

The 707 fulfilled that expectation. Boeing had not invented the jet airliner, nor had it entered the market without formidable competition. But it produced the aircraft that transformed jet travel from an experiment into a worldwide commercial enterprise. After more than two decades in Douglas' shadow, Boeing was no longer merely returning to the commercial-aircraft industry. It was taking command of it.

Boeing did not stand still with its new creation. In rapid succession, it introduced several versions of the 707, varying the aircraft's fuselage length, wingspan, engines, fuel capacity, and structural weight to meet the differing requirements of domestic and international airlines. Some models carried more passengers, while others offered the range needed for nonstop transcontinental and intercontinental service. Boeing appeared to understand that the airlines did not want a single standardized jetliner; they wanted a family of aircraft tailored to different routes and operating conditions.

Douglas initially followed a more conservative path. It introduced improved DC-8 variants, but many of the early changes centered on increased operating weights, greater fuel capacity, and newer engines rather than major alterations to the basic airframe. The company seemed confident that its longstanding relationships with the airlines and its reputation as the dominant builder of commercial transports would sustain the DC-8 against Boeing's newcomer.

That confidence proved costly. Having finally returned to the commercial market after more than two decades in Douglas' shadow, Boeing pursued airline requirements aggressively and continued accumulating orders. DC-8 sales, meanwhile, slowed sharply during the early 1960s, and by 1963 the program appeared to be losing momentum.

Douglas eventually responded with far more substantial changes. Beginning in the mid-1960s, it developed the stretched

Super 60 series, which offered significantly greater passenger capacity and restored much of the DC-8's competitiveness. The DC-8-61, -62, and -63 transformed the original design into some of the largest and most capable jetliners of their era. The improvements revived sales, but they came too late to overcome the lead Boeing had already established.

When commercial production ended, Boeing had built 878 aircraft in the 707 and closely related 720 series, while Douglas produced 556 DC-8s. The DC-8 was therefore far from a failure; it became a successful and exceptionally durable airplane. But the 707 had outsold it decisively and had established Boeing as the leading manufacturer of long-range commercial jets. Douglas delivered the last DC-8 in 1972, while commercial 707 production continued until 1978.

The 707 airframe also enjoyed an extended life in military production. Boeing adapted it for specialized aircraft such as the Air Force's E-3 Sentry airborne-warning platform and the Navy's E-6 Mercury communications aircraft. These military orders kept the underlying production line active long after commercial deliveries ended, with the final new 707 airframe completed in 1991. Including commercial and military versions, Boeing built slightly more than 1,000 aircraft based on the 707 design.

***

With jetliners becoming commonplace on long routes between major cities, airlines began looking for ways to bring the speed and comfort of jet travel to smaller cities and shorter routes. Boeing responded by developing the 707-020, a shortened version of the 707 designed for markets that could not support the larger aircraft's passenger capacity. Although it retained the 707's broad wings and four engines, the shorter aircraft had a substantially lower gross weight. It could operate from shorter runways at smaller airports and consumed less fuel than the full-size 707.

United Airlines became the first carrier to order the aircraft for domestic service. At United's request, Boeing gave it a new designation — the 720 — to create the impression that the airline was introducing an entirely new model rather than merely a shortened 707.

The 720, however, taught Boeing an important lesson about aircraft "shrinks." Shortening a fuselage while retaining the original structure, systems, wings, and engines did not necessarily produce an efficient airplane. The 720 still carried much of the weight and complexity designed for the larger 707, but spread those costs across fewer passengers. With four engines to purchase, fuel, and maintain, its profit margins were less attractive than those of its larger sibling.

Airlines could continue using smaller piston-powered aircraft on regional routes, but they wanted to bring jet service to those markets — and to do so profitably. That would require an entirely different airplane. Riding the success of the 707, Bill Allen sought to extend Boeing's lead in the jet age into another market segment. He directed the company's engineers to develop an aircraft that could operate from smaller airports while profitably serving routes that could not support a 707.

The economics were straightforward. An airline could make money on a route only if it carried enough paying passengers to cover the cost of operating the aircraft assigned to it. Flying a 707 on a short route with only 100 passengers could quickly turn the trip into a financial loss. Boeing therefore needed a smaller jet whose acquisition and operating costs matched the traffic available in these markets.

In 1960, Boeing began work on a clean-sheet design. But clean-sheet did not mean starting entirely from scratch. Reusing proven components could reduce the cost of design, engineering, testing, tooling, manufacturing, and certification. The engineers began by identifying what could be carried over from the successful

707. They retained its 148-inch-wide fuselage cross-section, which allowed the three-by-three seating arrangement airlines favored. This also permitted Boeing to reuse fuselage tooling, cabin fittings, and other interior components. Keeping the same cross-section further allowed the new aircraft to retain the 707's basic nose and cockpit configuration.

The remainder of the airplane would be substantially new. Boeing needed fewer engines because every additional engine increased the aircraft's purchase price, fuel consumption, and maintenance expense. The new design would also require an entirely new wing and control surfaces. The wing would be the key to operating efficiently from smaller airports. Boeing had already built considerable expertise in swept-wing design, drawing on German aerodynamic research captured after World War II and applying it first to the B-47 and B-52 bombers and then to the 707. For the new aircraft, its engineers would take another major step forward.

The challenge was to design a wing that allowed the airplane to take off from a relatively short runway, cruise quickly and efficiently in the thin upper atmosphere, and then fly slowly enough to land safely on another short runway. At maximum gross takeoff weight, the largest 707 variants required as much as 10,900 feet of runway. During the late 1950s and early 1960s, only the world's largest airports offered runways of that length. Although many airports were extending their runways to accommodate jet transports, smaller cities remained beyond the practical reach of aircraft such as the 707.

Boeing's engineers designed a thin, swept wing optimized for high-speed flight at altitude. To make that wing perform safely at low speeds, they equipped it with an unusually sophisticated system of high-lift devices along both its leading and trailing edges. When extended, the leading-edge slats and trailing-edge flaps effectively increased the wing's chord and curvature, producing substantially more lift at slower speeds. Once retracted, they restored the clean aerodynamic profile needed for efficient high-altitude cruise. Although slats and flaps were not new, Boeing's arrangement used more advanced designs and configurations than those found on earlier aircraft.

The engineers also installed spoilers on the upper surfaces of the wings. After touchdown, these panels rose into the airflow, increasing drag and dumping much of the lift still being generated by the wings. This placed more of the aircraft's weight on its landing gear, making the wheel brakes more effective.

With the trailing-edge flaps, leading-edge devices, and spoilers fully deployed, Boeing engineers joked that "the wing came apart." The description was understandable. Viewed from behind during landing, the opened structure contained gaps through which the runway underneath could be seen.

Because the new airplane would be smaller than the 707 and carry roughly two-thirds as many passengers, Boeing determined that it required only three engines. Fewer engines would reduce fuel and maintenance expenses and, because engines were among the costliest components of an airliner, substantially lower the aircraft's purchase price. However, the odd number of engines created a configuration challenge. Boeing placed all three at the rear of the aircraft — one on each side of the fuselage and the third embedded in the tail. Air reached the center engine through an intake above the fuselage and an S-shaped duct embedded in the fuselage leading to the engine.

To keep the horizontal stabilizer clear of the two side-mounted engines, Boeing placed it at the top of the vertical stabilizer, creating what became known as a T-tail. For airports with limited ground equipment, the engineers also incorporated a rear airstair that allowed passengers to board and disembark without a separate stairway.

After two years of design and development, Boeing rolled out the first 727 at its Renton, Washington, factory in 1962. Boeing skipped the 717 designation because it had previously assigned the number internally to the KC-135 program to distinguish the military tanker from the civilian 707.

If Boeing had ushered in the jet age with the 707, it stormed the industry with the 727. Airlines were no longer forced to choose between large jets for long-distance routes and slower propeller aircraft for shorter ones. They now had a jetliner capable of profitably serving smaller cities and regional markets. While the 707 could carry as many as 189 passengers over distances of 3,000 miles or more, the original 727 was designed to carry roughly 110 passengers on routes of up to 2,500 miles.

The new aircraft arrived none too soon. Sud Aviation of France had begun achieving international success with the Caravelle, a smaller twin-engine jet designed for short and medium-range routes. United Airlines ordered 20 Caravelles in 1961. The 727, however, was larger and offered greater range and passenger capacity while retaining the ability to operate from many of the same smaller airports.

During the certification program, Boeing sent the 727 on an around-the-world promotional tour to demonstrate its versatility and generate airline interest. Among its most impressive achievements was becoming the first commercial jetliner to operate from La Paz, Bolivia. At more than 13,000 feet above sea level, La Paz presented an exceptional challenge because the thin air reduced both engine performance and aerodynamic lift. The flight provided a powerful demonstration of the 727's short-field and high-altitude capabilities.

Orders soon followed. United and Eastern Air Lines became the launch customers, each ordering 40 aircraft. The 727 went on to dominate the medium-range market for much of the next two decades. As it had with the 707, Boeing developed several variants, the most successful being the stretched 727-200, which typically carried about 135 passengers in a two-class configuration. By the time production ended in 1984, Boeing had built more than 1,800 copies — a record for a commercial jetliner at the time.

Douglas, meanwhile, faced a growing competitive threat. Its DC-8 had been eclipsed in sales by the 707, and Boeing's 727 was now capturing another emerging segment of the jetliner market. Yet the cost of designing, engineering, certifying, and producing a direct competitor was escalating rapidly. Douglas feared that it could not recover those development costs quickly enough by challenging the well-established 727 head-on.

Instead, the company targeted an even smaller market: routes carrying up to about 90 passengers over distances of roughly 1,500 miles. Douglas approved development of a compact twin-engine jetliner that would become the DC-9. Its smaller size and simpler design promised even lower acquisition and operating costs than the 727.

Like the 727, the DC-9 mounted its engines on either side of the rear fuselage and placed the horizontal stabilizer atop the vertical tail in a T-tail configuration. But rather than using three engines, it needed only two. Its simplified systems also allowed it to be flown by two pilots, eliminating the flight engineer required aboard the 707, 727, and DC-8. The smaller crew, reduced fuel consumption, and lower maintenance requirements made the aircraft particularly economical on routes that could not support a larger jet.

The DC-9 also incorporated a built-in rear airstair, allowing passengers to board and disembark at airports without extensive ground facilities. In its general configuration, it resembled the French Caravelle, but it was a newer design with improved performance and operating economics for roughly the same passenger market.

The DC-9 entered commercial service with Delta Air Lines in December 1965 and quickly became a major success. Within three years, Douglas had accumulated more than 200 orders. By the time production of the original DC-9 family ended in 1982, nearly 1,000 had been built. That total did not include the later MD-80, MD-88, MD-90, and MD-95 families. Although marketed under new designations, each was an evolutionary development of the same basic DC-9 airframe. Douglas was once again a formidable competitor — and Bill Allen and the rest of Boeing took notice.

***

With the DC-9 becoming a runaway success, Boeing faced a difficult decision. The market was already crowded, and any competing aircraft would have to recover its development costs quickly and generate a profit without requiring an enormous production run. By 1965, Boeing's executives were engaged in a vigorous debate over whether the company should enter that market at all.

Boeing's engineers would have to develop an aircraft that could compete directly with the DC-9 while remaining profitable at a relatively low sales volume. The marketing department argued strongly in favor of proceeding. An airline that purchased a Boeing jet at the lower end of the market might later turn to Boeing as it expanded or replaced its larger aircraft. By drawing customers into Boeing's family of airplanes, training programs, spare parts, and support services, the company stood a better chance of retaining them for decades.

Although the proposed aircraft would be largely a clean-sheet design, Boeing again looked for proven components it could reuse. As it had with the 727, the company retained the basic fuselage cross-section, nose, and cockpit configuration of the 707. This reduced development, tooling, manufacturing, and certification costs while giving the new aircraft an immediate advantage over the DC-9. The 148-inch-wide fuselage used by the 707 and 727 accommodated six-abreast seating in a three-by-three arrangement, which airlines had come to prefer. The narrower DC-9 could seat only five passengers across in a two-by-three configuration. Boeing's wider cabin therefore offered greater capacity without requiring a proportionately longer aircraft.

Like the DC-9, the new Boeing would use only two engines, lowering its purchase price, fuel consumption, and maintenance costs. Its simpler systems would also allow it to operate with two pilots, eliminating the flight engineer required aboard the 707 and 727.

Unlike either the DC-9 or the 727, however, Boeing's new aircraft would carry its engines beneath the wings rather than at the rear of the fuselage. It was a decision made in 1965 that would reverberate through Boeing — and across the aviation world — more than half a century later.

***

The arrival of turbine power brought a quantum leap in aviation technology — not only in propulsion, but also in the aircraft structures, aerodynamic designs, and cockpit systems required for high-speed, high-altitude flight. These advances came at an enormous cost.

Gone were the days when a manufacturer could design and build an airplane in weeks or months, then recover its investment by selling only a few dozen copies. Jetliners were vastly more complex and expensive to engineer, test, certify, and manufacture. Except in unusual circumstances, no single airline order was large enough to make a new program profitable. Manufacturers needed commitments from several carriers — and eventually hundreds of aircraft sales — to justify launching a new design.

As a result, the commercial-aircraft sales cycle evolved during the postwar jet age. A manufacturer typically began by presenting a preliminary concept to selected airlines to gauge their interest. At this stage, the airplane existed mainly on paper, and even its external appearance might remain unsettled.

The proposal was essentially a series of "what-if" questions: Would the airline buy an aircraft capable of carrying a specified number of passengers over a certain range? What if it consumed a particular amount of fuel, cruised at a given speed, or offered takeoff performance 10 percent better than that of competing models?

Airline responses helped the manufacturer refine the aircraft's size, range, performance, and operating economics. If sufficient interest emerged, the proposal could be presented to the board of directors for permission to market the aircraft formally — but not yet to build it. The sales team would then begin searching for a launch customer.

As the term implied, the launch customer was the first airline to make a firm commitment to the new aircraft. Once that commitment was secured, and depending on the level of interest in the rest of the market, the manufacturer could officially launch the program, begin committing substantial company funds, finalize the design, and prepare for production.

But the launch customer was not necessarily the first airline willing to sign an order. The manufacturer usually sought an influential carrier whose endorsement might persuade others to follow. Its commitment would serve as the first falling domino, creating confidence in the aircraft and encouraging additional orders.

For the airline, becoming a launch customer carried both risks and rewards. The greatest risk was that it would be the first to encounter the inevitable teething problems of an entirely new aircraft. Aviation history was filled with new models that suffered technical faults during their introduction. At the most tragic extreme stood Britain's de Havilland Comet. More commonly, early problems produced delayed and canceled flights, disrupting schedules, increasing costs, and damaging an airline's reputation. The potential benefits, however, could be substantial. Manufacturers eager to secure a prestigious first customer often offered steep discounts. Those concessions compensated the airline for assuming the technical and financial risks associated with introducing an unproven aircraft.

A launch customer could also exert considerable influence over the final design. Knowing that its order was essential to the program, the airline might request features tailored to its routes, operating practices, or passenger requirements. The manufacturer then had to balance those demands against the need to keep the aircraft attractive to the broader market. A feature valuable to one airline could add weight, cost, or complexity that discouraged others from buying it.

This pattern — gauging airline interest, refining the concept, and then securing an influential launch customer — became firmly established during the jet age. The 707 provided an early example when Pan American pressed Boeing to widen the fuselage to accommodate six-abreast seating.

In 1964, as Boeing began developing its new short-range jet, its sales representatives approached airlines to determine their interest in an aircraft capable of carrying approximately 100 passengers while operating from airports with short runways and limited ground facilities. The competitive outlook was not especially promising. Douglas was already enjoying strong demand for the DC-9, and Boeing's marketing department estimated that the new aircraft might sell only about 250 copies during a production run of roughly ten years.

Boeing believed the program could nevertheless be profitable because the new aircraft would share major elements with the 707 and 727. That commonality would reduce development, tooling, manufacturing, training, and support costs. Company estimates suggested that Boeing could recover its investment after selling approximately 100 aircraft, leaving the remainder of the anticipated production run to generate a reasonable profit.

Boeing was not expecting another market-dominating success. Its objective was to complete its product line by offering airlines Boeing jets for short, medium, and long-range routes. The board authorized the program, and the search began for the airline whose commitment could bring the new airplane to life.

***

Lufthansa was a rapidly growing European flag carrier in the early 1960s. West Germany had emerged from the devastation of World War II and rebuilt its economy, although the country remained divided by the Cold War. After a dispute over the historic Lufthansa name, the West German airline retained it, while East Germany established Interflug as its national carrier.

Lufthansa had already ordered Boeing's 707, 720, and 727 for its busiest routes. Although it operated intercontinental services, much of its network connected cities within West Germany and across Western Europe. Many of those destinations were served by secondary airports with short runways, limited ground equipment, and relatively modest passenger demand. That made Lufthansa an ideal launch customer for Boeing's proposed small jetliner, now designated the 737.

By 1964, Lufthansa wanted to extend the speed, reliability, and passenger appeal of jet service to its thinner European routes. Boeing, meanwhile, had fallen behind Douglas and Sud Aviation in the small-jet market and was eager to fill the gap in its product line. Lufthansa had not yet selected an aircraft for those routes, and Boeing mounted an intensive campaign to persuade the globally recognized carrier to launch the 737.

As a prospective launch customer, Lufthansa expected favorable pricing, but it also sought changes to the aircraft itself. One of its most important requirements was that baggage and cargo could be loaded without specialized ground equipment. Many of the smaller airports it served lacked powered belt loaders and other cargo-handling machinery. Lufthansa did not want to purchase new equipment or finance airport upgrades merely to introduce the aircraft. Ground crews needed to be able to load baggage into the lower cargo holds while standing on the ramp. That requirement meant the fuselage had to remain close to the ground.

Joe Sutter, an increasingly prominent Boeing aeronautical engineer, had worked on the 367-80, 707, and 727 programs. In 1964, he was assigned to help develop the proposed 737. Early concepts resembled the 727, DC-9, and Caravelle, with two engines mounted near the rear of the fuselage. However, Sutter and his team instead considered placing the engines beneath the wings, as Boeing had done on the 707. Wing-mounted engines would be easier for mechanics to inspect and service, and this would be an additional selling point to not only Lufthansa but other airlines that would consider the plane. Conventional engine pylons, however, would require taller landing gear to provide adequate ground clearance. That, in turn, would raise the fuselage beyond the reach of baggage handlers standing on the ramp.

The engineers found a compromise. Rather than suspending the engines well below the wings on long pylons, they mounted them tightly against the wings' undersides. This allowed the 737 to retain a short landing gear configuration and a low fuselage while giving mechanics direct access to the engines without lifts or elevated work platforms.

The low-slung configuration became one of the 737's defining characteristics. It satisfied Lufthansa's requirement for simple ground handling and helped make the aircraft attractive to airlines serving airports with limited infrastructure.

In February 1965, Lufthansa ordered 21 aircraft, becoming the 737's launch customer. Boeing had secured the endorsement of a carrier with worldwide recognition, and additional orders soon followed. United Airlines subsequently ordered 40 aircraft but wanted greater passenger capacity. Boeing stretched the fuselage to meet United's requirement and designated the longer model the 737-200. Lufthansa's shorter version became the 737-100.

Other airlines began placing orders for what was nicknamed the "Baby Boeing," its short, broad fuselage giving it the appearance of a miniature 707. By 1970, Boeing had delivered more than 250 aircraft — roughly the total number the company had originally expected to sell during the program's entire production life — and orders were still coming in.

The basic configuration developed in 1964 had been shaped in large part by Lufthansa's operational needs. Its low fuselage and closely mounted engines proved highly practical for airlines of that era. But the same design choices would constrain Boeing's engineers decades later, with consequences that would reverberate throughout the company and the aviation world.

Unbridled Advancements

At the beginning of the 1960s, a person born at the turn of the 20th century could remember when horse-drawn vehicles were still common and steam locomotives represented the fastest practical form of transportation. Their first airplane ride might have been aboard a fragile machine built largely of wood and fabric, flying only a few hundred miles at an altitude of several thousand feet.

Six decades later, that same person could board a jetliner in New York and arrive in London roughly six hours later, having crossed 3,000 miles in comfort high above the clouds with more than 100 fellow passengers. They could also watch men strapped into small capsules atop enormous rockets and launched beyond Earth's atmosphere, while scientists and politicians predicted that space travel might soon become commonplace. The transformation was almost incomprehensible. In little more than half a century, transportation had advanced from horses and steam engines to intercontinental jetliners and spacecraft. Transportation had always influenced where people lived, worked, and conducted business. Aviation was now poised to accelerate those changes on an unprecedented scale.

The new decade was also marked by growing international tension. The United States and the Soviet Union competed for global influence across nearly every arena — military, political, scientific, industrial, and technological — while attempting to avoid a nuclear confrontation that could destroy both countries and much of the world.

The two superpowers engaged in a continuous contest to prove the superiority of their respective political and economic systems. Each achievement became evidence that one side was stronger, more advanced, and better equipped to shape humanity's future. Since direct war between nuclear powers was almost unthinkable, victory would be measured through influence, technological achievement, and national prestige.

By the end of the 1950s, that rivalry had opened a dramatic new front — the Space Race.

In October 1957, the Soviet Union placed Sputnik, the world's first artificial satellite, into orbit. The achievement electrified the world but shocked the United States. Americans had assumed that their country led the world in advanced science and technology. The realization that the Soviets had reached space first raised unsettling questions about American education, industrial capability, missile technology, and national security.

The United States responded by accelerating its own space program, but the Soviets continued accumulating historic firsts. In April 1961, Yuri Gagarin became the first human to orbit Earth. The United States launched Alan Shepard into space less than a month later, but his mission lasted only about 15 minutes and followed a suborbital trajectory. America would not place an astronaut into orbit until John Glenn's flight in February 1962, nearly eleven months after Gagarin's achievement. The Soviets appeared to be running away with both the technological victories and the international prestige that accompanied them.

On May 25, 1961, President John F. Kennedy addressed a joint session of Congress and committed the United States to landing a man on the Moon and returning him safely to Earth before the decade was over. The declaration established a clear national objective, but it also amounted to a challenge to the Soviet Union. The Soviets never formally accepted Kennedy's challenge, yet both sides understood that the Moon had become the ultimate prize. The nation that put boot prints on it first would claim not only a historic achievement but also symbolic leadership in science, technology, and the future exploration of space. It was a race neither side believed it could afford to lose, but only one could win.

For the United States to prevail, it would have to marshal the full extent of its industrial and engineering capabilities. Companies experienced in building the world's most advanced aircraft, engines, electronics, computers, guidance systems, and materials would have to apply their expertise to machines unlike anything previously constructed. The effort resembled the industrial mobilization of World War II. Although its purpose was different, its scale was only somewhat smaller. The technologies required to reach the Moon would draw heavily from aviation, missile development, electronics, computing, metallurgy, and precision manufacturing. The companies already producing sophisticated aircraft and weapons would play central roles.

Boeing stood at the forefront of that industrial force. The company had extensive experience designing large aircraft, strategic bombers, missiles, and complex aerospace systems. The race to the Moon promised Boeing substantial revenues, but perhaps even more valuable was the knowledge it would gain in propulsion, materials, systems integration, electronics, manufacturing, and project management. Boeing became one of NASA's principal industrial partners as the agency pursued the goal of placing American astronauts on the Moon before the end of 1969. Other major contractors included North American Aviation, McDonnell Aircraft, Grumman, Douglas Aircraft, General Motors, and IBM.

The effort extended far beyond those familiar names. Approximately 20,000 companies, universities, research institutions, and government organizations contributed to the Apollo program and the broader American space effort. At its peak, roughly 400,000 people were involved in the work required to send astronauts to the Moon. That extraordinary national undertaking reached its defining moment in July 1969.

Boeing won several lucrative NASA contracts during the race to the Moon, and its contributions became indispensable to the success of the Apollo program. The company built five Lunar Orbiter spacecraft, all launched successfully in 1966 and 1967, that photographed 99 percent of the Moon's surface and helped NASA identify suitable landing sites for Apollo.

Boeing was also awarded the contract to develop and build the Saturn V's enormous first stage, known as the S-IC. Thirteen Saturn V rockets were launched, including ten that carried astronauts. Nine of those crewed missions traveled to the Moon, and six — Apollo 11, 12, 14, 15, 16, and 17 — landed twelve men on its surface. Apollo 13 was intended to become the third lunar landing, but an oxygen-tank explosion in the North American Aviation-built service module forced the crew to circle the Moon and return to Earth.

Boeing also built the Lunar Roving Vehicle, the lightweight, battery-powered car that allowed astronauts to travel far beyond the distances they could have covered on foot. Carried aboard Apollo 15, 16, and 17, the rover greatly expanded the astronauts' range and enabled them to explore more of the lunar surface during the final three Apollo missions.

***

While Boeing was helping America race to the Moon, it still had an airliner business to run. On the heels of the runaway successes of the 707, 727, and 737, Bill Allen turned the company's attention toward what many believed would be the next great leap in commercial aviation: supersonic flight.

Throughout much of the 20th century, advances in aviation had been measured largely by increases in speed, capacity, and range. Jet power had already cut transatlantic travel times roughly in half, reducing a flight from New York to London from about twelve hours in a piston-powered airliner to six hours in a jet. Supersonic aircraft promised to halve that time again, making the same journey in approximately three hours.

Military requirements had often driven advances that later found civilian applications, and World War II had produced enormous gains in aircraft speed, size, range, propulsion, and structural design. Technologies developed to carry heavier bomb loads farther and faster could, in civilian service, be adapted to carry fare-paying passengers over long distances in less time. Recognizing that possibility, Allen had established a Boeing study group in the early 1950s to examine the feasibility of a commercial supersonic transport.

Boeing was not alone in looking toward that future. Britain and France, having entered the commercial jet age with their own passenger aircraft, were also exploring supersonic travel. Both began feasibility studies in the early 1950s, and by the beginning of the next decade each was moving toward development of a commercial supersonic transport.

But the obstacles were formidable, both technically and financially. Neither country possessed an aerospace industry nor the financial means capable of funding such an undertaking entirely through private capital, nor was either government eager to shoulder the full cost alone. But by combining their industrial and financial resources, Britain and France could make the project possible. Their partnership gave rise to one of the most iconic aircraft ever built — Concorde.

Despite being the world's largest aircraft manufacturer by the early 1960s, Boeing could not finance a comparable aircraft on its own. Development of an American supersonic transport would require substantial federal support. The more fundamental question, however, was whether the United States needed to produce a rival to Concorde at all.

Supporters envisioned a future in which passengers crossed oceans, continents, and even individual countries at supersonic speed. A New York executive might fly to London in the morning, attend a luncheon meeting, and return home in time to brief colleagues that same afternoon. The possibilities seemed intoxicating, especially when viewed against aviation's recent history. In little more than three decades, passenger aircraft had evolved from wood-and-fabric machines carrying a handful of people at little more than 120 miles per hour into sleek metal airliners powered by turbine engines and carrying more than a hundred passengers at near the speed of sound.

Opponents, however, questioned whether taxpayers should underwrite such an expensive commercial venture. Their argument was straightforward: if private investors were unwilling to finance a supersonic transport, perhaps the potential market was not large enough to justify one. It was a difficult argument to dismiss.

Supersonic speed would inevitably command a premium. The enormous cost of developing and producing such an aircraft would have to be recovered through higher fares. For a leisure traveler flying from New York to Paris, arriving in three hours rather than six might not justify paying 50 percent more — or perhaps considerably more — for a ticket. A few extra hours aboard a conventional jet would be a small sacrifice if the savings could instead be spent at the destination.

For business travelers, however, time was money. The wealthy and well-connected would also pay for the fastest and most prestigious form of transportation available, even when traveling for pleasure. In the early 1960s, business travelers and affluent passengers represented a much larger share of the flying public than they would in later decades. The question was whether that comparatively small market could generate enough revenue to justify the extraordinary development cost of a supersonic airliner.

The answer was almost certainly no. Government support would be essential, and for the moment, politicians opposed to committing taxpayer money to the project held the upper hand.

When the John F. Kennedy administration entered the White House in 1961, debate was already raging over whether the federal government should finance the development of a commercial supersonic transport. Kennedy appointed Najeeb Halaby as administrator of the Federal Aviation Administration. Born in the United States to a prominent Arab American family, Halaby had distinguished himself as a World War II aviator and, shortly after the war, became the first pilot to make a nonstop transcontinental flight in a jet aircraft. Deeply connected to the aviation and aerospace communities, he became an ardent advocate for an American supersonic transport, commonly known as an SST.

Through his political and industry connections, Halaby helped persuade the Kennedy administration to support government-funded development of an American SST. The argument was straightforward: Britain and France were moving ahead with a jointly funded supersonic airliner, and the United States risked appearing technologically inferior if Europe succeeded while America remained on the sidelines. If supersonic aircraft represented the future of commercial aviation, the United States could lose not only its leadership of the industry but also tens of thousands of aerospace jobs.

Such an outcome would have been politically difficult to accept. The United States was still recovering from the shock of the Soviet Union's Sputnik launches and remained determined to avoid another "Sputnik moment." Barely a decade earlier, Britain had gained an early lead in jet transportation with the de Havilland Comet, and only the aircraft's design failures had allowed American manufacturers to overtake it. Although many Europeans questioned whether an SST could ever be commercially viable, the project had become a matter of national pride in both Britain and France and enjoyed broad public support.

The Anglo-French Concorde project was formally announced in November 1962. The following May, while the United States was still debating whether to develop an SST at taxpayer expense, Pan American World Airways stunned the aviation industry by taking options on six Concordes. Pan Am joined Air France and British Overseas Airways Corporation as one of the proposed launch customers for the European aircraft.

The announcement sent shock waves through both the aviation industry and Washington. Congress moved quickly, authorizing the FAA to solicit proposals for preliminary designs from Boeing, Lockheed, and North American Aviation. Funding was also provided for the major engine manufacturers to study the powerplants required to propel the new aircraft beyond the sound barrier for sustained periods.

After reviewing the initial proposals, the government selected Boeing and Lockheed to proceed with detailed designs. Lockheed's L-2000 resembled Concorde but was considerably larger, with seating for slightly more than 200 passengers compared with approximately 100 aboard Concorde. Boeing's 2707 — its designation deliberately evoking the successful 707 — was more radical. It incorporated variable-geometry, or "swing," wings and was designed to carry as many as 300 passengers. Both American designs were also intended to fly faster than Concorde's Mach 2 cruising speed. Lockheed proposed a speed approaching Mach 3, while Boeing targeted Mach 2.7.

As part of the final competition, Boeing and Lockheed constructed full-scale mock-ups for evaluation in late 1966. In January 1967, Boeing's 2707 was declared the winner. Lockheed's L-2000 was faster and presented fewer technical risks because of its conventional fixed-wing design, but Boeing's more complicated swing-wing configuration promised several operational advantages.

With the wings swept forward, the 2707 could take off and land at speeds closer to those of conventional jetliners. It could therefore operate from many airports already served by large passenger aircraft without significantly disrupting existing traffic patterns. Once airborne, the wings could sweep backward for high-speed flight. Because their position could be optimized for different phases of flight, Boeing also argued that the aircraft would be more fuel-efficient than a fixed-wing design.

The 2707's greater passenger capacity offered another advantage. The cost of each flight could be distributed among more fare-paying passengers, theoretically producing either higher profits or lower ticket prices and making supersonic travel available to a broader segment of the public.

Boeing moved quickly into development. It expected to begin constructing the first prototype in 1969, conduct the initial flight tests by 1972, and place the aircraft into airline service by the middle of the decade.

***

While the SST debate intensified, another major opportunity emerged. The U.S. Air Force needed an aircraft capable of rapidly transporting outsized cargo around the world. Although the Soviet Union was steadily expanding its military capabilities, the United States remained the only nation able to project substantial power almost anywhere on the globe. Preserving that advantage required the ability to move tanks, trucks, helicopters, and large quantities of palletized cargo across intercontinental distances. Even the largest military transports then in service could not perform that mission efficiently or in sufficient volume.

In 1963, the Air Force invited Boeing, Douglas, and Lockheed to propose a new heavy transport. Among its requirements was a cargo opening approximately 17 feet wide and nearly 14 feet high. The aircraft also had to permit loading through both the front and rear of the fuselage, allowing vehicles and cargo to be driven or pushed through without reversing the loading process. The new transport was required to carry approximately 180,000 pounds of cargo — nearly three times the capacity of the C-141 Starlifter, the recently introduced Lockheed jet transport that formed the backbone of the Air Force's strategic airlift fleet.

The Air Force's specifications led all three manufacturers toward remarkably similar configurations. Each proposed a cavernous fuselage, wings mounted above the cargo compartment, and a cockpit positioned above the main deck. The elevated cockpit preserved the full length of the cargo compartment while permitting the aircraft's nose to open for front loading.

The differences were comparatively minor. Boeing and Douglas placed their horizontal stabilizers roughly in line with the upper fuselage, while Lockheed adopted a T-tail with the horizontal stabilizer mounted atop the vertical fin. Lockheed positioned its cockpit at the front of a raised spine extending along the fuselage, giving the aircraft an egg-shaped cross section. Boeing and Douglas instead used a pronounced cockpit bulge above the forward fuselage.

All three designs depended on an entirely new generation of engines. These turbofans were enormous compared with contemporary jet engines, with front fans measuring more than seven and a half feet in diameter rather than the roughly four-foot diameter common at the time. They were known as high-bypass turbofans because most of the air entering the engine flowed around, or bypassed, the combustion core. Much of the thrust therefore came from the large volume of air accelerated by the fan rather than from the hot exhaust leaving the turbine. The new engines promised nearly twice the thrust of existing powerplants while consuming less fuel and producing less noise.

Because the three aircraft offered broadly comparable capabilities, price became a decisive factor. In late 1965, the Air Force selected Lockheed, the lowest bidder. Its design entered production as the C-5 Galaxy and, after extensive modernization over subsequent decades, continues to serve as one of the Air Force's principal strategic transports.

The loss was disappointing but hardly devastating to Boeing. Its Seattle factories were already busy producing the 707 and 727 and preparing for the introduction of the 737. Even so, losing such a prominent government competition was unusual for a company that had enjoyed a close relationship with the military since its earliest years and had secured some of the nation's most important aircraft contracts during World War II and the early Cold War.

Yet the competition could scarcely have come at a more opportune time. Although Boeing lost the military contract, the design work it completed would soon help the company secure one of the greatest victories in the history of commercial aviation.

***

Around the same time that the Air Force was seeking a new heavy military transport, Pan Am president Juan Trippe approached Boeing about developing a much larger passenger aircraft. Air travel remained relatively expensive, but rising prosperity in the United States and Western Europe, combined with the speed and capacity of the new jetliners, was placing increasing strain on the world's major airports. Many had simply expanded facilities originally built during the 1920s and 1930s for much smaller piston-powered aircraft. Congestion was becoming a serious problem.

Trippe firmly believed that supersonic transports represented an important part of aviation's future and had already placed options on Concorde. Yet he also understood that regular SST service remained at least a decade away and, when it arrived, would initially be affordable only to business travelers and the wealthy. For decades, he believed, supersonic aircraft would operate alongside conventional subsonic jetliners. SSTs would serve travelers willing to pay a premium for speed, and conventional aircraft carrying everyone else.

SSTs would also suffer from limited range. Their enormous fuel consumption would confine them largely to direct routes between major population centers, such as New York or Washington and London or Paris. Passengers continuing to more distant destinations would have to transfer to another flight. That transfer could erase much of the SST's advantage. By the time a supersonic aircraft descended, landed, taxied to the terminal, unloaded its passengers, and transferred them to another flight, a long-range subsonic jet that remained continuously at altitude might arrive at nearly the same time. Even when the SST remained faster overall, the difference might not be great enough to offset the lower cost and greater convenience of a nonstop conventional flight.

Trippe had another motivation. A larger aircraft carrying more passengers could distribute its operating costs across more seats. Lower fares would make flying affordable to more people, generating still greater demand for air travel.

Bill Allen immediately recognized the implications. More passengers would mean more flights and, ultimately, more airplanes. Boeing's own studies showed that passenger traffic would continue to grow faster than new airports could be built or existing ones expanded. A much larger aircraft could carry more people without requiring a corresponding increase in the number of takeoffs, landings, and airport gates.

The question was how to create such a dramatic increase in capacity. Boeing had already stretched the 707 several times, and the latest iteration, the 707-320, was approaching the practical limits of the design. What was needed was not another incremental enlargement but a quantum leap — an aircraft capable of carrying so many more passengers that it could fundamentally reduce the cost of air travel.

Boeing had recently spent considerable time and money developing its unsuccessful proposal for the Air Force's giant transport competition. Rather than begin again from a blank sheet of paper, the company examined whether elements of that design could be adapted for commercial service.

Joe Sutter, the senior engineering manager who had worked on the 737 program, was placed in charge of the new project. Like the 737, the aircraft he was about to develop would become far more successful than anyone initially imagined. It also presented the kind of complex engineering challenge that Sutter relished. His team began by examining Boeing's military transport proposal to determine which features could be retained and which would have to be discarded or substantially redesigned.

One capability Boeing wanted to preserve was the ability to load cargo through the nose. Although the new aircraft would initially serve primarily as a passenger transport, many industry leaders believed that SSTs would eventually dominate long-distance passenger travel. Once that happened, they reasoned, very large subsonic aircraft would be relegated largely to carrying freight, which did not need to travel at supersonic speeds. To accommodate that anticipated future, Sutter's team retained the elevated cockpit used in Boeing's military transport design. Placing the flight deck above the main fuselage allowed the aircraft's nose to be hinged upward, providing direct access to the cargo deck if and when the design was retired from passenger service and started a new career as a freighter.

The engineers also considered whether the aircraft should have two full-length passenger decks or one very wide main deck. A double-deck arrangement promised even greater capacity, but it created serious boarding and evacuation problems. Passenger boarding bridges were only beginning to appear, and most equipped gates had just one. Unless airports installed two bridges at each gate, passengers destined for the upper deck would have to board through the main cabin and climb an internal stairway, slowing the boarding process.

Wing placement presented another challenge. Boeing's Air Force proposal had used a high-mounted wing so that the cargo floor could sit close to the ground. The configuration made it easier to load heavy equipment at remote airfields and kept the engines clear of dirt and debris on unimproved runways. Those advantages were unnecessary for a commercial airliner. A high wing would place the engines far above the ground, making routine inspection and maintenance more difficult and requiring specialized lifting equipment. Commercial aircraft flew almost every day and therefore required more frequent servicing than military transports.

Sutter's team considered mounting the wing midway up the fuselage, between the decks of a double-deck design, as well as placing it conventionally beneath the main cabin. Emergency evacuation favored the lower position. During a water landing, much of an aircraft's buoyancy comes from its wings. Passengers seated below them could therefore find themselves dangerously close to, or beneath, the waterline.

Boeing ultimately selected a single very wide passenger deck above a low-mounted wing. The configuration provided ample room for passengers while preserving the aircraft's future usefulness as a freighter capable of carrying large pallets and outsized cargo.

Each wing would carry two of the newly developed high-bypass turbofan engines. The aircraft was expected to seat approximately 365 passengers in the mixed first-class and tourist-class arrangement typical of the era — roughly twice the capacity of the 707-320. Boeing designated the new aircraft model as the 747.

Allen and his team presented the design to Trippe and Pan Am's leadership. They enthusiastically embraced it. In April 1966, Pan Am placed an order for 25 aircraft valued at more than $550 million, the largest commercial airplane order ever placed up until that time. The order formally launched the program, and Sutter's engineering team began the enormous task of designing, building, testing, and certifying an aircraft unlike anything the industry had previously attempted.

Douglas and Lockheed soon began developing widebody airliners of their own. Neither directly challenged the 747, because the market for aircraft of its size was unlikely to support two competing designs. Instead, both manufacturers targeted the next segment down — the high-density routes that required more capacity than existing narrow-body jets but not an aircraft as large as the 747.

Whether that market was large enough to sustain both companies was another question, but neither was willing to concede it. Douglas developed the DC-10, while Lockheed produced the L-1011 TriStar. Each used three high-bypass turbofans — one beneath each wing and a third mounted in or near the tail — and was designed to carry roughly 270 passengers over distances of approximately 3,000 miles.

The American aerospace industry had dominated commercial aviation throughout the postwar era and the first decade of the jet age. As the 1970s approached, it appeared poised to dominate the next generation as well, with widebody airliners and supersonic transports expected to shape aviation far into the future. From nearly every perspective, Boeing appeared positioned to lead across the entire market.

Bill Allen must have been pleased by the view from his Seattle office when he decided to step down as Boeing's president in 1968. He had led the company for more than two decades, beginning at the end of World War II, and had guided it through immense economic, political, and technological change.

When Allen took control of the company, Boeing depended primarily on military contracts for its revenues. By the end of his tenure, commercial aircraft generated the larger share of its business, and Boeing had become the world's leading commercial airplane manufacturer. Half a century after the company's founding, Bill Boeing's ambition to build aircraft primarily for civil transportation had finally been realized.

Under Allen, Boeing had evolved from a leading military aircraft manufacturer into the preeminent innovator in commercial aviation and space. No previous Boeing leader — and arguably no executive in the history of aviation — had overseen the development of so many consequential aircraft.

During his tenure, Boeing produced the Model 377 Stratocruiser, B-47 Stratojet, B-52 Stratofortress, KC-135 Stratotanker, and the 707, 727, 737, and soon-to-fly 747. With the exception of the B-47, examples or later derivatives of nearly all those aircraft would remain in operation well into the 21st century. Even the Model 377's military descendant, the KC-97, lived on through the heavily modified Super Guppy, which would continue carrying rocket stages and other outsized cargo into the 21st century.

The time seemed right for Allen to hand control of the company to a successor, although he would remain on Boeing's board for several more years. His replacement was Thornton A. "T." Wilson. Allen had spent years grooming Wilson for senior leadership and pushed for his selection as Boeing's next president.

Wilson had joined Boeing in 1943 after earning a graduate degree in aeronautical engineering. Assigned to the company's bomber programs, he contributed to the design of the B-47 and later served as project engineer on the B-52. Allen had selected him to lead Boeing's efforts on the Minuteman missile project. In 1964, again with Allen's support, he became head of corporate operations and planning, which put him on a direct path to senior leadership.

With Wilson's appointment as CEO, Boeing's leadership once again passed to an engineer who had risen through the company's ranks, much as it had under Claire Egtvedt and Philip Johnson before Allen. Wilson possessed the instincts of an innovative engineer, but also the resolve and confidence required to make difficult decisions.

Boeing could scarcely have chosen a more suitable leader, for the next several years would prove among the most consequential in the company's history.

Europe’s Answer to Europe’s Fears

By 1969, the political headwinds in the United States were turning against development of a supersonic transport. Boeing's SST program was falling behind schedule as engineers struggled with the immense technical challenges of pushing an airliner beyond the speed of sound. Meanwhile, the public was becoming increasingly alarmed about the aircraft's potential environmental effects. In addition to the disruption caused by sonic booms, scientists and environmental groups warned that SST emissions at high altitude might accelerate depletion of the ozone layer. Members of Congress began discussing whether to reduce — or eliminate — federal funding for the program.

Yet whether the SST survived or not, the future of commercial aviation still appeared firmly anchored in the United States. Boeing, Douglas, and Lockheed were all developing widebody airliners, programs that supported tens of thousands of jobs and built upon an already enormous American aerospace industrial base. At the same time, the United States was pulling ahead of the Soviet Union in the race to the Moon. The federal government had poured vast sums of taxpayer money into the aerospace industry to achieve President Kennedy's goal of landing astronauts on the lunar surface before the end of the decade. That investment would produce technological advances not only in aviation and spaceflight, but across numerous other industries.

By the late 1960s, the American aerospace industry dwarfed that of every other region of the world. Across the Atlantic, the Anglo-French Concorde program was delayed, massively over budget, and facing declining public enthusiasm. Although Europe retained a small lead in the development of a commercial SST, Boeing was advancing rapidly despite its own difficulties. More important, Boeing and Douglas jetliners already dominated airline fleets around the world, and Lockheed was now entering the commercial transport market as well.

European leaders began sounding the alarm. The enormous programs underway in the United States were giving American manufacturers an industrial capacity and technological capability that Europe might never be able to match. Boeing's 747, Douglas' DC-10, and Lockheed's L-1011 threatened to consolidate American supremacy in commercial aviation for the remainder of the century and even beyond.

Europe possessed a long and distinguished aviation history, along with skilled engineers and mature aircraft manufacturers. But the industry was fragmented along national lines, with individual companies serving domestic markets too small to support development programs on the scale of those underway in the United States. Unless Western Europe acted quickly, its aircraft manufacturers risked being reduced to second-tier suppliers for the American giants. Thousands of aerospace jobs — and perhaps the continent's independent aircraft industry itself — were at stake.

In 1967, the governments of Britain, France, and West Germany signed a memorandum of understanding to form a loose coalition that would develop and market a widebody airliner. The proposed consortium would combine the best capabilities of the three nations' aerospace companies and compete directly with the Americans.

Britain eventually withdrew. With Concorde's costs escalating and its commercial future increasingly doubtful, the British government was reluctant to commit additional taxpayer money to what the public feared might become another white elephant. France and West Germany forged ahead, and in 1969 they formally created Airbus Industrie.

Although Airbus was intended to compete commercially, its founding governments understood that it could not survive without financial assistance. The devastation of World War II, the smaller size of the European national economies, and the fragmentation of the continent's aerospace sector left its manufacturers without the financial resources of their American competitors. Therefore, Airbus became more than a commercial venture. It was also an instrument of industrial policy, created to preserve and advance the aerospace interests of Western Europe in general, and France and West Germany in particular.

At first, this government involvement caused little concern in the United States. American suppliers expected the Europeans to purchase many of their aircraft systems and components from established U.S. manufacturers. Boeing, Douglas, and Lockheed also regarded Airbus as little more than a minor entrant. How could a start-up threaten companies that had spent decades accumulating technical expertise, building worldwide service networks, cultivating airline executive relationships, and establishing reputations for quality and innovation?

Airbus had an enormous amount of ground to make up. By 1969, Boeing's 747 was already undergoing flight testing and was scheduled to enter airline service by 1970. Douglas and Lockheed were approximately a year behind, with deliveries of the DC-10 and L-1011 expected in 1971. Lockheed would later encounter delays when Rolls-Royce, the manufacturer of the L-1011's engines, fell into financial crisis and entered bankruptcy protection, but the aircraft would nevertheless enter service in 1972.

While its American competitors were flying prototypes or had already begun cutting metal, Airbus engineers were still confronting a blank sheet of paper. The odds appeared overwhelmingly against them.

The American manufacturers had been entrenched in the airlines of the noncommunist world since the end of World War II. Their sales organizations operated like well-oiled machines, and their executives had longstanding relationships with the world's major airlines. Their aircraft were admired by those who built, maintained, flew, and traveled aboard them. Their reputations for quality were largely beyond reproach and, unlike Britain's ill-fated

Comet, had not been devastated by a series of crashes caused by fundamental design defects.

The Americans also possessed worldwide service organizations. When one of their aircraft was grounded by a mechanical problem, trained engineers and mechanics were usually nearby. Spare parts were stocked in regional warehouses or could be delivered through established distribution networks. Airbus had no comparable infrastructure. Even if it could design an outstanding aircraft, airlines would hesitate to purchase it unless they were confident that Airbus could keep it flying.

One way Airbus sought to reduce that resistance was by competing on American terms. Although most of the world had adopted the metric system, the United States — the center of global aircraft manufacturing — continued to use imperial measurements. American airliners were designed in inches and feet, and their weights were expressed in pounds.

Airbus decided to use imperial measurements as well. Doing so made its aircraft easier to compare with American models, eliminated cumbersome conversions during sales presentations, and allowed the company to quickly tap into established supply chains for fasteners, sheet metal, subassemblies, and other standardized components.

The consortium also adopted English as its official working language. German-speaking engineers and French-speaking engineers would conduct their joint business in the language already dominant in international aviation. These decisions made Airbus more compatible with the industry it was attempting to enter.

Its most consequential decision, however, concerned the aircraft itself. Although the new Airbus would be broadly comparable in capacity to the Douglas DC-10 and Lockheed L-1011, it would use two engines rather than three.

The emerging widebody market was already crowded. Boeing's 747 occupied the upper end of the market. It was enormous even by comparison with the other widebodies under development, using four engines to carry approximately 365 passengers — or more, depending upon configuration — over long intercontinental routes.

Douglas and Lockheed deliberately avoided competing directly with it. Their DC-10 and L-1011 were roughly one-third smaller than the 747 by maximum takeoff weight and were intended to carry approximately 250 to 270 passengers in typical cabin configurations on heavily traveled domestic and shorter international routes. Each used three large turbofan engines and became known as a trijet. On the DC-10, the third engine was mounted above the fuselage through the vertical stabilizer. On the L-1011, it was installed within the rear fuselage and fed by an S-shaped intake positioned ahead of the tail, an arrangement reminiscent of Boeing's smaller 727.

The market for aircraft as large as the 747 was limited, which was one reason Douglas and Lockheed chose not to challenge it directly. The larger opportunity appeared to lie in carrying substantial numbers of passengers over shorter distances. North America, Western Europe, and parts of Asia all contained dense routes that could support widebody aircraft smaller than the 747. Airbus believed those markets might also support a third competitor.

To succeed, however, it needed to distinguish its aircraft from the two American trijets. From that challenge emerged both the genesis and the genius of Airbus.

The company designed its new A300 around only two of the powerful turbofan engines being developed for the new generation of widebody airliners. The design was intended to carry approximately 240 to 260 passengers on short to medium-haul routes. The result was an aircraft capable of serving many of the same routes as the American trijets, but with one fewer engine. Airbus was able to accomplish this because, although it would carry roughly the same number of passengers as its American competitors, the airframe was lighter because it was designed for a shorter range — it wouldn't need to carry as much fuel as its American competitors who planned to pitch their models to airlines who would deploy them on longer intercontinental routes.

That difference offered powerful economic advantages. One fewer engine meant lower fuel consumption and therefore a lower cost per available seat-mile — the industry measure of how much it costs to fly one passenger seat one mile. Because those costs were spread across roughly the same number of seats as aboard the competing trijets, the A300 could potentially generate profits more easily on the same routes.

The aircraft would also cost less to purchase. Engines were among the most expensive components of a jetliner, accounting for a substantial portion of its total acquisition cost. Eliminating one engine reduced not only the purchase price, but also the cost of maintenance, spare engines, replacement parts, and overhaul work. A twin-engine aircraft could also offer greater dispatch reliability. With one fewer engine to inspect, service, and potentially suffer a mechanical problem, there were fewer opportunities for an engine-related delay to prevent an on-time departure.

There was, however, one important disadvantage. Aviation regulations limited how far a twin-engine airliner could travel from a suitable airport while flying across oceans or other remote regions. Authorities in North America and Western Europe generally required twin-engine aircraft to remain within approximately one hour's flying time of an alternate airport in the event one engine had to be shut down. That was also another reason Douglas and Lockheed incorporated a third engine on their models, which was to not be restricted on long overwater routes by regulatory restrictions.

Although jet engines were far more reliable than the piston engines they replaced, regulators believed that the loss of one engine might indicate an increased risk — however remote — that the remaining engine could also fail. Aircraft with three or four engines were considered to possess greater redundancy and were therefore permitted to fly more direct routes farther from diversion airports.

Airbus was not deterred. Most of the routes it initially intended the A300 to serve were within North America and Europe and did not require prolonged flights over large bodies of water. On those routes, the limitations imposed upon twin-engine aircraft mattered little, while the economic advantages of operating one fewer engine mattered greatly.

All four manufacturers — Airbus, Boeing, Douglas, and Lockheed — were entering the widebody era at roughly the same time. Airbus' two-engine configuration had the potential to become a genuine game changer, particularly in the heavily traveled domestic and regional markets targeted by the DC-10 and L-1011.

Yet Airbus still faced a long and difficult struggle. In little more than a decade, the commercial aviation market had changed dramatically, and the United States had established itself as the global standard for advanced passenger aircraft. Even if Airbus produced an engineering marvel, airlines would inevitably compare it with the products, sales organizations, and support networks of the American manufacturers.

Airbus lacked both the experience and the worldwide infrastructure of its competitors. It was also entering the market approximately two years behind the DC-10 and L-1011. Douglas had begun developing the DC-10 in 1967, and Lockheed had launched the L-1011 soon afterward. Airbus did not begin detailed development of the A300 until 1970. That delay would become one of the greatest obstacles to the aircraft's commercial success.

Initial sales were weak. The first orders came largely from airlines associated with Airbus' sponsoring governments, including Air France and Lufthansa. Then, just as the A300 was approaching certification in 1973, the Arab oil embargo sent fuel prices soaring and triggered a recession that resulted in a severe downturn in aircraft sales.

The timing could scarcely have been worse. Between 1974 and 1977, Airbus failed to record a single new A300 sale. Production slowed to one aircraft every two months.

To keep the assembly line operating, Airbus began building "white tails" — aircraft produced without customers, painted white, and placed in storage in the hope that buyers would eventually be found. By early 1977, executives at the company's headquarters in Toulouse were openly discussing the possibility that A300 production might be halted permanently. Some believed only six to eight weeks of work remained.

Then, from across the Atlantic, came a lifeline.

Deregulation and How an American Astronaut Saved European Aviation

Beginning in 1938, the federal government — through the Civil Aeronautics Board, or CAB — regulated nearly every aspect of commercial aviation within the United States. The CAB treated air transportation much like a public utility. Competition was limited, and the government determined which routes an airline could serve and how much it could charge for a ticket.

The CAB's mandate was to promote both the growth and safety of air travel. It sought to ensure that airlines earned a reasonable return on the routes they operated, theoretically reducing the temptation to cut corners in pursuit of profit. It also required carriers to serve smaller or remote communities that could not otherwise support profitable air service.

Nearly every important airline decision required CAB approval, and the process was rarely quick or easy. The barriers to starting a new airline were almost insurmountable. Most of the major carriers operating during the 1970s had been established before the CAB was created and were effectively grandfathered into the regulated system. There were exceptions. Southwest Airlines, now one of the world's largest carriers, began operations in 1971, at the height of CAB regulation. It avoided federal oversight by operating exclusively within the state of Texas. Because it did not cross state lines, it remained outside the CAB's jurisdiction.

By the 1970s, however, the CAB had become a large and lumbering bureaucracy. Airline requests could take years to approve. In several notorious cases, applications to begin serving new routes remained unresolved for nearly a decade.

During its early decades, the CAB had largely accomplished what it had been created to do. It fostered a stable environment in which commercial aviation could grow and flourish. By the 1960s, however, many policymakers and economists believed it had outlived its usefulness and was becoming an impediment to the industry it had helped create. As with most government agencies, dismantling it would prove far more difficult than establishing it.

Several developments during the early 1970s forced Congress to reconsider whether the CAB could keep pace with the rapidly changing demands of air transportation.

The first was the extraordinary growth in passenger traffic. The number of people traveling by air had doubled since the beginning of the jet age, and the arrival of widebody aircraft promised another surge before the end of the decade. Their greater capacity and lower seat-mile costs were expected to make air travel affordable to more people and open additional markets. That growth would require airlines to adjust routes, fares, and flight frequencies far more quickly than the CAB's cumbersome approval process allowed. There was growing concern that the agency would become a bottleneck, preventing the industry from responding efficiently to changing market conditions.

The second development was the 1970 bankruptcy of the Penn Central Railroad, the largest corporate bankruptcy in American history up until that time. The railroad industry had been declining throughout the postwar period as air travel expanded and the interstate highway system allowed trucks to move freight more flexibly and, in many cases, at lower cost. Many observers believed the railroads might have adapted more effectively had they not been constrained by the Interstate Commerce Commission. Heavy regulation limited their ability to abandon unprofitable routes, change prices, and restructure their operations. Penn Central's collapse raised fears that the airline industry might eventually suffer the same fate.

The bankruptcy prompted Congress to examine the regulation of the railroads and, more broadly, the nation's entire transportation system. Increasing competition and improving efficiency became central objectives of that review.

The third development was the Arab oil embargo of 1973. The sudden restriction of petroleum supplies delivered a severe blow to the American economy and hit the airline industry especially hard. Fuel costs became an enormous burden on airline profit-and-loss statements, upending business models that had been built during an era of relatively inexpensive energy.

Airlines needed the freedom to alter schedules, routes, and fares quickly as operating conditions changed. Once again, the CAB's regulatory machinery hindered their ability to respond efficiently and protect their profitability.

Another consideration was the continuing cost of air travel. Although flying had become affordable to a broader portion of the population, it remained beyond the routine reach of many Americans. Air travel was still dominated by business travelers and the financially well-off. When the average family flew, it was often a rare event requiring considerable advance planning and saving.

Economists, members of Congress, and much of the public increasingly recognized the broader economic benefits of making air travel more accessible. Greater competition, they argued, would lower fares, expand service, and allow more individuals and businesses to take advantage of the speed and mobility aviation offered.

On October 24, 1978, President Jimmy Carter signed the Airline Deregulation Act into law. Its official long-title name was, *An Act to amend the Federal Aviation Act of 1958, to encourage, develop, and attain an air transportation system which relies on competitive market forces to determine the quality, variety, and price of air services, and for other purposes.* Its formal title described its purpose clearly.

The act did not eliminate federal oversight of aviation safety. Rather, it gradually removed the CAB's authority over fares, routes, and market entry. The agency itself was dissolved at the end of 1984, with its remaining responsibilities transferred to other departments.

The immediate consequence was that America's airlines suddenly found themselves operating in a vastly different economic environment. Although carriers had frequently complained about CAB interference, many had also benefited from the stability it provided. On approved routes, fares were structured to offer airlines the opportunity to earn a reasonable return. That did not guarantee that an airline would be profitable overall. The CAB could not control labor expenses, administrative overhead, poor management, or other indirect costs. Airlines could — and sometimes did — lose money despite operating within the regulated system. Nevertheless, regulation provided a degree of protection from direct price competition and unrestricted entry by new rivals.

Deregulation removed that protection. Airlines were now responsible for finding their own paths to profitability while redesigning and optimizing their route networks. They were like adult children who had spent their lives in a comfortable home governed by strict but protective parents, only to be suddenly kicked out the door and into the outside world and expected to survive on their own.

The Airline Deregulation Act did not emerge overnight. It was the culmination of years of debate, particularly after the Penn Central bankruptcy prompted Congress to examine the nation's transportation infrastructure. The oil embargo had further exposed the airline industry's vulnerability to rapidly changing economic conditions.

Most airlines had grown accustomed to the CAB and valued the stability it provided. Yet even they recognized that the industry needed greater flexibility if it was to remain healthy, competitive, and capable of surviving whatever challenges the future might bring.

By 1977, Eastern Air Lines carried one of the heaviest debt burdens in the American airline industry and depended on one of its most seasonal revenue streams. As its name suggested, Eastern had built its network primarily along the Eastern Seaboard and into

Florida and the Caribbean. By the early 1970s, it carried more passengers than any other airline in the Western world.

Eastern's traffic peaked during the winter, when residents of the Northeast and Midwest joined its regular business travelers in heading south for warmer climates. Routes between New York and Florida were among the busiest in its system. More than most major airlines, Eastern depended heavily on vacationers and holiday travelers, leaving it especially vulnerable to economic downturns that caused consumers to postpone discretionary travel.

By the spring of 1973 — months before the Arab oil embargo — the United States was already confronting an emerging energy shortage. Although the country at the time imported roughly 30 percent of the petroleum it consumed, the immediate problem was not simply a shortage of crude oil. Supplies of refined products, particularly gasoline and aviation fuel, were also tightening as demand outpaced domestic production and refinery capacity.

By the early 1970s, the American economy had outgrown much of the nation's petroleum infrastructure. At the same time, the Nixon administration's wage and price controls created uncertainty within the oil industry and discouraged investment in additional production and refining capacity. The controls had been imposed as part of Nixon's response to rising inflation and the economic disruption surrounding the United States' 1971 withdrawal from the Bretton Woods monetary system.

The shortages worsened during the peak summer travel season of 1973. Then, in October, Egypt and Syria attacked Israel, beginning the Yom Kippur War. After the United States provided military support to Israel, the Arab members of the Organization of Arab Petroleum Exporting Countries imposed an embargo against the United States and several other nations that supported Israel.

The embargo sharply reduced oil supplies and helped drive petroleum prices to unprecedented levels. What had begun as an energy shortage became a national economic emergency.

The automobile industry was among the sectors hit hardest. Detroit's assembly lines were producing predominantly large, heavy automobiles powered by equally large engines. Many achieved little more than eight to ten miles per gallon. As gasoline prices rose and shortages produced long lines at service stations, consumer demand shifted abruptly away from the gas-guzzling cars that dominated the American market. Because automobile manufacturing and its vast network of suppliers, dealers, lenders, and related industries represented a major portion of the American economy, the effects spread quickly. Consumers curtailed discretionary spending, and leisure and vacation travel were among the first expenses to be eliminated.

Few airlines were more exposed than Eastern. Its seasonal business model depended on a strong winter travel market, particularly to Florida and the Caribbean. The winter of 1973–1974 became one of the worst periods in the airline's history. The collapse in discretionary travel struck Eastern harder than most competitors, while its substantial debt — much of it incurred to purchase new jet aircraft — left it with little financial flexibility.

The recession that began in November 1973 was both deep and prolonged. Although the economy officially emerged from it in March 1975, the recovery did not restore the confidence or stability Americans had enjoyed during much of the postwar era. High inflation persisted and incomes stagnated — a painful combination that became known as "stagflation."

The oil crisis also delivered a psychological shock. For the first time in the postwar era, many Americans confronted the possibility that the nation's prosperity depended on resources and events beyond its control. Long gasoline lines, rising prices, unemployment, and turmoil in the Middle East created a sense of national vulnerability. The future no longer appeared as predictable or assured as it once had.

That uncertainty further discouraged consumer spending, prolonging the airline industry's difficulties even after the recession formally ended. Airlines faced higher fuel costs, weaker passenger demand, and a regulatory system that restricted their ability to adjust routes, schedules, and fares quickly enough to respond.

Amid that turmoil, Congress began reconsidering whether the regulatory structure governing American transportation was protecting the airlines or preventing them from adapting. The oil crisis did not by itself create the movement toward airline deregulation, which had been gathering strength for years, but it added urgency to the debate.

For the airlines, deregulation promised both freedom and danger. It would allow them to respond more rapidly to changing markets, but it would also remove many of the protections they had enjoyed under federal regulation. It was a double-edged sword — but one that eventually had to be drawn. The economic upheavals of the 1970s had made it increasingly clear that the status quo could not continue.

***

In December 1975, an American hero named Frank Borman took the reins of Eastern Air Lines as its chief executive officer. Borman was known throughout the world as one of the pioneering astronauts of the Space Race. He had flown into space twice during the 1960s, but it was his second mission that made him a legend.

In December 1968, Borman commanded Apollo 8, accompanied by James Lovell and William Anders. The three became the first human beings to escape Earth's gravitational embrace, travel through deep space, and orbit another world. At the time, Apollo 8 ranked among humanity's greatest technological achievements, eclipsed only by the Apollo 11 lunar landing seven months later.

Borman retired from NASA and the Air Force in 1970 and joined Eastern as senior vice president of operations. Although he had little conventional business experience, he was accustomed to making difficult decisions under intense pressure. It would not take long for Eastern to witness that quality firsthand.

Eastern was the launch customer for Lockheed's L-1011 TriStar, having ordered the new widebody jet during the late 1960s. The airline placed the first production L-1011 into service in April 1972.

Shortly before midnight on December 29 of that year, Eastern Flight 401, operated by an L-1011 delivered only four months earlier, crashed into the Florida Everglades while approaching Miami International Airport. Of the 176 people aboard, 101 were killed.

When Borman was awakened in the middle of the night by his operations staff informing him of the crash, he immediately secured the use of a helicopter in Miami where he was living and flew it to the crash site where he helped rescue survivors from the murky waters and get them to emergency responders. His highly visible response to the catastrophe reinforced his reputation as a decisive leader willing to place himself at the center of a crisis rather than direct events from a distant office.

The investigation ultimately determined that the aircraft itself had not caused the crash. The flight crew had become preoccupied with a malfunctioning landing-gear indicator light and failed to notice that the autopilot had disengaged from its altitude-hold mode. The aircraft gradually descended into the Everglades. It was the first fatal crash involving an L-1011, but it was attributed primarily to crew error rather than a failure of the new airplane.

Borman's conduct following the accident enhanced his standing within Eastern and helped propel him through its executive ranks. He became chief executive officer in December 1975 and chairman of the board the following year.

When Borman took over the reins of the company, he was looking at an ugly corporate landscape in front of him. Eastern had struggled to earn consistent profits and was carrying a formidable debt burden, much of it associated with the purchase of a large fleet of L-1011s and other new jet aircraft. It also faced high labor costs and increasingly strained relations with its unions. The energy crisis and the deep recession that followed had weakened demand for air travel just as fuel prices were rising dramatically.

Eastern was left operating large, expensive widebody aircraft on routes that could not always generate enough passengers to fill them. When traffic was strong, the L-1011 offered an efficient means of carrying large numbers of travelers. When traffic collapsed, however, its empty seats became costly liabilities, consuming expensive fuel without producing revenue.

The L-1011 had also arrived late. Unlike the Boeing 747 and McDonnell Douglas DC-10, which could be ordered with engines produced by established American manufacturers, the TriStar depended entirely on the new Rolls-Royce RB211 turbofan.

The RB211 promised impressive advances in thrust, efficiency, and noise reduction, but its development became a technological and financial ordeal. Costs escalated, deadlines slipped, and Rolls-Royce became insolvent in February 1971. Because the company and the engine program were considered strategically vital, the British government nationalized Rolls-Royce and provided the support necessary to complete development.

The crisis temporarily halted work on the TriStar and placed Lockheed itself in jeopardy. The L-1011 eventually entered service in April 1972, roughly a year behind its principal competitor, the DC-10. That delay gave McDonnell Douglas a crucial commercial advantage.

The RB211 experienced early service difficulties as Rolls-Royce worked to bring the technologically ambitious engine to full maturity. A series of modifications steadily improved its performance and reliability, and it eventually developed into one of the industry's most respected families of high-bypass turbofan engines. For Eastern, however, those later improvements could not eliminate the economic problem confronting it during the mid-1970s.

The airline had acquired an aircraft designed for an era of rapidly expanding passenger traffic and relatively affordable fuel. Instead, it found itself operating the TriStar amid recession, reduced discretionary travel, and historically high petroleum prices. Eastern could not consistently fill the aircraft, yet it still had to meet the debt payments incurred to purchase them.

By 1977, oil remained expensive, and continuing instability in the Middle East raised the possibility of another sudden price increase. Borman understood that Eastern needed an aircraft better matched to the changing market — one that could carry substantial passenger loads while offering lower operating costs and greater flexibility.

Frank Borman put the word out throughout the aircraft industry that Eastern was looking for a more economical airplane.

***

By 1977, Airbus had endured a sales drought lasting more than two years. Its production backlog had dwindled to only a handful of aircraft, and word was spreading that the assembly line might shut down within months — perhaps even weeks.

When Airbus executives learned that Eastern Airlines was searching for a new aircraft, they recognized the opportunity for what it was: possibly their last chance to save both the A300 and Airbus itself. Few sales campaigns in aviation history had ever carried such enormous stakes.

On the day of the meeting, Airbus presented the A300's performance capabilities and economic advantages to Eastern's executives. Frank Borman and his team were impressed but remained noncommittal. Their caution was understandable. The A300 was a new aircraft built by a relatively young manufacturer that lacked the record, infrastructure, and global reach of its American competitors.

Airbus then made an extraordinary offer. Four unsold "white-tail" A300s were sitting in Toulouse, and Eastern could lease them immediately for a six-month trial at no charge, apart from the costs of operating them. If the aircraft failed to meet Eastern's expectations, the airline could return them without further obligation. If they performed well, Eastern could purchase them at a substantial discount and place additional orders under a highly favorable, most-favored-customer discount pricing structure.

Frank Borman and his executive team took the bait and by the end of 1977 the four white tails, now emblazoned with the Eastern Airlines livery, were flown across the Atlantic to begin crew training and passenger service. After just four months and before the six-month trial period was over, Eastern had become so impressed with the A300s performance, which included a fuel burn a full 30 percent less than the L-1011's, that in April of 1978 it bought the four A300s it had been flying and ordered 19 more with options to purchase additional ones. Airbus was off life support.

The Eastern order could not have come at a more opportune moment for Airbus. When the Airline Deregulation Act was signed into law just six months later, there was a seismic shift in the airline industry in the U.S. With the industry deregulated, the legacy carriers had to quickly become competitive with regard to the routes they serviced and the prices they charged for each ticket. With a dozen new airlines getting ready to start operations, and dozens more in the wings waiting to secure investor backing and regulatory approval, it was going to become a cutthroat business. The demand for airplanes was going to increase exponentially. The new upstart carriers were going to need airplanes and the legacy carriers had to figure out how to best utilize their existing fleets or acquire newer and more efficient planes.

At the time, Airbus' executives couldn't appreciate how close they came to missing the golden egg that the Deregulation Act laid for them. Had Frank Borman and Eastern Airlines vacillated in their decision or delayed it by as much as two months, it's likely Airbus would have shuttered its factories in Toulouse and Hamburg and left the future of commercial aviation entirely in the hands of America's aviation giants.

And Frank Borman had no idea at the time how his decision to take Airbus' offer was tossing the company a lifeline that would fundamentally change the global aviation market in the coming decades. Before Eastern's order for the A300, Airbus had sold a mere 53 copies of the plane worldwide. However, in 1978 alone, it recorded 73 additional orders, including the ones from Eastern Airlines. Airbus' future was now secure.

Frank Borman's decision drew fierce criticism. How, detractors asked, could an American hero — an icon of the nation's aerospace power and technological prowess — purchase airliners from a foreign manufacturer? He faced condemnation from many quarters, including executives at rival American carriers.

But Borman was not being un-American. He was acting as a disciplined corporate executive and a free-market capitalist. He had a fiduciary duty to Eastern's board and shareholders to address the airline's mounting financial problems. Eastern was burdened with large, expensive aircraft that were poorly suited to many of its routes and to the economic conditions of the time. The A300 offered a means to right-size the fleet and reduce operating costs.

Nor was the purchase entirely foreign in economic terms. Many of the aircraft's components — including its most expensive systems, the engines and much of the avionics — were manufactured in the United States.

Meanwhile Back in Seattle

When Thornton Wilson took over as Boeing's chief executive in 1968, his honeymoon lasted barely a year. Pratt & Whitney's development of its new high-bypass turbofan had encountered technical problems, delaying the 747 and pushing its first revenue flight with Pan Am into early 1970. Although difficulties were inevitable given the unprecedented technology incorporated into the aircraft, the 747 program was also suffering from cost overruns and other production delays. Boeing had to dig deeper into its pockets, adding to an already heavy debt load.

Broader economic pressures were also beginning to weigh on the company. Having achieved its central objective of beating the Soviet Union to the Moon, the U.S. space program was retreating rapidly from its peak spending years. That decline directly affected Boeing, which had played a major role in the Saturn V program and numerous other space initiatives. An even more damaging blow came in 1971, when Congress terminated funding for the American SST. The decision eliminated a major anticipated source of revenue and abruptly dismantled a program around which Boeing had staked much of its future.

At the end of 1967, shortly before Bill Allen announced his retirement, Boeing recorded 332 commercial jet orders. Its record year had been 1965, when it received 417. By 1969, annual orders had fallen to 151. Two years later, Boeing recorded just 90 orders from airline customers.

The ordering frenzy of the 1960s — driven largely by airlines rushing to join the jet age — had created excess capacity throughout the industry. By the end of the decade, carriers were struggling to fill their aircraft profitably and began sharply reducing new orders. A recession that began in 1969 weakened passenger growth and further depressed demand. That same year also marked the peak of U.S. spending on the Vietnam War. By 1971, expenditures associated with the conflict had fallen to less than half their 1969 level, sending additional shock waves through the economy.

Boeing was the economic engine of Seattle and the surrounding region, as well as its largest employer. When Boeing sneezed, the rest of the region caught cold. But between 1969 and 1971, the company did far more than sneeze. It contracted the corporate equivalent of pneumonia and found itself on life support, staring at the possibility of bankruptcy.

Boeing had bet the company on the 747, spending more than $1 billion — approximately $9.6 billion in 2026 dollars — to bring it to market. Yet the orders it had received were nowhere near sufficient to recover that investment. Worse, some airlines that had already ordered the aircraft were beginning to get cold feet as the industry confronted an oversupply of passenger capacity.

The 747 was then the largest privately financed industrial project in American history. Its costs extended far beyond the aircraft itself. Boeing had built an enormous new factory at Everett and invested heavily in the tooling, equipment, infrastructure, and supplier network required to develop and produce the airplane. The stakes extended across the Pacific Northwest. A large share of the region's residents either worked for Boeing, worked for a company that depended on Boeing, or knew someone who did. Boeing helped define the region, and the region's economy was inseparable from Boeing.

Thornton Wilson was a realist. He made no attempt to sugarcoat the situation for the board, Wall Street, or the region's political leaders. Drastic measures were required, and he initiated a sweeping cost-cutting campaign. Between 1968 and 1971, Boeing's workforce plunged from a peak of approximately 142,000 employees to fewer than 39,000 — a reduction of more than 70 percent. The consequences were felt throughout the region and across the country.

At the height of the crisis, Seattle's unemployment rate reached approximately 13 percent, compared with a national rate of about 5 percent. The city's population declined by roughly

60,000, falling below 500,000. Boeing had gone in only a few years from the corporate equivalent of a superstar athlete with an apparently limitless future to a critically ill patient fighting for survival.

In a sense, Boeing had brought some of the crisis upon itself. Two decades of extraordinary success under Bill Allen had fostered a belief that the company could do no wrong — that its technical ambitions were limitless and that every new frontier was within its reach. The harsh realities that emerged after 1968 delivered a brutal corrective. A succession of crises, arriving within a remarkably short period, created a perfect storm that nearly overwhelmed the company. It is doubtful Boeing would have emerged intact without Thornton Wilson's discipline and realism at the helm.

After slashing the workforce and shedding excess capacity, Wilson and his leadership team had to position Boeing for an industry that had changed fundamentally. The company survived, but survival alone was not enough. It would have to reinvent itself.

Boeing continued exploring aircraft concepts that could address the void left by the SST's cancellation and the growing demand for greater fuel efficiency. Once the immediate crisis had passed, Wilson and the rest of Boeing's leadership took stock of their altered competitive environment.

By the end of 1972, Boeing's commercial order book had recovered to 170 aircraft. Orders rose modestly to 182 in 1973 and remained at roughly that level in 1974. The effects of the oil embargo eventually caught up with the industry, however, and orders fell to 117 in 1975 as airlines canceled aircraft they no longer needed and throttled back on orders for new ones. Orders began recovering again in 1976, but Boeing remained far below the peaks it had enjoyed during the mid-1960s. Something new would be needed to stimulate demand.

During the early 1960s, airlines had sought to bring the speed of jet travel to progressively smaller markets. By the 1970s, they wanted to bring the principal advantages of widebody aircraft — greater comfort and lower seat-mile costs — to smaller markets as well.

Within a few years of introducing the 747, Boeing studied several smaller derivatives. The original aircraft had been designed for high-capacity, long-range intercontinental service, a market that was not large enough to support multiple competing models. Boeing therefore examined the next segment down to determine whether it could enter it profitably.

Its engineers considered a three-engine derivative of the 747 — a shortened fuselage retaining the aircraft's familiar forward hump, with one engine beneath each wing and a third mounted in the tail, much like the Douglas DC-10 and Lockheed L-1011. Boeing's market analysts concluded that the segment was unlikely to support three competing aircraft, four if Airbus' A300 was included, and the concept was dead on arrival.

Douglas and Lockheed were left to battle for the middle of the widebody market, and their aircraft capably served airlines seeking to carry approximately 260 to 300 passengers on shorter, heavily traveled routes. But continuing advances in aerodynamics, structures, and engine technology during the decade after the first widebodies appeared created opportunities for a more efficient generation of aircraft. That was the opportunity Boeing intended to pursue with its next design.

***

Thornton A. Wilson had succeeded Bill Allen in 1968 and overseen the 747 as it entered service. He then steered Boeing through one of the darkest periods in its history. By the middle of the 1970s, with the Western economy slowly recovering from recession and the first oil shock, Wilson knew the company had to begin investing in its future — even though demand for new aircraft remained well below its earlier peaks.

Initially, Boeing concentrated much of its product-development effort on the aircraft that still offered the greatest promise: the 747. Its basic configuration provided ample room for growth. Within a year of entering airline service, Boeing introduced improvements that modestly increased capacity but, more important, extended the aircraft's range, allowing airlines to serve additional city pairs nonstop.

The original 747 already possessed one of the longest ranges of any passenger aircraft. Some airlines purchased it not primarily for its enormous capacity but for the routes its range made possible. New York to Athens, for example, was not necessarily dense enough to require an aircraft of the 747's size, but the airplane made it possible to fly the route nonstop.

Pan Am, Iran Air, and several other carriers wanted to operate even longer routes, including nonstop service between New York and Tehran. Boeing responded with a shortened version of the 747. The new aircraft retained the original model's basic wing, engine arrangement, and forward upper deck, but its fuselage was shortened by approximately 47 feet, from about 231 feet to 184 feet. Passenger capacity fell to roughly 276, but range and cruise performance improved substantially. Boeing designated the aircraft the 747SP, for "Special Performance."

Despite its difficult development, the 747 had firmly established itself as the icon of long-haul commercial aviation. It also faced no direct competitor, and the market appeared too limited to justify the enormous investment required to develop one. But that same limitation meant Boeing could not depend on the 747 alone.

Between 1971 and 1975, the aircraft accumulated only 96 orders, an average of fewer than 20 per year. During the first decade after its launch, it received approximately 301 total orders, most of them concentrated in the program's early years. If Boeing was to prosper again, it would need new products for larger and more broadly contested markets.

Wilson, like his predecessors, understood that innovation could not stop merely because the company had narrowly survived a financial crisis. The 707 was still winning occasional orders, but its age and design limitations were becoming increasingly apparent when compared with newer aircraft such as the 747, Douglas DC-10, Lockheed L-1011, and Airbus A300.

The 707 was also approaching the end of its competitive life. Wilson understood a principle that had guided Boeing before: it was better to make one's own product obsolete than allow a competitor to do it. In late 1972, Boeing began serious studies of a successor under the designation 7X7.

From the outset, the 7X7 was envisioned as a twin-aisle widebody aircraft smaller than the 747. In a broad concept, it would occupy a market similar to that of the Airbus A300. Airbus was struggling to sell the A300 during the early 1970s, but the problem was not necessarily the aircraft itself. Airbus was a new manufacturer without Boeing's industrial scale, reputation, support network, or established relationships with airlines. It had also introduced the aircraft at one of the worst moments the commercial aviation market had experienced.

The 707 was not the only Boeing aircraft beginning to show its age. The 727 had been designed only a few years later. Although it continued to attract orders at a healthier rate than the 707, Boeing could see that its eventual replacement would also have to be addressed.

In 1975, after United Airlines and other carriers expressed interest in an aircraft with greater capacity and better economics for routes then served by the 727, Boeing studied another stretch of the design. The original 727-100 typically accommodated about 106 passengers. Its successful stretched successor, the 727-200, carried approximately 155. A proposed 727-300 would have stretched the fuselage farther and accommodated roughly 169 passengers in a typical airline configuration. But the concept faced a fundamental problem: the 727's engines. Its three low-bypass turbofans were among the loudest in commercial service. Unlike newer high-bypass turbofans, which routed most incoming air around the engine core, the 727's engines forced a much larger proportion through the combustion section. The result was an intense exhaust noise, particularly during takeoff. Increasingly stringent federal noise standards made the aircraft's existing propulsion system an obstacle to any long-term derivative.

Even apart from the noise issue, airlines showed limited enthusiasm for another stretched 727. They wanted aircraft incorporating the latest technology, particularly the new generation of high-bypass engines that promised substantial improvements in fuel efficiency.

Boeing's engineers examined replacing the 727's rear-mounted engines with larger high-bypass turbofans. Their greater diameter and weight created difficult installation and structural problems. Engineers also considered placing two engines beneath the wings while retaining the aircraft's T-tail, but that configuration introduced complications of its own. None of the engineering problems were necessarily insurmountable. The more important question was whether solving them made economic sense. Commercial aircraft design modification is not simply a matter of producing a workable design. Every major alteration must undergo extensive engineering analysis, testing, production planning, and regulatory certification. At some point, the cost of modernizing an old airframe approaches the cost of developing a new one, while still producing an aircraft constrained by its original architecture.

That was the dilemma Boeing faced with the proposed 727 derivatives. A redesigned aircraft might cost nearly as much to develop as a clean-sheet model while delivering far fewer benefits.

There was another compelling reason to start over. Advances in propulsion were being matched by rapid progress in electronics and computer technology. The space program had demonstrated how computers could transform the control, guidance, and navigation of complex vehicles. Those technologies now promised greater precision, efficiency, and safety in commercial aviation. It was a no-brainer. Boeing began studying a clean-sheet successor to the 727. The new project was designated the 7N7.

From the outset, the 7N7 was conceived as a narrow-body aircraft with a single aisle and six-abreast seating. It would use high-bypass turbofan engines, improved aerodynamics, lighter materials, and a new generation of cockpit avionics. Together, those advances were expected to produce a dramatic improvement over the 727. Boeing projected operating costs per passenger seat-mile approximately 30 percent lower than those of the older aircraft — a gain similar to what the 7X7 was expected to achieve over the 707.

Boeing now had two major development programs running concurrently: one to replace the 707 and another to succeed the 727. Whatever their final configurations, both aircraft would have to reach the market within roughly the same period.

The financial implications were daunting. Boeing's most recent clean-sheet program, the 747, had nearly bankrupted the company. Developing two aircraft simultaneously, only a few years after that crisis, seemed almost impossibly ambitious. The only practical solution was to design them not as wholly separate products, but as members of a coordinated family.

Wilson directed the 7X7 and 7N7 teams to share resources and, wherever practical, common systems. The aircraft would use similar cockpit layouts, instruments, avionics, electrical and hydraulic components, and interior features. Commonality would reduce duplicated engineering effort, lower development costs, simplify production, and lessen program risk.

It would also provide airlines with a powerful economic incentive to operate both models. Closely related cockpits and operating procedures would reduce the time required for pilots to transition from one aircraft to the other. Rather than maintaining completely separate pilot pools, an airline operating both models could deploy crews more flexibly, reducing training expenses and improving workforce utilization.

As the two programs evolved, one of the most consequential decisions was to eliminate the flight engineer's station. Large airliners had routinely carried flight engineers since the emergence of complex four-engine aircraft in the late 1930s. Their job was to monitor and manage the engines, fuel, hydraulics, electrical equipment, pressurization, and other aircraft systems.

That requirement had produced increasingly crowded cockpits filled with gauges, switches, and warning lights. New digital systems could now perform much of that monitoring automatically. Instead of requiring a flight engineer to scan dozens of instruments continuously, computers could track system performance and alert the pilots only when a condition required attention.

The new flight decks would also replace many traditional round gauges with electronic displays. Information could be organized and presented when needed rather than permanently occupying space on the instrument panel. The result would be a cleaner two-pilot cockpit, reduced crew workload, and improved situational awareness. Airbus was pursuing similar advances for later versions of the A300.

By the late 1970s, the commercial market was improving and the global economy had emerged from the recession that had marked the middle of the decade. Airline deregulation in the United States promised new routes and greater competition, while many aircraft designed during the 1950s and early 1960s were approaching obsolescence. Another major replacement cycle appeared to be forming.

Within approximately one month during the summer of 1978, Wilson and Boeing's board authorized development and marketing of both new aircraft. The larger 7X7, intended to succeed the 707, became the 767. The smaller 7N7, designed to replace the 727, became the 757.

The two aircraft were expected to enter airline service within months of each other. Because the 767 studies had begun earlier and were further advanced, it would enter service first.

Developing two clean-sheet commercial aircraft concurrently for different market segments was an extraordinary gamble. Boeing had never attempted anything comparable. Yet Thornton Wilson had already guided the company through a financial upheaval without precedent in its history. He had stabilized the enterprise, preserved its engineering capabilities, and laid the foundation for renewed growth. Boeing's can-do engineering culture had survived.

Under Wilson, Boeing had emerged from the turmoil of the early 1970s leaner, more disciplined, and positioned to capitalize on the next expansion of commercial aviation. The same could not be said of its two closest American rivals.

***

Douglas' reluctance — and, some would argue, lack of foresight — to develop new leading-edge aircraft allowed its 1960s product line, the DC-8 and DC-9 families, to stagnate. Its decision to plunge into the widebody market with the DC-10 then strained the company financially to the breaking point. In 1967, Douglas merged with the McDonnell Aircraft Corporation, a major defense contractor best known for advanced military aircraft such as the F-4 Phantom, then in widespread service with the U.S. Navy, Air Force, and Marine Corps, as well as several allied nations.

The merger provided Douglas with the financial support it needed to survive. But the company's decision to compete directly with Lockheed in a widebody market that many believed could not support two rival aircraft meant that the expected revenues and profits from the DC-10 would not meet expectations.

Lockheed was in an even more precarious position. McDonnell Douglas could at least continue generating revenue from the DC-8 and DC-9, even as both designs aged and annual sales declined. Lockheed, by contrast, had no other commercial aircraft in production. Its future in the airline market depended entirely on the L-1011.

Although Lockheed remained one of the nation's largest defense contractors, it was drowning in debt. Delays in the L-1011 program — many of them caused by problems with the aircraft's Rolls-Royce engines — drove costs sharply higher. Lockheed also suffered massive losses on the Air Force's C-5 Galaxy program after structural cracking appeared in the wings and the company was required to help correct the problem. It lost another anticipated source of revenue when the U.S. Army canceled a major helicopter program, while its shipbuilding operations encountered a host of additional difficulties.

In 1971, Lockheed sought a federal loan guarantee, warning that its survival depended on obtaining approximately $1.5 billion in financing from private creditors. The Nixon administration concluded that Lockheed was too important to the nation's defense industrial base to be allowed to fail and sent legislation to Congress authorizing the guarantee.

The issue was further complicated by the collapse of Rolls-Royce, the sole engine supplier for the L-1011. The British company entered receivership, contributing to delays that brought the L-1011 to market roughly a year after the DC-10. The British government agreed to rescue Rolls-Royce and continue development of the RB211 engine, but its commitment was tied to assurances that Lockheed would also receive financial support. Without both, the L-1011 program could not survive.

The proposed loan guarantee — approximately $12.5 billion in 2026 dollars — provoked a bitter debate in Congress. Critics argued that the government should not rescue a private company from the consequences of poor management and excessive risk. Supporters warned that Lockheed's collapse would eliminate thousands of jobs, damage the defense industry, and jeopardize the L-1011 program. The Senate ultimately approved the measure only after Vice President Spiro Agnew cast the tie-breaking vote.

The loan guarantee allowed Lockheed to survive, but the company soon became engulfed in an international bribery scandal. Investigations revealed that Lockheed had made or offered improper payments in several countries to influence aircraft purchases. The most notorious case involved Japan, where payments were funneled to political figures and intermediaries to help secure All Nippon Airways' selection of the L-1011.

The 1970s were an unforgiving period for commercial aircraft manufacturers. Yet among the three major American producers, Boeing was clearly the best positioned for the future. McDonnell Douglas had survived only through merger and was locked in a costly struggle with Lockheed. Lockheed had required federal assistance to avoid collapse and remained dependent on a single commercial aircraft program.

Boeing had endured its own near-death experience, but under Thornton Wilson's disciplined leadership it emerged from the turmoil comparatively intact. More important, it retained the financial capacity, engineering talent, and strategic vision needed to move forward. As the industry entered a new era, Boeing was once again in the driver's seat.

Let’s Buy a Jetliner

As painful as the 1970s had been, by the decade's end it was clear that Boeing had finally put the worst behind it. By 1978, its balance sheet was far stronger than it had been at the beginning of the decade, thanks in large part to the cash reserves Thornton Wilson had painstakingly rebuilt to approximately $2.5 billion — about $13 billion in 2026 dollars. That same year, Boeing booked a record 485 commercial-airplane orders, finally surpassing the previous record it had set in 1965. Nearly one-quarter of those orders were for the new 757 and 767 families, both launched that year. The timing of the two programs could scarcely have been more propitious because just as Boeing regained its financial footing, the 757 and 767 provided the company with the products it needed to capitalize on a rapidly strengthening commercial-aircraft market.

In July 1978, perhaps taking a cue from Boeing's decision to develop two closely related aircraft for different market segments, Airbus launched a follow-on to the A300 called the A310. The new aircraft was not merely a modified A300 but a complementary model designed to broaden Airbus' product line. Much like Boeing's new 757 and 767, the A300 and A310 would share numerous components and design features.

The A310 retained the A300's widebody fuselage cross-section and was offered with the same families of engines, but it was substantially shorter and designed to carry roughly 220 passengers. Its smaller capacity, combined with aerodynamic and structural improvements, gave it considerably greater range than the original A300. More importantly, the A310 was intended to compete directly with Boeing's 767.

Like the 767, the A310 would also feature a modern, computer-based flight deck designed for a two-pilot crew. Airbus subsequently planned to incorporate much of the A310's cockpit technology into later versions of the A300, giving the two aircraft a common flight deck and a high degree of crew commonality — an approach similar to the one Boeing adopted with the 757 and 767.

With the launch of the A310, Airbus now offered more than a single aircraft design for its airline customers. And by the end of 1979, Airbus, that wanna-be player in the airline manufacturing sandbox, had garnered 256 orders for its A300 and A310 widebody models in total since their respective launches.

One little-noticed fact about Airbus' 256 orders was that the new European consortium had already surpassed one of America's established Big Three manufacturers in the widebody market. In only ten years, Airbus had secured more orders for its A300 and A310 than Lockheed had accumulated for the L-1011. Airbus had launched its first widebody after Boeing, McDonnell Douglas, and Lockheed had introduced theirs, and it had brought the A300 to market later than all three. Yet within a decade of the consortium's formation, it had already overtaken one of them in total widebody orders. Airbus was no longer merely an ambitious European experiment; it had become a credible competitor capable of taking business from the most powerful aircraft manufacturers in the world.

The movement toward airline deregulation, culminating in the Airline Deregulation Act of 1978, also helped reshape the commercial-aircraft market. America's legacy carriers understood that decades of regulation by the Civil Aeronautics Board had left many of them burdened with inefficient route structures, inflated costs, and fleets poorly suited to the more competitive environment that was approaching. As airlines prepared for a world in which fares, routes, and market entry would increasingly be determined by competition rather than government allocation, they began reassessing the size, composition, and operating economics of their fleets. That broader industry realignment was one factor behind the surge in aircraft orders recorded by Boeing and the other manufacturers in 1978.

But oil prices were once again heading skyward because of events unfolding in the Middle East. The Iranian revolution, which began in 1978 and culminated in early 1979 with the overthrow by radical Islamists of the Western friendly Shah, resulted in disruptions to global oil supplies. Iran was the world's second largest producer of oil at that time and the political turmoil disrupted the country's oil production capabilities. And, although high oil prices would naturally create a demand for the new generation of aircraft available, it also would result in strong headwinds for the economy. In 1979 Boeing and other aircraft manufacturers would see a drop in orders from the previous year. Inflation was persistently high throughout the 1970s and it showed no signs of abating. By 1980, another recession would hit the U.S. economy slowing orders for planes even further.

Then came a major shake-up in the commercial aircraft industry. In 1981, Lockheed announced that it would shut down the L-1011 production line after completing its remaining orders, with the final aircraft scheduled for delivery in 1985. Having suffered enormous losses on the program and after having delivered only 250 TriStars, Lockheed would withdraw permanently from the commercial airliner business and concentrate instead on its defense operations.

There were no celebrations at McDonnell Douglas over the departure of its principal rival. The company was still licking the financial wounds inflicted by its ferocious battle with Lockheed for the middle segment of the widebody market. The contest would become a classic business-school case study of what happens when two competing products enter a market large enough to support only one profitably. The lesson was simple — both companies lose.

Lockheed had unquestionably lost the contest first, but McDonnell Douglas had hardly emerged victorious. It had been severely weakened and would struggle for years to remain competitive in the commercial aircraft business. The damage had come not only from fighting Lockheed for the same limited piece of the market, but also from costly design compromises intended to make the DC-10 more attractive to airlines. McDonnell Douglas would never fully recover. It would survive for only another decade and a half.

The field of major Western commercial aircraft manufacturers had now narrowed to three: Airbus, Boeing, and McDonnell Douglas. And among them, Boeing was unquestionably the elephant in the room, at least for now.

***

The economic slowdown that began in 1979 foreshadowed the recurring cycle of booms and busts that would become a defining feature of the commercial aviation industry over the following decades. The pattern had first emerged in the late 1950s, when airlines around the world rushed to replace their propeller-driven fleets with jets. That surge was followed by a steep decline in orders as a slowing economy and excess capacity left airlines with more aircraft than they needed.

The cycle repeated itself in the late 1960s and early 1970s with the arrival of a new generation of widebody aircraft. An improving economy, rising passenger demand, and eventually airline deregulation helped fuel another ordering boom. But that expansion was soon followed by another trough as economic conditions deteriorated and airlines once again sharply curtailed their purchases.

Commercial aircraft orders became a kind of bellwether for the broader economy, particularly in the United States. Boeing's importance to the national industrial base was not unlike that of Detroit's automakers: hundreds of thousands of jobs, suppliers, communities, and related industries were tightly intertwined with its fortunes. When airlines stopped buying airplanes, the effects reverberated far beyond the factory floor.

Ordering a commercial jetliner required years of advance planning. An airline placing an order for new aircraft might not receive its first delivery for anywhere from eighteen to thirty-six months, and sometimes longer. Manufacturers had to make production decisions long before they could know with certainty what the economic environment would look like when those airplanes finally rolled off the assembly line.

Building a jetliner also depended on a vast network of subcontractors and suppliers that furnished everything from engines and avionics to landing gear, structural components, and cabin equipment. Many of those contracts were negotiated around expected production volumes, with higher volumes generally reducing the cost of each component and lower volumes driving unit costs upward. Once those commitments had been made, they could not always be increased, reduced, or canceled without substantial cost.

A company such as Boeing had to anticipate where the markets served by its airline customers would be several years into the future. It then had to align the flow of parts, materials, and labor into its production lines in Seattle and Renton with that forecast. The process demanded enormous amounts of market analysis, economic forecasting, and production planning, but it also required a considerable measure of educated guesswork.

When an airline placed an order for new aircraft, the contract negotiated with Boeing contained stringent provisions governing the obligations of both parties. Boeing needed assurance that the airline would follow through on its commitment, because much could change during the two or more years between the signing of an order and the delivery of the finished aircraft. By reserving production capacity for one customer, Boeing was also limiting its ability to offer those delivery positions to another.

The airline was generally required to make an initial deposit when the contract was signed, securing delivery positions on Boeing's production line. That deposit was typically followed by a series of pre-delivery payments as the aircraft moved closer to production and completion. Together, those payments could represent a substantial portion of the aircraft's purchase price and helped Boeing finance the materials, labor, and supplier commitments required to build it.

The remaining balance was normally paid when the completed aircraft was delivered and formally accepted by the airline. The payment structure gave Boeing some protection against a customer canceling its order after production resources had already been committed, while also giving the airline contractual protections if Boeing failed to meet agreed delivery dates, performance guarantees, or other terms of the purchase agreement.

In addition to firm orders, airlines could also negotiate options to purchase additional aircraft. A firm order was governed by a detailed purchase agreement specifying such matters as price, financing, delivery positions, aircraft configuration, payment schedules, and the obligations of both the airline and the manufacturer. Although firm orders could later be amended, deferred, or canceled under certain circumstances, they represented a binding commitment by the airline to purchase the aircraft. An option, by contrast, gave the airline the contractual right — but not the obligation — to purchase additional aircraft under terms negotiated in advance. It functioned as a form of reservation. An airline might, for example, require thirty new jetliners, twenty to replace aging aircraft approaching the end of their service lives and another ten to support expansion into new markets. Because the airline could not know whether economic conditions two or three years later would justify that expansion, it might place a firm order for twenty aircraft while securing options for ten more.

Those options could reserve future delivery positions and often preserve previously negotiated pricing terms, subject to escalation clauses and other contractual adjustments. They generally required a much smaller financial commitment than a firm order, sometimes in the form of an option fee rather than a substantial deposit. The contract would also establish deadlines by which the airline had to decide whether to exercise each option.

If the airline elected to proceed, the option would be converted into a firm order. At that point, the airline would become responsible for the required deposits, pre-delivery payments, and other contractual obligations governing the original firm aircraft. If it declined to exercise the option by the specified deadline, the manufacturer could release the reserved production position to another customer. Because so much could change during the two or more years between an aircraft order and its scheduled delivery, purchase agreements generally contained detailed provisions governing cancellation, default, and termination. Under certain circumstances, an airline could cancel all or part of a firm order, although doing so could carry substantial financial consequences.

Depending on the terms of the contract, the airline might forfeit its initial deposits and pre-delivery payments, pay agreed damages, or negotiate a settlement with the manufacturer. The closer an aircraft was to production or delivery, the greater the potential cost of cancellation, because the manufacturer might already have committed labor, materials, supplier capacity, and valuable production slots to that customer's airplanes.

Those payments were intended to compensate the manufacturer for at least part of the financial damage caused by the cancellation. A manufacturer like Boeing could remain contractually obligated to suppliers for components and services ordered for the aircraft, while also facing the challenge of finding another customer willing to accept the vacated delivery position, and possibly an aircraft already built or configured to the original buyer's specifications.

For its part, the airline would seek to negotiate the most favorable terms possible, chief among them being a substantial discount from the manufacturer's published price. By the 1980s, catalog prices had become little more than reference points from which negotiations began. The actual price paid by an airline could be 30 to 40 percent below the published figure, and sometimes even lower, depending on the size of the order, the importance of the customer, market conditions, and the intensity of competition among manufacturers.

Large, well-established airlines generally possessed the greatest bargaining power because their orders were substantial and their endorsement carried considerable influence within the industry. But even a start-up carrier purchasing only a handful of aircraft could secure meaningful discounts, financing assistance, training support, spare parts, maintenance arrangements, and other favorable concessions. A manufacturer eager to place a new model with an additional customer might be willing to offer especially generous terms.

Winning an aircraft order carried value beyond the immediate revenue it generated. Each announced sale served as a highly visible endorsement of the manufacturer and its product. The entire airline industry would take notice, particularly when a respected carrier selected one aircraft over a competing model. Even an order from a new entrant could add momentum to a sales campaign, fill open production positions, and signal that another customer had placed its confidence in the airplane. In that sense, every order was not merely a commercial transaction, but another vote of confidence in the manufacturer's product.

Price was only one of many contractual terms open to negotiation. A large, financially secure, or "blue-chip" airline, could often negotiate a smaller initial deposit to secure production positions, more favorable pre-delivery payment schedules, greater flexibility in deferring aircraft, and less onerous cancellation provisions. It might also obtain concessions involving financing, spare parts, training, maintenance support, performance guarantees, or the right to substitute one aircraft model for another.

The balance of negotiating power depended heavily on where the industry stood in its ordering cycle. During a boom, when production lines were full and delivery positions were scarce, manufacturers held the advantage and could insist on larger deposits, tighter deadlines, and stricter cancellation terms. During a downturn, when orders were scarce and open production slots multiplied, the advantage shifted to the airlines, which could demand deeper discounts and more generous contractual protections.

That recurring swing between scarcity and excess capacity was fast becoming the bane of the commercial aircraft manufacturing business. Manufacturers had to make long-term commitments to workers, suppliers, factories, and production rates, yet the bargaining power behind those commitments could reverse abruptly with the economic cycle.

When the recession ended in 1983, it gave way to one of the longest periods of sustained economic expansion the United States had experienced up to that time. By then, the airline industry was five years into deregulation, and the upheaval it had unleashed was becoming unmistakable. Some of the start-up carriers created in its wake had already disappeared, while others had proved to be little more than flashes in the pan, failing almost as quickly as they had entered the market.

Deregulation had also set off a wave of mergers, acquisitions, and corporate restructurings among the established airlines. Carriers that only a few years earlier had been regarded as the blue-chip stalwarts of the industry were now struggling to adapt to an operating environment in which routes, fares, and competition were increasingly determined by the market rather than by federal regulators.

Among the most vulnerable were Pan American, Eastern Air Lines, and Trans World Airlines. Each entered the deregulated era burdened by some combination of high operating costs, rigid route structures, aging fleets, heavy debt, and deeply entrenched management practices. Although their financial conditions differed, all three would struggle to adjust to the new airline world order. The red ink spreading across their balance sheets was an early warning that even the most celebrated names in commercial aviation were no longer assured of survival.

***

Although no one could have known it at the time, 1983 marked the beginning of one of the longest and most vigorous economic expansions in modern American history. As the economy gathered momentum, airline passenger traffic began rising rapidly with it. Carriers could see from the pace and direction of the recovery that their existing fleets might soon lack the capacity needed to accommodate growing demand, particularly on their busiest routes.

Airlines began planning for expansion once again. Aircraft that had seemed unnecessary during the recession were suddenly needed to add capacity, replace aging equipment, and open new routes. The commercial aircraft industry was entering another upward turn in its relentless cycle of boom and bust, and orders began climbing once more.

Around the same time, the Reagan administration, which had taken office in 1981, was reconsidering the future and proper role of the Export-Import Bank of the United States, commonly known as the Ex-Im Bank. Established by Franklin Roosevelt in 1934, the Bank was an independent agency of the federal government created to facilitate the export of American-made goods and services by providing loans, loan guarantees, and export-credit insurance.

The Ex-Im Bank was not intended to displace private lenders. Instead, its charter required it to supplement rather than compete with commercial financing. It generally became involved when private credit was unavailable on reasonable terms, when political or commercial risks discouraged conventional lenders, or when a foreign government was offering subsidized financing to support a competing exporter. In such cases, the Bank could provide or guarantee financing that allowed an overseas customer to complete the purchase of American products.

The Bank's importance tended to increase when private credit tightened or when foreign export-credit agencies offered more favorable financing to buyers. Commercial aircraft were especially well suited to its mission. Jetliners were enormously expensive, their financing extended over many years, and many prospective customers were often foreign airlines that could not always obtain sufficient commercial credit on competitive terms.

As the world's airlines entered the jet age, Boeing, McDonnell Douglas, and Lockheed became among the most visible beneficiaries of Ex-Im financing. The Bank's support could make the difference between an overseas airline buying an American jetliner and selecting a competing aircraft whose sale was backed by another government. Ex-Im financing consequently became not merely a source of credit, but an important weapon in the increasingly fierce international contest for commercial-aircraft orders.

As the recession took hold in the early 1980s, Boeing lobbied vigorously to preserve the charter and lending authority of the Export-Import Bank. It was during this period that the early salvos were fired in what would become a perennial transatlantic dispute between Boeing and Airbus over government support for commercial aircraft.

Boeing argued that the Ex-Im Bank was essential to maintaining a level competitive playing field. Airbus was backed by the governments of France, West Germany, Britain, and Spain, which had helped finance the development of its aircraft through government loans and other forms of support. Because much of that financing was repayable only if an aircraft program achieved commercial success, Airbus and its industrial partners bore less financial risk than they would have faced had development been financed entirely through conventional private capital.

That support, Boeing maintained, gave Airbus greater freedom to absorb development costs, offer aggressive pricing, and pursue strategically important orders on terms that a wholly commercial enterprise might have found difficult to justify. European export-credit agencies could also help finance Airbus aircraft purchased by foreign airlines, adding another government-supported instrument to the European sales effort.

Boeing argued that it needed a comparable lever of its own. The Ex-Im Bank did not subsidize the development or manufacture of Boeing aircraft. Instead, it could provide loans, guarantees, or insurance to help qualified foreign airlines finance their purchases when private credit was unavailable, prohibitively expensive, or placed Boeing at a disadvantage against government-supported foreign competitors. To Boeing, preserving the Bank was not a request for preferential treatment, but a necessary defense against a competitor whose sponsoring governments were already deeply involved in supporting both the development and export of its airplanes.

This was one of the opening volleys in a heated debate that would continue for the next three decades before finally coming to a head in the early 21st century. Boeing portrayed Airbus as the beneficiary of government largesse and itself as a private enterprise forced to compete against the treasuries of Europe. The argument contained considerable truth, but it also required Boeing to overlook the extent to which its own rise had been intertwined with the United States government.

Boeing had benefited enormously from American military spending throughout its history. Its development of the B-47 and B-52 gave the company invaluable experience in swept-wing aerodynamics, jet propulsion, pressurized structures, high-speed flight, and the design and manufacture of large jet aircraft. The government did not simply hand Boeing the plans or the money to build the 707 as Boeing invested heavily in the Dash 80 prototype and accepted a substantial commercial risk of its own. But the technical expertise, engineering workforce, manufacturing capacity, and institutional knowledge that made the gamble possible had been cultivated in large part through military contracts financed by the American taxpayer.

Nor was the military Boeing's only government partner. The company participated extensively in NASA programs and the Space Race, benefiting from federally financed research and development across aeronautics, missiles, spacecraft, and launch vehicles. It was also selected to develop the airframe for the United States government's supersonic transport program, an ambitious effort that consumed years of publicly funded research before Congress terminated it in 1971.

None of this was precisely equivalent to the European launch aid provided to Airbus. Boeing's defense and space contracts generally paid the company to develop products and capabilities the United States government itself wanted, whereas Airbus launch aid was directed specifically toward the creation of commercial aircraft. Nevertheless, Boeing's claim that it had prospered entirely through the workings of the private market was difficult to sustain. Like Airbus, it had grown within an aerospace industry in which government financing, government research, and national strategic interests had always played a central role.

Boeing countered that whatever benefits it had once derived from government-financed military and space programs belonged largely to the past. Its commercial aircraft division, the company argued, now had to compete on its own merits against a European consortium whose governments were directly underwriting the development of its airplanes. More immediately, Boeing faced the prospect of introducing both the 757 and 767 into commercial service in the midst of a severe recession and one of the deepest downturns the airline industry had experienced.

The warning signs were already appearing. In early 1982, American Airlines, suffering heavy losses, canceled its firm order for fifteen 757s. The cancellation was particularly alarming because the 757 was still months away from entering service and had attracted only a limited number of customers. Boeing feared that American's decision might trigger a cascading effect, encouraging other financially distressed airlines to cancel, defer, or reconsider their own commitments.

At the same time, Airbus' new A310 was emerging as a formidable competitor for airlines seeking an aircraft smaller than the original A300 but capable of operating medium- and long-range routes. Brazil's VASP placed orders for Airbus aircraft, including the A310, while Singapore Airlines selected both the A310 and the Boeing 757 for direct operational evaluation. Singapore eventually concluded that the wider A310 better suited its regional network and disposed of its small 757 fleet in favor of the Airbus model.

For Boeing, developments such as these underscored the competitive threat. The company was investing heavily in two new airplanes just as demand was weakening, customers were retrenching, and Airbus was demonstrating that it could win strategically important sales far beyond its European home market. Boeing feared that without competitive export financing, additional foreign orders might slip away, not necessarily because its airplanes were technically inferior, but because Airbus and its sponsoring governments could offer a more attractive overall financial package.

Boeing claimed that these foreign sales losses reflected an uneven competitive playing field. In its view, the government support behind Airbus allowed the European manufacturer—and the export-credit agencies supporting its customers—to offer financing terms that Boeing could not match through ordinary commercial channels. The contest, Boeing argued, was therefore being decided not solely by the merits of the airplanes, but by the financial packages accompanying them.

Airbus rejected that characterization. It maintained that airlines were selecting its aircraft because they were competitive on their own merits. The A310, Airbus argued, was every bit as technologically advanced as Boeing's 767 and, for some missions, better suited to an airline's operating requirements. Its victories were the result of range, capacity, efficiency, cockpit technology, and overall economics and not merely government support.

The disagreement would become a familiar refrain in the rivalry between the two manufacturers. When Airbus won an important order, Boeing would often point to subsidies and preferential financing. Airbus answered that Boeing was using government support as an excuse for losing a fair commercial competition. Each side's argument contained an element of truth, and the line between technical merit, pricing, financing, and government assistance was rarely as clear as either company claimed.

McDonnell Douglas, for its part, remained largely on the sidelines of the emerging Boeing-Airbus dispute, at least for the time being. The company had its own interest in preserving American export financing and shared Boeing's concerns about government support for Airbus, but its immediate problems lay closer to home.

With Lockheed withdrawing from the commercial aircraft business, McDonnell Douglas sought to extract as many additional sales as possible from the DC-10 and recover at least some of the enormous investment consumed by its bruising battle with the L-1011. Yet Lockheed's departure did not automatically deliver the market to McDonnell Douglas. Demand for large trijets was weakening, the DC-10's reputation had been damaged by a series of highly publicized accidents, and the economic downturn was forcing the company to reduce production.

At the same time, McDonnell Douglas decided to concentrate its resources on the MD-80, the stretched and modernized derivative of the DC-9 that offered airlines improved fuel efficiency, greater capacity, and lower noise without requiring the company to finance an entirely new aircraft project. Rather than matching Boeing's gamble on the all-new 757 and 767, McDonnell Douglas pursued a more cautious strategy built around incremental improvements to existing designs.

For the time being, the fiercest arguments over subsidies and export financing would be waged primarily between Boeing and Airbus. McDonnell Douglas was still a major commercial-aircraft manufacturer, but it was increasingly focused on defending the product lines it already possessed rather than financing the next generation of airplanes it would need to remain competitive.

The 1980s would prove to be a record-setting decade for the commercial aircraft manufacturers. The long economic expansion generated sustained growth in passenger traffic, but the industry was also approaching a major fleet-replacement cycle. According to estimates circulating at the time, as much as 40 percent of the world's commercial jet fleet would need to be replaced by the end of the decade.

By the mid-1980s, an entire generation of jetliners delivered during the great ordering boom of the 1960s was approaching twenty years or more of service. These aircraft were not necessarily reaching the end of their service lives because with proper maintenance, many could continue flying for years. But they were approaching the age at which airlines traditionally began weighing the economics of replacement. Aircraft delivered in the 1960s were increasingly expensive to maintain and generally consumed more fuel, produced more noise, and required more flight crews than the newer generation of jetliners entering the market. They also faced increased reliability issues. Rising passenger demand gave airlines an additional incentive to replace them with aircraft that offered greater capacity, improved reliability, and substantially better operating economics.

The manufacturers stood to benefit from two powerful forces working simultaneously, which were that airlines needed additional aircraft to accommodate rapid traffic growth, while much of the existing fleet was becoming economically obsolete. Beginning in the middle of the decade, the airplanes delivered during the first great wave of the Jet Age would gradually begin giving way to the generation represented by the Boeing 757 and 767, the Airbus A310 and A320, and the McDonnell Douglas MD-80 and DC-10.

As the decade progressed, Boeing consolidated its commercial product line around four aircraft families: the 737, 747, 757, and 767. Production of the commercial 707 had ended in 1978, while the last 727 rolled off the assembly line in 1984. The numerical sequence of Boeing's model designations, however, did not reflect the size of the airplanes nor the market segments they served. Arranged from smallest to largest, Boeing's lineup was the 737, 757, 767, and 747.

Viewed from that perspective, a conspicuous gap existed between the 767 and the much larger 747. Boeing had no aircraft specifically designed to occupy the high-capacity market between its largest twin-engine widebody and its four-engine jumbo jet. That segment was occupied principally by the McDonnell Douglas DC-10, which had been designed to offer airlines an aircraft smaller and less costly than the 747 while still providing substantially more capacity than the narrow-body jets beneath it. The DC-10 was expressly valued by airlines as a step-up in capacity between smaller aircraft and the 747.

Across the three surviving Western manufacturers, the commercial aircraft market could therefore be represented — roughly — by the following progression: the Boeing 737 and McDonnell Douglas MD-80 at the lower end; the Boeing 757 above them; the Boeing 767 and Airbus A310 in the smaller widebody category; the larger Airbus A300; the McDonnell Douglas DC-10; and finally the Boeing 747 at the top.

The boundaries were not precise. Airlines selected aircraft according to a combination of passenger capacity, range, cargo volume, runway performance, and operating economics, and several models overlapped one another. The A310 was a shortened, lower-capacity, longer-range derivative of the A300, while the DC-

10 was offered in both domestic and intercontinental versions. Nevertheless, the broad hierarchy illustrated Boeing's strategic span, but also an interesting gap. And that was it possessed an airplane for nearly every major segment except the one lying between the 767 and the 747.

Boeing clearly possessed the broadest product portfolio, with aircraft serving more capacity levels and market segments than either of its competitors. From the short-haul 737 to the long-range 747, Boeing could offer airlines a family of aircraft capable of meeting nearly every major requirement. That strategic advantage was not lost on Airbus.

A Difference of Philosophy

In April 1972, at the height of the Cold War, the United States and the Soviet Union agreed to conduct a joint space mission as part of the broader policy of détente. The strategy sought to ease tensions and promote peaceful coexistence between two nations that continued to regard each other as adversaries, while reducing the risk of a nuclear confrontation that could have catastrophic global consequences.

That agreement culminated in the Apollo-Soyuz Test Project in July 1975, when an American Apollo spacecraft docked in orbit with a Soviet Soyuz spacecraft. Three American astronauts and two Soviet cosmonauts spent two days conducting joint experiments and demonstrating that the rival superpowers could cooperate in space. The mission was widely regarded as a tremendous success and served as a symbolic release valve for tensions between the two nations.

What was interesting about the mission was the fact that these two enemies were forced to showcase every facet and detail of their spaceships to the other side during the training and run-up to the mission. For each nation, their spaceships epitomized the pinnacle of their respective technological capabilities and know-how. They would in effect be opening up their "kimonos" for the other to look inside. And what was the reaction of each crew's view of the other side's machine?

For the American astronauts, they were struck by the simplicity of the Soyuz spacecraft. Although it could be flown manually, its design centered on automated operation and remote control from the ground. The cosmonauts were expected to make only limited control inputs during flight, typically in the final stages of a precise maneuver such as docking with another spacecraft.

On the other hand, the Soviet cosmonauts were astonished by the complexity of the Apollo spacecraft. To them, its intricate systems created an inherent danger — the more complex the spacecraft, the more opportunities there were for something to go wrong. What they found even harder to comprehend, however, was the extraordinary level of trust placed in the American astronauts who would fly it. Unlike Soyuz, which relied heavily on direction and control from the ground, Apollo was designed to be operated primarily by the astronauts aboard the spacecraft.

What the Russians struggled with was the level of independence and decision-making that was given to their American counterparts. These spacecraft were at the time the most sophisticated pieces of machinery ever developed, costing a substantial amount of each nation's talent and treasure. How could the Americans leave so much of value in the hands of a fallible human being that was prone to make mistakes and, sometimes, exercise bad judgment?

Each nation's form of government, culture, and history shaped how it designed and developed its space program. Russia had been governed autocratically for much of its history, and generations of its citizens had been conditioned to respect authority rather than challenge it. Decisions about how a system would operate — or, in this case, how a spacecraft would be flown — were made by a higher authority. If someone in authority said that a certain button had to be pushed at a particular time, that button would be pushed without question.

On the other hand, the Americans were steeped in a culture that revered freedom and rewarded initiative and self-reliance. Questioning authority was de rigueur in American society. If someone in authority said a certain button had to be pushed at a certain time, no one would bat an eye if the person that was supposed to push the button asked "Why?"

The Soviet Union was rooted in the political traditions of the Old World, where centralized authority had shaped much of European history. For centuries, monarchs, aristocracies, churches, and state institutions exercised power that ordinary citizens had little ability — and often little right — to challenge that authority. Although the Soviet Union replaced monarchy with communism, it preserved the underlying expectation that authority would direct society and provide solutions from above. Within that tradition, granting any one individual unrestricted control could seem less like empowerment and more like an unnecessary risk.

***

In 1984, Airbus officially launched a new single-aisle aircraft program aimed at the short and medium-haul market dominated by Boeing's 737 and McDonnell Douglas' MD-80 series. It was the fastest-growing segment of commercial aviation and was expected to expand dramatically, making it an obvious target for Airbus.

By then, Airbus had earned a strong reputation for building safe and dependable widebody aircraft, but its A300 and A310 remained largely niche products. Eastern Airlines' landmark order had helped secure the company's future and establish its credibility, yet Airbus still ranked a distant third behind Boeing and McDonnell Douglas. Its customers were primarily smaller and national carriers outside North America, then by far the world's largest commercial aviation market.

Although Airbus offered two models, the A300 and A310 were closely related aircraft adapted for different missions. Unless the company broadened its product line, it risked remaining a one-product manufacturer and a perpetual niche competitor.

Boeing's 737, although recently re-engined and stretched for better performance, was a design that was already over twenty years old. McDonnell Douglas' MD-80 series was from the same mold as Boeing's 737: re-engined and stretched (and renamed from its original DC-9 nomenclature) for better performance but incorporating a design that was also over twenty years old. Both of these aircraft dominated the short-to-medium haul markets around the globe, and they incorporated design elements that were meant to address the aviation needs of an era that didn't exist in air travel anymore.

Airbus' leadership decided to take advantage of the American manufacturers' seeming contentment to sit on their laurels and take the safe path of sticking with refreshed but old designs for their customers' needs. Airbus saw an opportunity, an opening to step through and take a leap forward. With the short-to-medium haul market expected to experience explosive growth in the coming years due to airline deregulation, the two dominant American manufacturers didn't have any immediate plans to address this growth beyond just selling their current designs. They both viewed this market as a safe arena for the two players that occupied it.

Sensing an opportunity, Airbus decided to develop an all-new clean-sheet design to serve this market. And in order to become a viable player, the new airplane design would incorporate the latest technological advancements of aviation in the past 20 years. In keeping with its established naming nomenclature, this new model would follow the A300 and A310 and be called the A320.

The greatest advancements in aviation since the early 1960s when the two dominant airplane models were developed for the short-to-medium haul markets were in avionics — the cockpit instruments that allowed the pilots to fly the plane — and in powerplant technology. Design engineering advancements had also been made in refining the wings of airplanes for more efficiency, as had manufacturing techniques to incorporate new alloys and construction methods to make airplanes lighter and less costly to manufacture, as well as less costly maintain. But avionics and powerplants were where the big advancements had been made.

Airbus' new A320 model was going to leverage a variant of the CFM56 engine, the same one used on Boeing's upgraded 737-300, -400, and -500 series, although it would eventually offer an alternative engine, the V2500, manufactured by International Aero Engines. But Airbus really needed to differentiate its offering beyond just another "me-too" airplane. It had to bring to market a real game changer, and this is where the company in fact did bring forth a real differentiator.

For starters, Airbus was going to design the new model around the latest in avionics. Unlike the Boeing and McDonnell Douglas offerings which were retrofitted with some, but not all, of the newer avionics capabilities, the A320 would be designed from the ground up around an entirely new "all-glass" cockpit, in reference to the amount of space taken up by glass monitors. But Airbus took it one step further.

With the advent of crewed spaceflight in the early 1960s, American and Soviet engineers equipped their spacecraft with hand controllers that transmitted electrical commands to their guidance and control systems. Those commands activated small external thrusters that controlled the spacecraft's pitch, yaw, roll, and movement through space.

During the late 1960s and early 1970s, aircraft designers applied similar principles to a new generation of combat aircraft, most notably the F-16. Rather than connecting the pilot's control stick directly to the aircraft's control surfaces through a complex network of cables, pulleys, and mechanical linkages, the F-16 used electrical wires to transmit the pilot's commands to flight-control computers. Those computers interpreted the inputs and instructed hydraulic actuators to move the control surfaces on the wings and tail. The resulting fly-by-wire system reduced weight and mechanical complexity while allowing the computer to make continuous corrections that enhanced the aircraft's stability and maneuverability.

Some of these new fighters, most notably the F-16, were deliberately designed with relaxed aerodynamic stability. That instability made the aircraft faster, more responsive, and far more maneuverable than a conventionally stable design. It also made the aircraft extremely difficult, if not impossible, for a pilot to control through direct mechanical inputs alone.

Fly-by-wire solved that problem. The flight-control computers continuously interpreted the pilot's commands and compared them to the aircraft's flight characteristics and made dozens of minute adjustments each second to the aircraft's control surfaces. Those corrections kept the aircraft stable throughout even the most aggressive maneuvers, performing a task no human pilot could accomplish manually.

This new approach became known as fly-by-wire, and it offered several important advantages. In the cockpit, it eliminated the need for a large control column or yoke whose movement had to be transmitted through mechanical linkages. Elsewhere in the aircraft, it reduced the need for reinforced passageways carrying cables, pulleys, and control rods. Replacing those heavy mechanical components with lightweight electrical wiring also reduced weight and freed valuable internal space. Just as importantly, fly-by-wire lowered maintenance demands because there were fewer cables, pulleys, and linkages to inspect, lubricate, adjust, and replace as they stretched or wore through normal use.

There is another important advantage to fly-by-wire: because the pilot's control inputs pass through a computer before reaching the control surfaces, the system can interpret and modify those inputs to produce the desired response.

A coordinated turn provides a useful example. In a properly coordinated turn, the aircraft banks and turns without slipping or skidding sideways. The pilot must balance aileron, rudder, and elevator inputs, and mastering that coordination has traditionally been considered a mark of good flying skill. Too much or too little control input, however, can produce lateral forces and an imperfect, less efficient turn. A fly-by-wire system can be programmed to coordinate those inputs automatically. The pilot commands the turn, while the flight-control computers make the precise, continuous adjustments needed to keep the aircraft properly balanced throughout the maneuver.

Perhaps the most consequential advantage of fly-by-wire, or FBW, was that designers could build safeguards into the flight-control software to prevent pilots from exceeding an aircraft's structural limits or operating outside its approved flight envelope. The system could restrict dangerously steep banks, excessive pitch attitudes, extreme angles of attack, or other control inputs that might lead to structural failure or loss of control.

Aviation history offered numerous examples of pilots, whether through panic, disorientation, or loss of situational awareness, pushing aircraft beyond their limits with catastrophic results. Fly-by-wire promised to reduce that danger by allowing the computer to intervene before a pilot's commands placed the aircraft at unacceptable risk. It did not make the A320 literally crash-proof, but it allowed Airbus to present the aircraft as substantially more resistant to certain forms of pilot-induced loss of control.

Airbus was not about to let that advantage go unnoticed. In the years leading up to the A320's introduction, the company carefully cultivated the message that its new aircraft would be a technological game changer — a generational leap beyond anything Boeing or McDonnell Douglas then had to offer. Airbus prominently touted the safety features of its new aircraft, often using stall protection as a prime example.

A wing stalls when its angle of attack — the angle at which the wing meets the oncoming airflow — becomes too great. An excessive angle-of-attack will cause the airflow over the upper surface of the wing to separate and will result in a wing stall — the wing will stop producing lift.

Commercial aircraft already incorporated devices intended to warn pilots of an approaching stall. A stick shaker feature on some aircraft would rapidly vibrate the control column in front of the pilots, while some aircraft also employed a stick pusher that forced the nose downward if the pilot failed to respond. Even then, however, a determined — or panicked — pilot could overpower the system and continue pulling back on the controls.

The A320 presented a fundamentally different approach. Under normal flight-control law written into its software, its computers continuously monitored the aircraft's angle of attack and intervened before the wing reached a full aerodynamic stall. As the aircraft approached the protected limit, the system would alter how it would respond to the pilot's control commands, preventing the pilot from commanding an excessive angle of attack, and could automatically apply maximum thrust to restore energy. In effect, the pilot could pull the control fully aft, but the computers — not the pilot alone — determined how far the aircraft would pitch up.

Airbus presented this as one of the defining advantages of the A320. Rather than merely warning the crew that the aircraft was approaching danger, the flight-control system was designed to prevent the pilot from crossing certain aerodynamic boundaries in the first place. This wasn't an absolute capability. In the event of systems failures, the plane's fly-by-wire architecture was designed to bring the pilot back into the loop. But for most flights where there was no systems degradation, the plane's fly-by-wire system held sway over what could and could not be done in the cockpit.

From Airbus' perspective, this was a logical next step in aircraft safety. Rather than allow a pilot to command a maneuver that might exceed the aircraft's aerodynamic or structural limits, the flight-control system would prevent the airplane from entering an unsafe condition. Airbus' engineers and software designers believed that those limits could be defined more reliably in advance than judged by a pilot under pressure in the cockpit.

The philosophy reflected a fundamental conviction, which was that in certain flight regimes, the aircraft's designers — armed with extensive engineering data, testing, and computer modeling — were better positioned to determine the airplane's safe limits than a pilot reacting in the moment.

Many within the aviation community nevertheless balked at Airbus' planned fly-by-wire system. The old saying, "If it ain't Boeing, I'm not going," seemed to take on new relevance. Many commercial pilots were uneasy with the idea of surrendering even part of their authority to a computer, particularly one capable of rejecting a cockpit command. The concern was less about what the A320 would ordinarily allow pilots to do than about the psychological barrier of being overruled by software. Engineers and programmers could anticipate thousands of flight conditions, pilots reasoned, but they could never guarantee that every possible circumstance had been foreseen. There might always be one unanticipated scenario in which the system responded unexpectedly — or prevented a pilot from taking the action needed.

Airbus countered that fly-by-wire would make the A320 safer, more efficient, and easier to fly than conventional airliners. That last advantage would have a strong appeal for airlines. Pilot training was expensive, and an aircraft that simplified cockpit procedures and eased the transition between models could reduce the cost and time required to qualify crews on the A320.

The A320's fly-by-wire system introduced another novel cockpit feature that many pilots found unsettling: the elimination of the traditional control column and yoke. Most postwar commercial aircraft placed a control column between each pilot's knees, topped by a yoke that could be turned to command roll and pushed or pulled to control pitch. The two pilots' controls were mechanically interconnected, so when one pilot moved his yoke, the other moved with it. This allowed each pilot to see and feel the other's control inputs.

Traditional controls also provided some degree of tactile feedback about the aircraft's response, whether through direct mechanical forces or artificial-feel systems. Pilots could sense increasing resistance as aerodynamic loads grew and could immediately recognize when the other pilot was making a control input.

The A320 dispensed with the control columns entirely. In their place, Airbus installed small sidesticks beside each pilot, similar in concept to those used in modern fighter aircraft. The sticks were spring-loaded to return to center, but because they sent electronic commands to the flight-control computers, they provided no direct tactile indication of aerodynamic forces. Nor were the two sidesticks mechanically linked — movement of one did not cause the other to move. Airbus considered such linkage unnecessary because cockpit procedures required one pilot to fly while the other monitored. When control was transferred, the pilots were expected to announce the exchange clearly. To some pilots, however, the absence of interconnected controls removed an important source of situational awareness, particularly if both pilots made inputs simultaneously even if they weren't supposed to.

The sidestick also reinforced a broader unease about the pilot's role in a highly computerized aircraft. To skeptics, it resembled a video-game controller and seemed to confirm that the flight computers, rather than the crew, held ultimate authority. A joke began circulating among pilots that said the cockpit of the future would contain only a pilot and a dog: the pilot's role was to monitor the plane's computers, while the dog role was to bite him if he tried to touch anything.

Airbus worked hard to overcome that skepticism. Because fly-by-wire eliminated the cables, pulleys, and leverage requirements associated with conventional controls, the company argued, a large control column was no longer necessary. Removing it also cleared the space in front of each pilot, improved access to the instrument panel, and allowed the installation of a retractable table for charts, paperwork, or meals during cruise.

To many traditional pilots, however, the control column was more than a mechanical device. It symbolized their status as the airplane's final authority. Replacing it with a small electronic sidestick seemed tangible proof that their role was being diminished.

Airbus' claims for the A320 nearly backfired during the aircraft's first years in service, when several highly publicized crashes raised doubts about its new technology. The most famous occurred at an air show in Habsheim, France, in June 1988. An Air France A320 carrying 136 passengers and crew was performing a demonstration flight before thousands of spectators. Its pilots were among the airline's most experienced A320 aviators and were involved in introducing the new aircraft into Air France's fleet.

During a low, slow pass over the airfield, the A320 descended to roughly thirty feet above the ground. Instead of climbing away near the end of the runway, it continued toward a forest of trees off the end of the runway and plowed into them and came to rest on the ground before eventually exploding in a fireball. Video of the accident was broadcast around the world. Incredibly, most aboard escaped, but three passengers died in the ensuing fire.

The official investigation concluded that the aircraft had been flown too low and too slowly, outside of the plane's normal operating envelope, and too late for the engines to accelerate before the airplane reached the trees. Regardless of the precise chain of events, however, the images appeared to contradict Airbus' assurances that the A320's computerized protections would prevent the aircraft from entering a dangerous flight condition.

The accident badly damaged the A320's reputation, and two further crashes intensified the scrutiny. In February 1990, an Indian Airlines A320 crashed short of the runway while approaching Bangalore, killing 92 of the 146 people aboard. In January 1992, an Air Inter A320 struck a mountain near Strasbourg during a nighttime approach, killing 87.

Investigators identified crew actions and cockpit decision-making as central factors in all three accidents, but the findings also exposed weaknesses in training, operating procedures, cockpit presentation, and the interaction between pilots and the aircraft's automated systems. Critics remained unconvinced by Airbus' claim that the A320 was inherently safer than conventional aircraft. Pilots had long been trained to respond instinctively to changing conditions, and skeptics argued that placing computers between the pilot and the control surfaces could create confusion or a mismatch between pilot intent and aircraft response.

The investigations led to changes in training, operating guidance, and elements of the A320's cockpit and automated systems. Combined with the aircraft's improving safety record in the following years, those measures gradually restored confidence in the design.

***

Although Boeing and McDonnell Douglas kept an eye on Airbus, neither initially regarded the newcomer or its planned A320 as a serious threat. By the end of 1984, the year Airbus formally launched the program, Boeing had received 141 orders for the 737 and McDonnell Douglas 128 for the competing MD-80. Airbus had secured only fourteen A320 orders, all from European carriers.

The disparity extended across their full product lines. Boeing recorded 182 commercial-aircraft orders that year, while McDonnell Douglas received 135. Airbus received only thirty-five in total. From the perspective of the two established American manufacturers, the European consortium still appeared to be a minor player.

Of the two, McDonnell Douglas had greater reason for concern. Its commercial lineup consisted principally of the MD-80, derived from the DC-9, and the widebody DC-10. Airbus already offered two widebodies — the A300 and A310 — that competed in many of the same markets with the DC-10. The A320 would now challenge the MD-80 as well, placing nearly the entirety of McDonnell Douglas' product line under competitive pressure.

What should have concerned both American manufacturers was Airbus' government backing. The consortium had been created with direct public support to preserve Europe's commercial-aircraft industry, and the A300 and A310 had been launched with substantial taxpayer-funded investment. The A320 would benefit from similar financing.

That support fundamentally altered the risks of developing a new aircraft. Commercial-airplane programs required billions of dollars in up-front investment, often many years before deliveries generated meaningful revenue. Government financing allowed Airbus to pursue those programs without bearing the full financial exposure faced by Boeing and McDonnell Douglas. Without such backing, the A300 might never have attracted conventional financing. Airbus had no production history or established customer base and was attempting to enter a market dominated by experienced American manufacturers. Private lenders would have considered the venture extraordinarily risky. Government investment made it possible for Airbus to launch the A300, follow it with the A310, and establish itself as a credible competitor.

By the time of the A320, however, Airbus was entering a much larger market from a stronger position. It had an established product line, proven manufacturing experience, and growing credibility among the world's airlines. It might have sought conventional financing for the new program, but as an enterprise created and sustained by its sponsoring governments, it again turned to them for support.

Airbus launched the A320 with considerable fanfare, promoting it as the future of short and medium-haul aviation and emphasizing its advanced safety features. The company had quietly studied the concept since the late 1970s and had worked closely with Delta Air Lines in hopes that an order from a prominent North American carrier would validate the program. Delta ultimately declined to buy the A320 at the time, but its participation gave

Airbus valuable insight into the requirements of major airlines and helped shape the new aircraft's design.

In addition to its all–fly-by-wire cockpit, Airbus gave the A320 a cabin nearly ten inches wider than the 737's. Like Boeing's 737, it used a single aisle with three seats on each side, but the added width provided more shoulder room and greater passenger comfort. The MD-80, by comparison, had a narrower fuselage with a two-by-three seating arrangement.

The wider fuselage added weight, but it also gave the A320 greater versatility. Its cargo hold could accommodate standardized containers, while baggage and freight aboard the 737 and MD-80 generally had to be hand loaded individually by ground crews.

Air France became the A320's launch customer, and other airlines soon followed. The aircraft made its first flight in 1987 and entered flight testing and certification with more than 200 orders already secured. The program proceeded smoothly, but the airplane still faced considerable skepticism, particularly from pilots wary of its computerized flight controls.

***

John Leahy joined Airbus in 1985 as head of sales, charged principally with breaking into the stubbornly resistant North American market. Before joining Airbus, he had sold light aircraft for Piper, giving him extensive experience in aviation sales but little exposure to the world of large commercial jetliners.

In 1986, Leahy approached Northwest Airlines, which operated an aging mixture of Boeing and McDonnell Douglas aircraft. Although Northwest was regarded as one of the industry's more efficient carriers, many of its airplanes were expensive to fuel and maintain. Its recent acquisition of Republic Airlines had also expanded its network, increasing its need for newer, more efficient aircraft.

Leahy approached Northwest's executive leadership and promoted the A320 as offering operating costs as much as 50 percent lower than some of Northwest's existing aircraft, while providing the roughly 150-seat capacity the airline wanted. Boeing offered the popular 737-300, but it was too small and burdened by a delivery backlog of more than three years. The larger 737-400 offered greater capacity but lacked the range Northwest was looking for. McDonnell Douglas' offerings were not compelling at all.

Northwest's executives nevertheless remained wary of ordering an unproven airplane from a manufacturer that had existed for less than two decades. Leahy answered their skepticism with a proposal that became legendary within the industry. He offered Northwest the discounted per-aircraft price normally associated with a 100-plane order while requiring the airline to commit to only ten A320s. So confident was he that Northwest would be satisfied that he also reserved ninety additional delivery positions for the carrier to exercise later.

The gamble succeeded. Northwest became the first North American airline to operate the A320, placing it into service in 1989. Leahy's confidence proved well founded — pleased with the aircraft's performance, Northwest eventually exercised the remaining ninety delivery positions.

It would not be Leahy's last sales coup in North America. In 1992, United Airlines stunned the industry by ordering fifty A320s. United had long been regarded as an almost exclusively Boeing customer, with historical ties stretching back to Boeing Air Transport six decades earlier. The decision was so unexpected that some members of Congress questioned whether Airbus' government-backed financing had enabled it to undercut Boeing.

United acknowledged that price played an important role, but it was not the only consideration. The A320's performance made it particularly well suited to operations from United's hot-and-high Denver hub. Nevertheless, the order intensified accusations that Airbus' government support gave it an unfair competitive advantage — charges that would grow louder as the company gained ground against Boeing and McDonnell Douglas.

By the end of 1993, when Airbus delivered United's first A320, the European manufacturer held only about 18 percent of the commercial-aircraft market. Yet it had already surpassed McDonnell Douglas and established itself as the industry's second-largest manufacturer.

Leahy, meanwhile, would become the public face of Airbus sales for the next three decades. He directed sales campaigns responsible for aircraft orders valued collectively at more than a trillion dollars, building a record unmatched in commercial aviation and earning a reputation as the industry's greatest salesman.

Reaching the Zenith

Thornton Wilson had led Boeing through a turbulent decade and positioned it to face the future with confidence. He ensured that the 747 succeeded despite its troubled development, then made the bold decision to launch the 757 and 767 concurrently. Boeing also secured a major defense contract in 1980 to develop and produce a new generation of air-launched cruise missiles. Designed to strike targets from beyond the reach of enemy air defenses, these weapons would later prove highly effective at penetrating and overwhelming sophisticated defense networks.

When Wilson stepped down in 1986, Frank Shrontz became Boeing's sixth chief executive. Shrontz had joined the company in 1958 after earning both a law degree and an MBA and serving three years in the U.S. Army. Like the legendary Bill Allen, he was not an engineer — only the second Boeing CEO without an engineering background. Shrontz instead brought a strong aptitude for administration and program management. He spent much of his career overseeing contracts and corporate administration, briefly leaving Boeing to serve as an undersecretary of defense during the Ford administration. After Ford lost the 1976 election, Shrontz returned to Boeing in 1977 and became production manager for the 707, 727, and 737 programs.

When Shrontz assumed production oversight, all three aircraft were mature designs nearing the later stages of their commercial lives. The 757 and 767 were being developed to replace the 727 and 707, but Boeing had no new aircraft planned for the market segment served by the smaller 737. Positioned below the 727 and the forthcoming 757, the 737 competed directly with the McDonnell Douglas DC-9. Over nearly two decades, the DC-9 and the 737 had become the workhorses of short-haul airline service around the world.

In 1980, McDonnell Douglas introduced an enlarged derivative of the DC-9 under the new designation MD-80. The fuselage was stretched to carry more passengers, and the aircraft received updated engines—though not the new high-bypass turbofans that offered greater fuel efficiency and performance. Its principal advantage was improved operating economics: by spreading each flight's costs across more passengers, airlines could reduce their cost per seat-mile. The engines, although based on older technology, also proved exceptionally durable and required less maintenance than their predecessors.

Despite the new name, the MD-80 remained fundamentally a DC-9. McDonnell Douglas certified it with the FAA as the "DC-9 (MD-80)," allowing the company to treat it as a derivative rather than an entirely new aircraft and thereby avoid much of the expense and complexity of a clean-sheet certification program.

McDonnell Douglas wanted to avoid the cost of a clean-sheet aircraft, but Boeing could not afford one either. The company had already invested heavily in the still-developing 757 and 767 programs. Yet the market served by the 737 and the DC-9 — now the MD-80 — was becoming increasingly important. Airline deregulation had sharply reduced fares and stimulated demand, particularly in North America, the world's largest aviation market. Much of that growth came from relatively short trips: visiting family, meeting clients, or traveling a few states away. Journeys that once required a full day or more by car could now be completed quickly and affordably aboard a Boeing 737 or a McDonnell Douglas DC-9 or MD-80.

The original 737-100 and -200 series — later known collectively as the 737 Original — had far exceeded Boeing's expectations. When the aircraft was launched in 1964, the company estimated a market of only 250 airplanes and viewed the design primarily as a way to fill a gap in its product line. Yet two decades later, the Original series was still selling. By 1984, however, the design was showing its age, and orders were showing signs of slowing as McDonnell Douglas gained market share with the newer MD-80.

While overseeing 737 production, Frank Shrontz received approval for Boeing's engineers to follow McDonnell Douglas' lead and modernize the existing design. Their principal focus was replacing the 737's older low-bypass turbofans with a new generation of smaller, more efficient high-bypass engines being developed by CFM International, a partnership between GE Aviation and France's Snecma. CFM had developed an engine called the CFM56 partly in hopes of winning a U.S. Air Force contract to re-engine its aging fleet of four-engined KC-135 tankers. Boeing's engineers explored adapting a modified version of that engine, the CFM56-3, to the twin-engined 737.

The new engines promised substantial improvements in fuel economy and performance without requiring an extensive redesign of the wing. Their additional power also allowed Boeing to stretch the 737-200's fuselage, increasing typical two-class capacity from about 102 passengers to 126. Designated the 737-300, the new model made its first flight in 1984 and entered service with launch customer USAir in early 1985. It proved enormously successful as air travel surged throughout the 1980s.

In 1985, seeking to capitalize on booming air travel and the popularity of the 737-300, Boeing stretched the design again. The resulting 737-400 carried about 147 passengers in a typical two-class configuration. Boeing also developed a shortened version, the 737-500, with a fuselage closer in size to that of the earlier 737-200.

By the time production of the 737 Classic series ended around the turn of the century, Boeing had delivered 1,988 aircraft. The 737-300 was the most successful variant, with 1,113 built. Combined with the 813 first-generation 737 Originals, total 737 production had reached 2,801 aircraft, making it the most-produced commercial jetliner in history up until that time — a remarkable achievement for an airplane Boeing had initially expected to sell only 250 copies.

***

By 1980, the 747 had become a mainstay of many airline fleets. Improved versions of the McDonnell Douglas DC-10 and Lockheed L-1011 had begun to encroach on the lower end of its market, but neither could match the 747's overall capacity and range. Boeing further tweaked the design that year with the 747-300, whose principal improvement was a stretched upper deck that increased passenger capacity. Even so, the design risked falling behind as newer technologies became available.

Airline executives had closely followed the development of Boeing's 757 and 767 and Airbus' A310, all of which featured glass cockpits, computerized flight systems, and two-person flight crews. Their modern flight-management systems automated many cockpit functions, improving safety and efficiency — critical advantages during an era of high fuel prices. Airlines began pressing Boeing to incorporate similar technology into the 747.

In 1985, shortly before Thornton Wilson stepped down, Boeing formally launched the 747-400. The new model featured a redesigned glass cockpit that eliminated the need for a flight engineer. Boeing also increased the wingspan by seventeen feet, yet the use of new alloys and construction techniques made the wing nearly three tons lighter than those of earlier versions. Combined with more efficient engines and additional fuel capacity, these improvements gave the 747-400 a range of more than 7,500 nautical miles and seating for about 416 passengers in a typical three-class configuration. It remained the world's largest passenger jetliner and retained its familiar title: the "Queen of the Skies."

The timing of the 747-400 could scarcely have been better. By the time of its entry into service in 1989, many airlines were preparing to replace their aging 747s, the oldest of which were approaching twenty years in service and becoming increasingly expensive to maintain. The -400's greater range and capacity also made new routes economically viable, further expanding its appeal.

Demand was strong from the outset. By the time the aircraft made its first flight in 1988, Boeing had received more than 100 orders. The 747-400 entered commercial service with Northwest Airlines in early 1989, and Boeing ultimately delivered nearly 700 aircraft across the series. It became by far the most successful version of the 747.

***

When Frank Shrontz took the helm in 1986, Boeing was benefiting from a surge in passenger travel driven by the economic expansion that had begun in 1983 and by airline deregulation. The company was increasing production across its commercial-aircraft lines, but ramping up output was neither quick nor simple. Many components required long lead times, and when airlines placed orders simultaneously during economic booms, delivery waits could grow from the normal eighteen to twenty-four months to as long as three years.

Managing production during the 1980s therefore became a delicate balancing act between manufacturers and their airline customers. Deregulation had made the industry more volatile. Some airlines that had secured early production slots were no longer ready to accept their aircraft on schedule, while others needed airplanes as quickly as possible. A secondary market emerged in which airlines sold or transferred delivery positions to carriers seeking earlier access. Some also surrendered valuable slots to raise cash.

Manufacturers generally welcomed these arrangements. They wanted completed aircraft delivered promptly because the largest portion of the purchase price was typically paid only upon delivery — often years after the original order had been placed.

When Shrontz took over in 1986, Boeing received nearly 350 orders across its commercial-aircraft lineup. Orders rose to 366 in 1987, then nearly doubled to a record 657 in 1988. With the economy booming and demand showing no sign of slowing, Boeing faced intense pressure to increase production. Shrontz had a formidable challenge on his hands.

Aircraft components required long manufacturing lead times, and if delivery schedules stretched much further, Boeing risked losing customers to its competitors. By the end of the 1980s, the company was producing only about 200 aircraft annually. The nearly 1,400 orders Boeing accumulated between 1986 and 1988 represented roughly seven years' worth of production.

Boeing needed to expand its manufacturing capacity, but before spending heavily on new facilities, it first had to extract as much efficiency as possible from its existing operations. Shrontz launched a broad effort to modernize the company's culture and production methods. Managers attended seminars and courses on contemporary management practices, manufacturing techniques, and supply-chain coordination. In the past, Boeing had largely ridden out the industry's recurring cycles of expansion and contraction. It could no longer afford to do so. Deregulation and the booming economy had created an opportunity for substantial growth, but Boeing could capture it only if it could build and deliver aircraft fast enough to meet demand.

Notwithstanding the brutal downturn of the early 1970s, Boeing had traditionally weathered the industry's recurring cycles of expansion and contraction. The unprecedented order boom of the late 1980s, however, demanded a different approach. Boeing had to learn not only to build airplanes faster, but also to withstand the inevitable downturn once the boom ended. Shrontz intended that transformation to be his legacy. Deregulation had ushered commercial aviation into a new era, yet Boeing was still relying on business practices rooted in the old one.

In 1989, Boeing shattered every previous sales record when airlines ordered 891 commercial jets, nearly half of them for the new 737 Classic series. Although the surge had been anticipated, accommodating it proved painful. A missing component at the beginning of an assembly line could disrupt every stage that followed. Aircraft were rushed out of the factory and parked on the tarmac awaiting parts or unfinished work — a practice known as "traveled work." Some customers received aircraft that still required additional corrections, or rework.

In one highly visible example, Boeing delivered a new 747-400 to British Airways with wrinkled fuselage skin caused by improperly fitted sheet metal. The defect did not pose an immediate safety concern, but it was unsightly and briefly tarnished Boeing's reputation after attracting attention in the aviation press.

It was, nevertheless, the kind of problem most chief executives would prefer — too many orders rather than too few. Even so, the strain was causing considerable anxiety within Boeing's executive suite. The July 17, 1989, issue of *Fortune* captured the dilemma. Its cover featured a confident Frank Shrontz standing before a 747 beneath the headline, "*Can Boeing Survive Prosperity?*"

***

During the 1980s boom, Boeing not only expanded the 737 family but also enhanced its successful 767. Although originally conceived as a medium-haul aircraft, the 767 had been engineered with enough structural capacity to accommodate higher maximum takeoff weights in future versions. Boeing's engineers exploited that growth potential by increasing its fuel capacity and extending its range sufficiently for transoceanic service. On heavily traveled routes such as New York to London, airlines could profitably deploy a 747 because demand was sufficient to fill its more than 400 seats. A route such as Atlanta to Barcelona, however, could not support an aircraft that large on a daily basis. The extended-range 767, with roughly half the capacity and operating costs of a 747, made such routes economically viable.

The significance of the 767 extended beyond serving smaller markets. Deregulation was fragmenting the airline industry into an increasing number of direct city-pair routes. A decade earlier, a passenger traveling from Atlanta to Barcelona might have flown first to New York, then to London, and finally to Barcelona. An extended-range 767 allowed an airline to offer the same journey nonstop, saving passengers hours of travel time. Although conceived primarily as a technologically advanced successor to the 707, the 767 entered service at precisely the right moment. It gave airlines an efficient aircraft with the range to bypass traditional gateway airports and connect smaller international markets directly.

Regulatory changes were also reshaping the industry. Before Boeing could operate the twin-engine 767 on long transoceanic routes, however, it had to overcome existing safety restrictions. At the beginning of the 1980s, regulations generally prohibited twin-engine aircraft from flying routes that took them more than sixty minutes from a suitable diversion airport, whether over water or remote stretches of land, in case one engine had to be shut down in flight.

Trijets such as the DC-10 and L-1011, and four-engine aircraft such as the 747, were not subject to the same restrictions. Regulators reasoned that if a twin-engine aircraft lost one engine, it would be left with no further margin for an additional engine failure because the remaining engine would become its sole source of propulsion. By the 1980s, however, the new generation of high-bypass turbofans had demonstrated exceptional reliability. Even when one engine had to be shut down in flight, there had been no cases in which the remaining engine subsequently failed for an unrelated mechanical reason. Dual-engine power loss had occurred, such as on Air Canada Flight 143, the famous "Gimli Glider," when the aircraft exhausted its fuel because too little had been loaded during refueling. But in a fuel-exhaustion scenario, having more than two engines would have made no difference because every engine would ultimately have shut down.

Shrontz and his team recognized that market fragmentation was creating new opportunities for long-range twin-engine aircraft.

Europe's Joint Aviation Authorities had already permitted the Airbus A300 and A310 to operate as far as ninety minutes from a suitable diversion airport at single-engine cruising speed, increasing their appeal among international carriers.

The FAA moved more cautiously but followed in 1985, approving the Boeing 767 for extended-range twin-engine operations. Its initial authorization allowed flights up to 120 minutes from a suitable diversion airport — thirty minutes beyond the limit then granted to the Airbus aircraft. The A300 and A310 soon received comparable approval. In aviation parlance, the new standards became known as ETOPS, an acronym commonly rendered as Extended-range Twin-engine Operational Performance Standards. It has also been referred to as Extended Twin-engine Operations. Within aviation circles, the acronym is sometimes jokingly said to stand for "Engines Turn Or Passengers Swim."

Although the FAA could certify an aircraft-engine combination for ETOPS, an airline also had to earn operational approval. The carrier had to demonstrate that it could operate and maintain the aircraft reliably, including through an established record of trouble-free service. ETOPS approval therefore depended not only on the aircraft's design and engine reliability, but also on the airline's maintenance practices, training, and operational procedures.

ETOPS aircraft were subject to additional maintenance safeguards intended to prevent a single error from disabling both engines. In general, the same maintenance task could not be performed on both engines during the same maintenance visit when an error could affect both powerplants.

The danger of such common-cause failures had been demonstrated by Eastern Air Lines Flight 855 in 1983. The Lockheed L-1011 was flying from Miami to Nassau when all three engines began losing oil after maintenance personnel failed to install O-ring seals on the magnetic chip detectors in each engine.

Two engines shut down, and the third failed shortly afterward. The crew eventually restarted one engine and returned safely to Miami, touching down with only that engine producing power. The incident became a classic example of how repeating the same maintenance error across multiple engines could place an aircraft in grave danger.

To receive ETOPS-120 approval, an airline initially had to accumulate at least 20,000 flight hours with the aircraft type without an in-flight engine shutdown. Although that figure may sound daunting, an airline operating ten 767s for an average of 1,500 hours each annually could reach the threshold in little more than a year.

In-flight shutdowns were already rare. Turbine engines were exceptionally reliable, largely because they had far fewer moving parts than the piston engines they replaced. When a shutdown did occur, the cause was often not the turbine itself but an ancillary component such as a fuel or oil pump. The three major engine manufacturers — General Electric, Pratt & Whitney, and Rolls-Royce — continually improved these systems, sometimes adding redundant components to reduce the possibility of an engine shutdown still further.

ETOPS opened an entirely new realm of airline operations for twin-engine aircraft, an advance pioneered in part by the emerging European manufacturer Airbus. Deregulation had encouraged airlines to operate more efficiently, and long-range twin-engine aircraft offered a powerful way to do so. They could serve intercontinental routes at substantially lower operating costs than comparable three- or four-engine aircraft, making thinner nonstop routes economically viable.

A large jet such as the 747 was expensive to operate and generally had to be filled, or nearly filled, to earn a profit. Yet many airlines initially bought the aircraft as much for its range as for its capacity. A route such as Chicago to Tokyo could be flown nonstop only by the 747 because, at the time, no other commercial airliner had sufficient range. Even when an airline could not fill every seat, it could charge higher fares because passengers had few alternatives.

A long-range twin-engine aircraft changed that equation. It could fly many of the same routes with fewer seats and substantially lower operating costs, allowing airlines to reduce fares. Lower prices stimulated demand, which could support more frequent service. A route previously flown only several times a week or once a day might eventually support multiple daily departures, giving travelers greater flexibility and convenience.

Boeing had designed the 767 with considerable room for growth. In rapid succession, its engineers increased the maximum gross takeoff weight of the 767-200, allowing it to carry more fuel and extending its range by roughly 2,500 nautical miles. Designated the 767-200ER, for "Extended Range," it could serve trans-Pacific routes. Boeing then stretched the fuselage to create the 767-300, increasing passenger capacity by approximately 20 percent, and subsequently raised that model's takeoff weight and fuel capacity to produce the 767-300ER.

During the order boom of the 1980s, airlines bought the 767 in large numbers. Its combination of range, lower operating costs, and ETOPS approval helped usher in a new era of long-distance air travel. By the end of the decade, the 767 had become the most widely used aircraft on the heavily traveled North Atlantic routes, displacing the 747, DC-10, and L-1011 as the preferred choice for many services.

***

By 1990, the first generation of widebody jetliners was beginning to age. Aircraft such as the 747, A300, DC-10, and L-1011 had helped bring air travel to the masses and fueled the greatest expansion in commercial aviation history. Under Frank Shrontz, Boeing had addressed the aging of its early 747 fleet with the highly successful 747-400, which airlines were ordering in large numbers as a replacement. But the earliest DC-10s, L-1011s, and A300s were also getting older. With Lockheed having withdrawn from commercial aviation, Boeing, McDonnell Douglas, and Airbus each saw an opportunity to capture the coming replacement market.

The three manufacturers approached that opportunity differently. Boeing, true to its culture and history, chose the boldest path.

The 767 had become a major success, and Boeing's engineers initially explored ways to enlarge it even further. One proposal, internally designated the 767-X, placed a partial upper deck over the rear half of the fuselage, producing an ungainly double-deck configuration. Airlines shown the design reacted with little enthusiasm. They wanted more than a larger 767. Having embraced the economics of twin-engine aircraft, they wanted a wider fuselage, greater capacity, and enough range to fly fully loaded across continents and oceans.

Shrontz and his team briefly considered a trijet version of the 767, but that would have amounted to little more than an improved successor to the DC-10 and L-1011. Boeing did not want merely to compete with McDonnell Douglas. It wanted to change the market. Therefore, the company began studying an entirely new aircraft: a supersized twin-engine jet larger than any then in service. It would be a clean-sheet design incorporating the latest advances in avionics, aerodynamics, and materials. The initial version would carry roughly 300 passengers over intercontinental distances, filling the gap between the 767, with about 250 seats, and the 747-400, with more than 400. Like the 767, it would also be designed with room to grow, allowing future versions to carry more passengers, more fuel, or both.

Boeing's new twin would require engines of unprecedented power. The aircraft was expected to exceed 550,000 pounds at maximum takeoff weight and carry more than 300 passengers, placing its capabilities close to those of the first-generation 747 — yet it would rely on only two engines.

Pratt & Whitney already had a potential starting point in its PW4000 series. The existing engine produced no more than about 68,000 pounds of thrust, but the company believed it could enlarge and modify the design to deliver as much as 99,000 pounds for Boeing's new aircraft. Rolls-Royce also committed to developing an engine for the program. Derived from the smaller Trent 700, the new Trent 800 would eventually produce up to approximately 95,000 pounds of thrust.

General Electric, meanwhile, was developing an all-new engine designed initially to produce at least 90,000 pounds of thrust, with future versions potentially exceeding 110,000 pounds. GE's design incorporated more new technology than the offerings from Pratt & Whitney and Rolls-Royce, which were based more heavily on existing engines, but it also carried greater development risk.

All three manufacturers recognized the direction in which commercial aviation was moving. Compared with trijets and four-engine aircraft, large twinjets offered airlines lower acquisition, maintenance, and operating costs. The future belonged to two engines.

***

At a maximum takeoff weight of roughly 550,000 pounds and with seating for about 300 passengers, the new aircraft would be enormous — yet powered by only two engines. Based on the performance promised by engine manufacturers, even modest growth beyond the initial design suggested that the aircraft, now designated the 777, could begin encroaching on the lower end of the 747's market.

Airlines typically deployed their 747s between major gateway cities. Many passengers flying to London, for example, were not ending their journeys there but connecting onward to other destinations. The 767 and Airbus A300 and A310 had enabled more direct service between smaller city pairs, but their capacity and range remained limited. The 777 promised to open additional nonstop routes while also allowing airlines to serve heavily traveled markets far more efficiently with only two engines.

Boeing understood that the 777 could cannibalize some 747 sales, particularly if the new design eventually grew into larger variants. But that was preferable to surrendering the market to a competitor. Even the largest 777s then envisioned would leave a role for the 747, though a diminished one, while ensuring that Boeing continued to dominate the large-aircraft segment.

Like the 757 and 767 before it, the 777 was intended to represent a major advance in commercial aviation. Having seen those earlier gambles pay off, Frank Shrontz urged his team to consider every opportunity to make the aircraft distinctive. McDonnell Douglas was developing an improved DC-10 called the MD-11, while Airbus was simultaneously developing two related widebodies: the twin-engine A330 for medium to long-haul service and the four-engine A340 for long-haul and ultra long-haul routes. The 777 therefore had to do more than compete. It had to change the market.

Aircraft manufacturers had traditionally developed new designs by presenting concepts to prospective customers and gauging their interest. With the 777, Boeing went further. Once airlines demonstrated strong demand for a high-capacity, long-range twinjet, Boeing invited several of them to participate directly in shaping the aircraft. The goal was not to compile separate wish lists, which in the past had produced costly airline-specific variations. Instead, Boeing encouraged the participating carriers to identify features they broadly shared, allowing the company to incorporate common operational requirements into a single, more versatile design.

Boeing invited eight prominent airlines that had expressed interest in the 777 to participate directly in its development: All

Nippon Airways, American Airlines, British Airways, Cathay Pacific, Delta Air Lines, Japan Airlines, Qantas, and United Airlines. Each sent representatives from its engineering, maintenance, flight operations, and cabin-service groups to work with Boeing's design teams. Together, they sought to define a set of best-in-class features that would appeal broadly across the airline industry. Boeing called this collaborative approach "Working Together" and promoted it heavily in trade publications and to the public.

The 777 would also transform the way Boeing designed an airplane. Traditionally, engineers first developed an aircraft's overall configuration according to its performance requirements. Scale models were then built and tested in wind tunnels, where airflow over the wings, fuselage, engine nacelles, and tail surfaces could be studied. Engineers repeatedly modified and retested the models until the aircraft's aerodynamic shape was optimized. The process could consume hundreds or even thousands of wind-tunnel hours and generate enormous quantities of data that had to be analyzed manually.

Once the aircraft's external shape was established, engineers created detailed drawings for every structural component, from the fuselage skin and underlying frames to the placement of individual fasteners. These drawings were prepared on paper, both at reduced scale for integration with surrounding structure plans and at full scale when precise dimensions were required for manufacturing.

A new aircraft's completed design could comprise thousands of blueprint sheets. Each had to be cataloged, copied, distributed, revised, and made available to Boeing's engineering and manufacturing teams, as well as to regulators responsible for reviewing and certifying the design.

Once an aircraft's design was finalized, manufacturers traditionally built a full-scale mock-up. Usually constructed from inexpensive materials such as wood, the mock-up reproduced the aircraft's principal structure and exterior shape. Engineers used it to verify that components — including fuselage panels, landing gear, wiring harnesses, plumbing, and pumps — would fit within their specified tolerances without interfering with one another. It also helped production planners determine the sequence in which parts should be installed most efficiently and economically. The mock-up was often the last opportunity to identify conflicts that had escaped notice on the drawing board.

Designing an aircraft, testing it in wind tunnels, producing thousands of engineering drawings, and constructing a full-scale mock-up was enormously laborious, time-consuming, and expensive. Thousands of engineers might be assigned to the development program. Under Frank Shrontz, Boeing's 777 team set out to change that process.

Computer technology had advanced dramatically since Boeing developed the 757 and 767 a decade earlier. The 777 team concluded that sophisticated computer-aided design, or CAD, software could be used to design every component of the aircraft digitally. Instead of relying on vast quantities of paper drawings, engineers could create and integrate the entire airplane within a computerized three-dimensional model.

Advances in computational fluid dynamics also allowed engineers to simulate airflow over the aircraft's surfaces. Computer models could display how air would move around the fuselage, wings, engine nacelles, and tail, reducing the need for repeated wind-tunnel testing. Physical testing remained necessary to validate the computer's predictions, particularly because inaccurate assumptions or data could produce misleading results, but the number of wind-tunnel hours could be substantially reduced.

Boeing's engineering leaders became confident enough in the digital model to propose eliminating the traditional full-scale mock-up altogether — one of the riskiest decisions made during the 777's development. Senior executives approved the proposal only in part. Boeing would still construct a full-scale mock-up of the forward fuselage, including the nose and cockpit.

That section was among the most difficult parts of the aircraft to design and assemble. Its complex contours required numerous differently shaped structural components, while its interior contained a dense concentration of avionics, wiring, controls, and other systems. If the nose and cockpit could be assembled precisely as depicted in the digital model, Boeing reasoned, it would demonstrate that the remainder of the aircraft could also be built without a full-scale mock-up.

Boeing's engineers had two more innovations in mind for the 777. The first was the incorporation of a digital fly-by-wire flight-control system. Taking a cue from Airbus, Boeing would place computers between the pilots' controls and the aircraft's primary control surfaces, replacing the direct mechanical control architecture traditionally used on Boeing commercial airplanes. But Boeing chose to implement fly-by-wire differently from Airbus in several important respects.

The 777 retained the traditional control columns and yokes in front of the pilots rather than adopting the sidestick controllers used on Airbus' planes. The two pilots' control columns were also mechanically linked, so movement by one pilot could be seen and felt by the other. Boeing further designed the controls to provide artificial force feedback, preserving much of the tactile sensation pilots associated with conventional flight controls.

Most significantly, Boeing adopted a different philosophy toward flight-envelope protection. The 777's computers could resist control inputs that would take the aircraft beyond certain aerodynamic limits, but those protections were generally "soft" rather than absolute. A pilot who deliberately applied sufficient control force could override them. Boeing's philosophy remained that, although computers could assist and protect the pilot, ultimate authority over the airplane would remain with the person at the controls.

The other innovation the company had in mind was related to the 777's wings. By the early 1990s, airports around the world were becoming increasingly congested, and gates capable of accommodating large aircraft were scarce. With a wingspan of just over 200 feet — comparable to that of the 747 — the 777 would compete for the limited number of widebody gates at major hubs.

Borrowing from carrier-based naval aircraft, Boeing proposed optional wingtips that could fold upward while the airplane taxied or was parked. The reduced span would allow the 777 to use gates designed for narrower aircraft such as the DC-10, whose wingspan was approximately 165 feet.

The concept was eventually abandoned for two principal reasons. First, the folding mechanism added considerable weight — always a liability in aircraft design. Even customers ordering fixed wings would have received aircraft carrying some of the underlying structural provisions, allowing Boeing to maintain production commonality. Second, airlines and airports were already planning to expand or reconfigure gates to accommodate the larger aircraft expected during the coming boom in air travel. Even so, the proposal demonstrated the ambition and ingenuity Boeing brought to the 777 program.

Although the 777 exceeded its initial development budget — not unusual for an entirely new aircraft — its development proceeded timely. Boeing intended the airplane to enter service with ETOPS approval from its first delivery to United Airlines and with a dispatch reliability rate exceeding 98 percent.

Traditionally, new aircraft accumulated the service experience required for ETOPS only after entering airline operations. Dispatch reliability above 98 percent also usually took time to achieve as manufacturers and airlines resolved the inevitable problems associated with a new model. By promising both from the outset, Boeing sought to assure airlines that the 777 could be integrated into their fleets without the prolonged teething troubles normally associated with an all-new aircraft.

***

The airline downturn Frank Shrontz and Boeing had anticipated arrived in 1990. The 1980s had produced the longest U.S. economic expansion since World War II, but rising inflation and higher interest rates began slowing consumer spending by the end of the decade. Economic growth had already weakened when Iraq invaded Kuwait in August 1990, abruptly threatening the stability of the Persian Gulf and the world's oil supply.

Oil prices surged, and the prospect of a wider war unsettled consumers and businesses alike. Saddam Hussein's assurances that Iraq would advance no farther carried little credibility. His military was among the world's largest at the time, Iraq had already used chemical weapons against its own populace, and there were widespread fears that it was pursuing nuclear weapons.

The combination of higher fuel prices, economic uncertainty, and fears of terrorism sent air travel into a steep decline. Boeing, McDonnell Douglas, and Airbus watched the order boom of the late 1980s come to an abrupt end just as each was spending heavily to develop a new generation of widebody aircraft. Although the recession lasted only from July 1990 until March 1991, air travel would take roughly two years to recover to its late-1980s level.

The early 1990s nevertheless marked a turning point not only for aerospace but for the world economy. The Soviet Union collapsed in 1991, leaving the United States as the sole superpower and opening vast regions of Eastern Europe and the former Soviet bloc to international trade. At the same time, rapid advances in computing and communications promised to transform how people lived and worked.

For more than four decades after World War II, the world had been divided broadly between the capitalist West, led by the United States, and the socialist East, led by the Soviet Union. Trade between the two blocs had remained limited, while the threat of nuclear war overshadowed international affairs. The Soviet collapse dissolved that division and fostered hopes for a new global order built on capitalism, expanding trade, and greater economic interdependence. Commercial aviation stood to benefit enormously. Air travel had long reflected the health of the broader economy. It expanded with growth and contracted quickly when confidence weakened. The opening of previously restricted markets, combined with rising global trade, promised an unprecedented increase in passenger and cargo traffic.

Before that growth could be captured, however, the world's commercial aircraft manufacturers had to decide what kind of airplanes this new era would require.

The Little Plane That Could

When Airbus launched the A320, Boeing still regarded the European manufacturer as a niche competitor and paid relatively little attention to the performance claims coming out of Toulouse. But when Northwest Airlines and then United Airlines rejected the 737 in favor of the A320, Boeing took notice. Airbus was already making rapid inroads with foreign carriers despite negative publicity surrounding the several early A320 crashes, but the Northwest and United orders sounded an alarm in Seattle.

Boeing could no longer ignore that the A320 represented the next generation of commercial aviation. It took advantage of the newest engines and incorporated advanced avionics intended to reduce pilot workload and, in theory, make the aircraft safer and easier to operate.

The Boeing 737 and McDonnell Douglas DC-9 and MD-80 families had brought jet service to smaller markets on a broad scale, but their designs imposed limitations. Airlines seeking to serve larger markets or fly longer routes profitably with those aircraft often had few practical options. Airbus approached the A320 differently. It was intended not only to replace aging 737s, DC-9s, and MD-80s on shorter routes, but also to give airlines the flexibility to serve denser and longer routes profitably.

Like Boeing, Airbus would offer its narrow-body family in several fuselage lengths. The A319 seated approximately 134 passengers, the original A320 about 164, and the A321 roughly 185. Airbus later introduced the smaller A318, nicknamed the "Baby Bus," which seated approximately 107. Such figures were generally based on approximate two-class configurations and allowed airlines to compare competing aircraft.

When United ordered the A320 in 1992, Boeing began seriously debating whether it needed an entirely new airplane to compete. The 737 design was approaching 30 years old. In an era of rapid advances in engines, avionics, materials, and aerodynamics, commercial aircraft designs could appear antiquated after only two decades.

The 737 had survived largely because Boeing re-engined and updated it during the early 1980s, producing the 737-300, -400, and -500 series. These "Classic" models gave the airplane a new lease on life, but beneath the improvements remained a design rooted in the early 1960s. A clean-sheet replacement would allow Boeing not merely to match the A320 but potentially surpass it, limiting Airbus' ability to capture additional market share. Yet developing an entirely new airplane posed three major challenges.

The first was engineering capacity. Boeing and much of its supplier network were already immersed in developing the 777. Finding enough engineers to launch another major program would be difficult, at least in the short term. Boeing had developed the 757 and 767 concurrently a decade earlier, but those airplanes were conceived as related programs. They shared design features, technologies, and engineering resources. A 737 replacement would be a separate undertaking requiring its own dedicated workforce.

The second challenge was time. An all-new airplane would probably take most of the decade to develop and certify. During those same years, airlines around the world would begin retiring large numbers of older 737s, DC-9s, and MD-80s. While Boeing developed its replacement, Airbus could capture much of that market with an airplane already in production.

The third and greatest consideration was money. Boeing had budgeted approximately $5 billion to develop the 777, an amount equal to nearly 20 percent of the company's total value when the program was launched. Those costs were already rising. By 1992, Frank Shrontz and his executive team concluded that launching another clean-sheet airplane could place enormous strain on Boeing's finances.

There was also the question of recovering the investment. The billions spent designing an airplane from the ground up would ultimately have to be incorporated into the price paid by airline customers. That could place a new Boeing narrow-body at a disadvantage against the A320, whose development had been supported by Airbus' sponsoring governments.

Therefore, Boeing considered another option: modernizing the 737 yet again.

From a capacity standpoint, the 737 fuselage was broadly comparable to that of the A320. It would require some stretching to achieve direct parity but lengthening a fuselage by inserting additional sections — known in the industry as "plugs" — ahead of and behind the wing was a relatively straightforward process.

Performance presented a more difficult problem. The A320's wing had been designed as an integral part of an entirely new airplane and offered excellent lift, range, and payload capability. Boeing's engineers examined enlarging the 737's wing, horizontal stabilizer, and vertical tail. Combined with newer, more powerful, and more fuel-efficient engines, those changes could make a third-generation 737 a formidable competitor.

Updating the fuselage, wings, tail surfaces, and engines would still cost billions, but far less than developing an entirely new airplane. Boeing could also retain the 737's existing type certificate. Pilots already qualified on earlier 737s would not require the extensive training normally associated with moving to a different aircraft type. Once qualified on one version of the 737, they could transition to another with comparatively limited additional instruction. That would become one of Boeing's strongest selling points for this new iteration of the 737.

The updated aircraft received the official designation 737 Next Generation, or 737 NG. Even so, the clean-sheet A320 retained several inherent advantages. Its fuselage was slightly wider, providing additional cabin space, and slightly deeper, allowing containerized cargo to be carried beneath the passenger floor. Standardized containers could accelerate the loading and unloading of baggage and freight, reducing turnaround times and operating costs.

The 737 NG would retain its short landing gear, a legacy of Lufthansa's original 1965 requirement that baggage be loaded manually by ground crews without specialized equipment. Because the basic fuselage would remain unchanged and could not accommodate standard cargo containers, Boeing saw little reason to raise the airplane farther off the ground.

The larger CFM56-7 engines planned for the 737 NG could still be fitted beneath the new wing, but their placement required further adjustment. As Boeing had done with the CFM56-3 on the 737 Classic, engineers positioned the engines slightly forward and higher. The nacelles also retained a flattened lower profile to preserve adequate ground clearance, including during crosswind landings in which one main landing gear might touch down before the other.

The A320 did offer fly-by-wire flight controls. Whatever reservations some pilots had about placing computers between themselves and an airplane's control surfaces, the technology reduced weight and mechanical complexity while offering potential maintenance savings.

Here, however, Boeing could turn the 737's age to its advantage. By the early 1990s, the 737 had become the most widely used jetliner in history and formed the backbone of many airline fleets around the world. Carriers replacing older 737s with the 737 NG would pay primarily for the airplanes themselves. Because the NG retained the same type certificate, airlines could avoid much of the cost of retraining pilots, as well as the cost associated for integrating a new type in their fleet and maintaining separate pilot pools for each type.

Maintenance technicians and ground-service personnel would also require relatively little additional instruction. Existing spare-parts inventories, support equipment, maintenance procedures, and supplier relationships could remain largely intact because the 737 NG would share many systems and service requirements with earlier models.

Replacing an established 737 fleet with the A320 would be far more disruptive. An airline would have to train pilots, mechanics, dispatchers, and ground crews; acquire new tools and support equipment; establish new spare-parts inventories; and build new supplier and maintenance relationships. For an airline operating dozens, and in some cases, hundreds of 737s, those transition costs could reach hundreds of millions of dollars beyond the purchase price of the aircraft.

The competing sales arguments were straightforward. Airbus could ask: Why remain tied to an aircraft designed in the early 1960s when a modern airplane offered greater efficiency, lower maintenance costs, advanced technology, reduced pilot workload, and the flexibility to operate profitably across a wider range of routes?

Boeing could respond: Why incur the expense and disruption of introducing an entirely different aircraft when the 737 NG offered comparable performance in a type your pilots, mechanics, and ground crews already knew, and your operating infrastructure was built around?

Airbus could, however, blunt the cost of switching by heavily discounting the A320, a tactic it would employ to great effect in the years ahead. And as Airbus gained market share, Boeing's complaints that its European rival enjoyed an unfair advantage through government financial support would grow steadily louder.

***

Although the 737 Classic remained a reliable cash generator, Frank Shrontz recognized the threat posed by the A320. He pressed for — and in November 1993 received — approval to develop the 737 Next Generation. The first 737 NG would make its maiden flight three and a half years later.

Boeing ultimately offered four principal variants, listed in order of increasing passenger capacity: the 737-600, -700, -800, and

-900. Each was positioned against a roughly comparable member of the Airbus family: the A318, A319, A320, and A321.

The A320 and 737 NG would soon enter a series of fiercely contested sales campaigns that helped determine the fortunes of both manufacturers. As global air travel expanded, the two aircraft families became the workhorses of airlines around the world. No other commercial aircraft would be produced in comparable numbers.

Both manufacturers continually refined their airplanes, improving range, fuel efficiency, payload, and operating economics. Although originally designed primarily for short and medium-haul service, later variants became capable of flying much longer routes. As airline networks evolved, some carriers even began deploying them on thin transatlantic routes that had once been the exclusive domain of larger widebody aircraft.

Jack Welch

John Francis "Jack" Welch Jr. was probably the most consequential CEO in Boeing's history — only Jack Welch was never a Boeing CEO.

Welch served as CEO of General Electric, or GE, from 1981 to 2001. Born in Massachusetts in 1935, he was an only child and completed his formal education at the age of 24, earning a master's degree and a PhD in chemical engineering from the University of Illinois at Urbana-Champaign. He immediately joined GE, where he spent the remainder of his professional career. With a deep disdain for bureaucracy, Welch built a reputation for eliminating layers of management, simplifying operations, and relentlessly cutting costs. He rose rapidly through GE's ranks and, in 1981, became the youngest CEO in the company's history.

Welch's reputation as arguably one of the most successful CEOs in American corporate history could be chalked up to the fact that when he took the reins at GE, the company's market valuation was slightly more than $12 billion — about $44 billion in 2026 dollars. By the time he retired twenty years later, in 2001, GE's value was estimated at more than $420 billion, or approximately $792 billion in 2026 dollars. Under Welch's leadership, the company completed more than 600 acquisitions, many of them in emerging markets, while simultaneously divesting underperforming business units.

But it was Welch's leadership style that drove much of GE's enormous growth during his tenure. He became known not only for cost-cutting, but also for his laser-like focus on performance. He implemented a policy that rewarded the company's top 20 percent of managers with eye-popping bonuses and stock options while automatically firing the lowest-performing 10 percent, regardless of any improvements or innovative ideas they may have introduced to pave the way for future growth.

Welch constantly scrutinized the numbers. If GE was not number one or number two in a particular market, it came as no surprise if he replaced the division's leadership or divested the business entirely. His propensity to fire people while retaining the real estate and other assets associated with an underperforming division earned him the moniker "Neutron Jack." The nickname referred to the neutron bomb then under development during the Cold War, a weapon designed to kill people with neutron radiation while leaving buildings and equipment intact.

Welch believed that the most important concerns for any company should be its products and customers because of the direct correlation between superior products, satisfied customers, and the ultimate measure by which any publicly traded company was judged: shareholder value. In other words, creating good products and services led to satisfied customers, which produced market-share gains and ultimately increased demand for both the company's products and its stock.

At some point during Welch's tenure, however, it could be argued that the relationship became inverted, and delivering shareholder value became the first priority. There have always been different ways to increase shareholder value, and one of the most effective is to improve profit margins regardless of whatever else may be happening within the organization.

Improving profit margins meant reducing operating costs. It meant running the company more efficiently. Jack Welch was a connoisseur of efficiency. If costs within a division were too high, he excelled at lowering them. And if they could not be lowered, then, well, sell the business to someone who could operate it better than GE.

In 1989, Jack Welch and a handpicked entourage embarked on a business trip to India. At the time, India had the second-largest population in the world, with more than 850 million people, ranking just behind China. The country portended a potential gold mine as an untapped market. Its burgeoning population, although still largely mired in poverty, was on the cusp of benefiting from a loosening of economic policies and trade restrictions under the leadership of its then-Prime Minister, Rajiv Gandhi.

The goal of the trip was to meet with Indian government officials and convince them to buy more GE products, particularly its engines and other manufactured goods. During one meeting, a minister in Gandhi's government boldly asked Welch whether, if India agreed to purchase the products GE was pitching, GE would in turn help jump-start the country's nascent technology sector, which India was keenly focused on developing and exporting.

The question caught Welch and his entourage off guard, but it also intrigued them. Welch wanted to know how GE could help India's technology sector, and the answer proved equally intriguing.

India's cost of living was a fraction of that in the United States, which meant a software programmer in India could be paid a fraction of what an American programmer earned. English was also recognized as an official language. Most Indians educated beyond the equivalent of high school typically spoke fluent English, meaning there would be few language barriers to overcome.

GE could assign some of the computer code being written each day in its American offices to Indian technology companies, which could perform the work at a fraction of the cost. At the time, GE relied heavily on programmers to write the code needed to operate the machines and systems it sold across numerous industries around the world.

Welch immediately understood the implications of the proposition. He was all too happy to agree because lowering the cost of doing business was always foremost on his mind, especially when it enhanced the ever-vaunted goal of shareholder value.

Besides, the work under consideration did not involve GE's most cutting-edge programming. It involved more mundane, lower-level assignments — writing code modifications, maintaining existing programs, and conducting quality testing. It was an easy proposition for Welch to accept, but one that would have far-reaching consequences for American industry.

Shortly after Welch returned from the watershed trip, GE sent a delegation back to India to lay the groundwork for integrating the country's developing technology sector into GE's software operations. What had originally been envisioned as the allocation of simple programming tasks — the low-hanging fruit — quickly grew legs.

GE's initial delegation, followed by others, identified a host of additional areas in which Indian technology companies could support GE's businesses in the United States. The arrangement quickly evolved into a symbiotic relationship. GE transferred ever more internal work to India, and India delivered it at a fraction of what it would otherwise have cost.

The outsourcing boom at an industrial scale was born, and it would change the face of American capitalism and culture for the remainder of the century and well into the next. Outsourcing fit perfectly within Jack Welch's ecosystem. If someone else could do something better and for less money, then let them do it.

GE's division leaders, under constant pressure from Welch to improve shareholder value, began identifying increasingly sophisticated programming work that could be outsourced to India. Soon, entire IT organizations within GE were decimated as the jobs they once performed moved overseas. On more than a few occasions, the American employees losing their jobs were required to train the Indian workers who would replace them.

Eventually, those same division leaders began looking beyond information technology for other tasks that could be outsourced to improve profit margins. Customer-support services, long considered a core function of any organization, soon began heading to India. An Indian worker, paid pennies on the dollar, was now answering customer inquiries about GE products.

Other companies sat up and took notice. CEOs across the landscape of American business began emulating Jack Welch because GE had become a Wall Street sensation under his leadership.

Companies began looking to India for a wide range of organizational support, including IT services, customer support, sales support, and even internal finance and accounting functions — the nerve centers of corporate America. Engineering services soon followed. After all, why pay $100,000 or more for an engineer fresh out of MIT when similar engineering capabilities and expertise could be obtained from an Indian engineer earning $10,000? Often, those same Indian engineers had studied at MIT or other prestigious American engineering schools before taking their newly minted degrees back home.

The industrial outsourcing phenomenon that began under Jack Welch would snowball, reaching into nearly every facet of American corporations and, eventually, the average American household, as tens of thousands of jobs migrated to a distant land.

By the early 2000s, outsourcing had become a major political issue, forcing politicians to address the growing number of American jobs being transferred to regions of the world where labor was cheaper. From the corporate standpoint, however, C-suite executives could not ignore the fact that they could substantially improve profit margins by shifting jobs from high-cost onshore locations to lower-cost offshore markets such as India.

Outsourcing came to have an almost narcotic effect on corporate leaders. Like a narcotic, it produced a quick high that was both euphoric and addictive because it delivered a direct and immediate improvement to the corporate bottom line.

***

Of course, outsourcing was not a new phenomenon. It had been practiced since our ancestors first began to walk upright. Outsourcing is really nothing more than specialization — the practice of allowing someone else to perform a task because they are better skilled at it than you are.

At one point in human history, everyone possessed essentially the same knowledge of the world around them — what was edible and what was not, how to hunt, how to gather food, and how to survive. The skills required for daily life were performed by nearly everyone and, for the most part, performed equally. But as human behavior became more sophisticated, individuals began to specialize, and communities increasingly relied on different people to perform different tasks.

The earliest form of outsourcing probably emerged when our ancestors began using tools. A nomadic group of hunter-gatherers may have included someone who had become particularly skilled at shaping spear points from flint. Perhaps he had a keen eye for finding the best stone or had perfected the technique of attaching the spear point securely to the shaft. Whatever the reason, others in the group may have asked him to make or repair their spears. In return, he may have received additional food, animal skins, or some other form of compensation.

Projecting that simple example across the millennia, it is easy to see how specialization became the forerunner of modern outsourcing. As human society grew more complex, people increasingly depended on others whose skills exceeded their own.

In the modern era, a business is simply an entity that specializes in doing something specific — obtaining raw materials, transporting them, manufacturing a product, distributing it, selling it, maintaining it, or providing some other service. When a consumer or another business purchases that product or service, it is turning to a specialized entity for a capability that was outsourced long ago.

Modern outsourcing, however, has created an ecosystem in which even a business that specializes in a particular product or service can outsource some of the very capabilities that allow it to function as a specialist.

The modern outsourcing phenomenon would not have grown so quickly or penetrated so deeply into American industry had it not been for two epochal events that occurred around the time of Jack Welch's 1989 trip to India.

The first was the collapse of the Eastern Bloc, which opened the door to the implementation and dramatic expansion of free-market systems across much of the globe. The free-market system held that markets should be free to trade with one another and that prices should be determined by supply and demand rather than government restrictions.

The second epochal event was the advent of the Internet. The Internet became commercially viable by 1992, just three years after Welch's trip to India, and grew explosively in the years that followed. It had the effect of removing barriers of time and distance and allowed information to be shared in ways unimaginable even to many of the people who had helped create it.

Information could now be transmitted instantaneously and seamlessly. A manager in the United States could pull up a spreadsheet on a computer screen and review it in real time with a counterpart half a world away, as though they were sitting beside each other in the same room. The outsourcing boom would not have grown as quickly, nor penetrated as deeply across the industrial landscape, without the capabilities offered by the Internet and the World Wide Web.

Before the Internet, outsourcing had been largely confined to manufacturing, with the textile industry serving as the most obvious example, although other industries had also moved production processes overseas. Sending manufacturing work abroad, however, required establishing an autonomous operation on a distant shore — one capable of functioning largely on its own while the parent company exercised oversight at arm's length because of barriers of time and distance.

The Internet changed that dynamic. Service and support functions quickly followed manufacturing offshore because managers at the parent company could now interact constantly and in real time with workers thousands of miles away. The work could be supervised almost as though it were still being performed inside the company's own offices. Outsourcing gradually became almost routine, with little consideration given to the consequences of transferring processes from across an organization to outside suppliers.

The most obvious, but often least considered, consequence was that once a process was outsourced, the knowledge associated with it — and often the infrastructure and support capabilities required to perform it — went out the door as well. Bringing that work back in-house could require a herculean effort at enormous cost.

Yet few executives gave much consideration to that possibility. The savings associated with outsourcing internal processes and the jobs attached to them were simply too great to ignore. Even when outsourcing did not produce dramatic savings — such as when work was transferred to an onshore supplier that could perform it more efficiently — it offered another important advantage. The risks, capital requirements, and operational burdens associated with a new project could be shifted from the company to the supplier.

Jack Welch did not invent outsourcing. What he did was open the door to its use on a scale and in ways that went far beyond what most people had imagined. Some would argue that the consequences weakened America's standing as an economic and industrial powerhouse and diminished its role as the engine of global commerce.

Into the Wilderness

Philip Murray Condit was born in Berkeley, California, in 1941 and grew up with a passion for aviation. He earned his private pilot's license at eighteen, followed by a bachelor's degree in mechanical engineering from the University of California, Berkeley, in 1963 and a master's degree in aeronautical engineering from Princeton University in 1965.

He could scarcely have chosen a more exciting moment to join Boeing. By 1965, the company was involved in nearly every major segment of the rapidly expanding aerospace industry. Its factories were producing the 707 and 727, while engineers were developing the 737 and beginning discussions with Pan American Airways over what would become the 747 jumbo jet. Boeing was also intensifying work on the 2707, America's proposed supersonic transport, and building the first stage of the Saturn V rocket that would carry astronauts toward the Moon.

The 2707 program, on which Boeing was staking much of its future, needed talented young engineers. Condit joined the company as an aerodynamics engineer and quickly earned a reputation for exceptional technical ability. During his first year at Boeing, he also obtained a patent for a concept known as the "Sailwing," a folding-wing design intended for small aircraft.

Within three years, Condit was promoted to a lead engineering position on the 747 program, where he applied experience gained on the SST to refining the jumbo jet's design and performance. He subsequently moved through a series of management positions and survived the massive layoffs that accompanied Boeing's financial crisis of the early 1970s. During that period, he returned to school and earned a master's degree in management from the Massachusetts Institute of Technology. With an established reputation as an aeronautical engineer and formal training in management, Condit appeared well positioned to rise through Boeing's leadership ranks.

In 1976, he was placed in charge of production for the 707, 727, and 737 programs. The assignment gave him valuable experience in aircraft program management and made him widely known throughout Boeing. Those three models formed the backbone of the company's commercial business, even as the 707 and 727 were beginning to show their age. It was working on those programs that put him into close proximity to Frank Shrontz, and the two developed a solid professional relationship.

Condit's major breakthrough came in 1978, when he was named chief project engineer for the new 757 and later became the program's director of engineering. He was effectively given responsibility for ensuring that the aircraft met its technical and performance objectives — a role ideally suited to an engineer whose abilities had become renowned throughout the company. It was a highly visible assignment, and Condit carried it out with confidence and distinction.

The 757's exceptional performance and commercial success further enhanced Condit's reputation. He understood both how airplanes were designed and how work moved through Boeing's vast engineering organization. He knew what had to be done, which people could do it, and how to produce an aircraft that met the needs of airline customers.

In many respects, Condit embodied the engineering culture upon which Boeing had been founded and built. As he rose through the company, however, his responsibilities extended far beyond designing airplanes. He proved to be an accomplished engineer who understood how to make an aircraft perform. Whether those abilities would translate into equally effective corporate leadership was another matter.

***

Leadership scholars identify numerous traits that distinguish effective executives from ineffective ones, but most can be traced to a small group of essentials: trustworthiness, courage, decisiveness, self-awareness, vision, integrity, and passion. Other qualities matter, but these are among those most consistently found in leaders who command the respect of their subordinates, peers, and superiors. Nor are such qualities confined to the workplace. Judgment, discipline, and integrity usually reveal themselves across the broader course of a leader's life.

Viewed across the breadth of his career, Philip Condit appeared to possess some of these qualities only unevenly. Before he became chief executive, Boeing had been led by a succession of forceful personalities, most of whom were known for their decisiveness. Once they chose a course, the organization was expected to follow it. Their leadership could be autocratic, but their authority rested on demonstrated competence, personal discipline, and an unwavering commitment to the company.

Condit did not fit that mold. He was less commanding, less confrontational, and far more inclined to seek consensus. Issuing blunt directives was not his style. Consensus-building can be an important executive skill, particularly within an organization as technically complex as Boeing, but it can become a liability when it reflects an unwillingness to make difficult choices or challenge powerful constituencies. Condit increasingly acquired a reputation for detachment and indecision — a leader more comfortable accommodating competing interests than imposing a clear direction.

His personal use of corporate privileges also raised questions about judgment and restraint. It was later reported that he lived for an extended period in a Boeing suite at Seattle's Four Seasons Hotel, accommodations ordinarily reserved for visiting executives and important guests. According to those accounts, the suite was remodeled to suit his preferences at company expense. Whether considered individually significant or not, such conduct contributed to an image of an executive who had grown comfortable treating corporate resources as personal entitlements.

As Condit accumulated authority, aspects of his personal conduct also became a source of concern within and outside the company. He was married four times, including marriages to a Boeing receptionist and, reportedly, to a first cousin, and he acquired a reputation for relationships with women employed by the company. Private behavior does not necessarily determine professional competence, but when personal relationships intersect with the workplace, they can raise legitimate questions about judgment, conflicts of interest, and the responsible exercise of executive power.

Condit had risen through Boeing as a gifted engineer and capable program manager. By the time he reached the top, however, he wielded an authority far broader than anything required to design or manage an airplane. The position demanded institutional judgment, moral discipline, strategic clarity, and the willingness to resist both personal temptation and organizational pressure. Those were qualities he did not always appear to possess in equal measure.

Perhaps Condit had been promoted beyond his capabilities — or, more precisely, into a role for which his technical experience and temperament had not prepared him. He had mastered the skills required to advance through Boeing's engineering hierarchy but appeared less equipped for the burdens of corporate leadership. The larger question was how Boeing — a company with a storied history, an enviable reputation, and a commanding position in global aerospace — had allowed someone so seemingly unprepared for those responsibilities to reach the chief executive's office.

***

Philip Condit's rise into Boeing's executive suite must be viewed against the geopolitical upheaval transforming the world around him. By 1991, the international order that had existed since the end of World War II was collapsing. The Soviet Union and its Eastern European bloc had disintegrated, ending more than four decades of confrontation between East and West. In their place emerged the promise of a new global order built on expanding trade, freer travel, and greater cooperation among nations.

Many world leaders regarded the end of the Cold War as a turning point in human history. The exercise of raw power and the need to "carry a big stick," as Theodore Roosevelt had famously advised nearly a century earlier, seemed less relevant in a world no longer divided between two hostile superpowers. Dialogue, diplomacy, and economic interdependence were expected to replace confrontation. Democracy appeared to be spreading across the globe, and consensus — the quality that most clearly defined Philip Condit's leadership style — was suddenly in vogue.

In October 1990, Boeing's board approved development of the 777. With the success of the 757 and 767 still fresh in the company's memory, Condit was a natural choice to lead the new program. He had played important roles in both aircraft and was appointed general manager of the 777 program. Condit embraced a collaborative development philosophy that Boeing promoted under the phrase "Working Together." Airlines, suppliers, engineers, production workers, and other stakeholders would participate in shaping the aircraft from its earliest stages.

The concept reflected Condit's consensus-driven approach. Rather than develop the airplane behind closed doors and present it to customers as a finished product, Boeing invited its airline customers into the design process. "Working Together" became more than a marketing slogan. It embodied both the culture of the 777 program and Condit's preferred method of leadership.

At the same time, the end of the Cold War was producing what became known as the "peace dividend." With the United States emerging as the world's only superpower and the threat of nuclear confrontation receding, many assumed that enormous defense budgets would no longer be necessary. Government spending could be redirected toward domestic and more peaceful purposes.

For the defense industry, however, the peace dividend carried ominous implications. The flow of Pentagon money that had sustained a vast network of aerospace and defense contractors was expected to contract sharply. Companies that had prospered during the Cold War now faced shrinking markets, intensified competition, and pressure to consolidate. A major realignment of the American aerospace industry appeared inevitable.

It was against this backdrop that Frank Shrontz selected Condit as his likely successor. In 1992, with Shrontz's strong recommendation, Boeing's board promoted Condit to president of the company, placing him next in line for the chief executive's office. An up-and-coming senior engineer named Alan Mulally subsequently assumed a larger leadership role on the 777 program.

Shrontz's decision would reverberate through Boeing for years, though perhaps not in the way he intended. Even highly capable leaders occasionally make consequential misjudgments. At the time, however, the choice of Condit appeared logical. Boeing was entering an era in which cooperation, international partnerships, industry consolidation, and complex negotiations would become increasingly important. A consensus builder may have seemed better suited to that environment than a forceful executive whose decisions might leave organizational casualties in their wake.

Condit's elevation reflected more than his technical accomplishments. His temperament appeared aligned with the spirit of the era. The post–Cold War world seemed to favor collaboration over confrontation, integration over rivalry, and consensus over command. Boeing's board had reason to believe that Condit represented the kind of leader the company would need to navigate those changes.

In 1996, Frank Shrontz retired as Boeing's chief executive, although he remained chairman of the board. The reins of the company passed to Condit, whose less autocratic and more collaborative style differed markedly from that of every one of his predecessors. He was preparing to lead Boeing through a global environment rich with opportunity but also fraught with danger — an environment that would punish any leader who failed to understand how its many moving parts were connected.

On April 29, 1996, Philip Murray Condit became just the seventh chief executive officer in the 80-year history of The Boeing Company. He had reached the pinnacle of his profession, the leader of one of the largest and most prestigious companies in the world. He now led one of America's most important corporations and wielded enormous influence over the future of global aerospace.

The question was whether the leadership style that had carried him to Boeing's summit would be strong enough to keep him there and to move the company forward into the new world order.

***

Upon taking control of Boeing, Philip Condit set out to leave his mark on the company. Rather than viewing himself merely as the steward of an institution that had become the gold standard of aerospace, he appeared determined to become an agent of transformation. Every chief executive possesses a measure of ego, but Condit may have envisioned himself reshaping Boeing on a scale not seen since the era of Bill Allen. He began building support for an ambitious remaking of the world's largest and most successful aerospace company.

The timing seemed extraordinarily favorable. The world was entering a new global era, but unlike earlier geopolitical transformations, this one appeared to promise prosperity rather than destruction. Trade barriers were falling, former adversaries were opening their markets, and international commerce seemed poised to expand at a rate without precedent. Greater trade would require the rapid movement of people and goods, much of it by air.

The wealth generated by that expansion would also allow millions more people to travel for business and leisure.

In 1996, the year Condit became chief executive, Boeing generated more than $22.5 billion in revenue and ranked twenty-eighth on the Fortune 500. It captured approximately 64 percent of the commercial-aircraft market and recorded 719 gross orders, one of the strongest sales years in its history. Global economic growth was accelerating, demand for air travel was rising, and Boeing stood in an enviable position to benefit.

Condit's central task appeared straightforward: preserve Boeing's strengths and capitalize on the expanding market. The company was like a magnificent racing vessel that had already proven itself faster and more capable than its competitors. As Condit took the helm, it entered a new race with a powerful wind at its back and its nearest rival trailing behind.

A prudent captain would have maintained course, kept the ship in peak condition, and ensured that the crew remained focused on extracting every advantage from the favorable conditions. The sails needed to be properly set, the hull maintained, and the crew disciplined and unified.

However, Condit chose a different course. Rather than preserve and refine the vessel that had carried Boeing to the front of the fleet, he began altering its structure, replacing its rigging, and redefining its purpose. He sought to transform a proven racing ship into a different kind of enterprise — and to do so while it was underway at sea.

***

The collapse of the Soviet Union and the Eastern Bloc created widespread expectations of a "peace dividend." With the threat of nuclear confrontation receding, defense spending would continue, but at a slower pace and, many hoped, on a reduced scale. The relentless development of new weapons and technologies needed to stay ahead of the Soviet Union would diminish, freeing vast sums for domestic and more peaceful purposes.

The resulting contraction triggered a rapid consolidation of the American defense industry. With fewer major weapons programs to compete for, the need for dozens of large defense contractors disappeared. Consolidation became essential to survival. In 1991, the United States had more than fifty prime defense contractors. Within several years, that number would fall to only a handful.

Philip Condit had spent his entire professional career at Boeing and had witnessed firsthand the cyclical nature of the commercial-aircraft business. He had survived the company's devastating downturn of the early 1970s through a combination of ability, timing, and perhaps good fortune. Although commercial aviation was entering a period of extraordinary growth when he became chief executive in 1996, Condit focused heavily on protecting Boeing from the downturns that would inevitably follow. His solution was diversification — specifically, expanding the company's access to government-funded defense programs.

Instead of concentrating Boeing's resources on selling commercial airplanes and exploiting an unprecedented surge in demand, Condit chose to play defense, both figuratively and literally. The irony appeared to escape both him and Boeing's board. By strengthening the commercial-airplane business and building a deep backlog of orders, Boeing might have created its own buffer against the next recession, even after allowing for the cancellations that typically accompanied an economic downturn.

Condit nevertheless deserves credit for recognizing the vulnerability created by Boeing's growing dependence on commercial aviation. For much of its history, the company had prospered through government contracts. Bombers, missiles, spacecraft, and other military and space programs had helped transform Boeing into an aerospace giant. But with the decline of its major bomber programs, the end of Apollo, and the cancellation of the SST, commercial airplanes had come to dominate the company.

In 1965, the year Condit joined Boeing, the company generated approximately $2.2 billion in revenue, with about half coming from commercial aircraft. By 1996, Boeing's revenue had risen to approximately $22.7 billion, of which $16.1 billion — more than 70 percent — came from the commercial-airplane business. Boeing had become increasingly dependent on a market Condit knew to be highly cyclical.

There may also have been another factor driving his search for alternative revenue. Condit was watching Airbus gain strength in Toulouse, and he may have viewed its rise with growing concern. Rather than respond by intensifying Boeing's traditional advantages in engineering, product development, and innovation, he sought a strategic backstop. Defense revenues could insulate the company not only from commercial downturns, but also from market share lost to Airbus.

This was a fundamentally defensive strategy. Instead of identifying Airbus' weaknesses and exploiting them, Condit sought protection from the consequences of its advance. Rather than double down on the engineering culture that had made Boeing the world's dominant aircraft manufacturer, he looked for other businesses that could offset the possibility that Boeing might no longer dominate commercial aviation.

The competitive trend was already troubling. In 1995, the final full year of Frank Shrontz's tenure as chief executive, Boeing accounted for roughly three-quarters of the large commercial-aircraft market, while Airbus held about 15 percent and McDonnell Douglas approximately 10 percent. By the end of 1996, Boeing's share had fallen to about 60 percent, while Airbus had risen to roughly 35 percent. McDonnell Douglas had slipped to approximately 5 percent, and speculation was growing that it might abandon the commercial-aircraft business altogether.

Airbus was gaining ground rapidly, and most of that growth was coming at Boeing's expense. McDonnell Douglas had become too weak to pose a serious competitive threat. Condit therefore faced a strategic choice: confront Airbus directly by investing aggressively in Boeing's commercial products, or reduce the company's exposure by expanding elsewhere. He chose the latter.

Condit believed he had to act quickly. What followed was counterintuitive, and the consequences would reverberate through Boeing and the aerospace industry long after he left the chief executive's office — and not in the way he intended.

***

When Airbus decided in 1984 to develop the A320 and expand its market reach, it was not pinning its future on a single airplane. The company intended to build an entire family of aircraft that would allow it to compete directly with Boeing's entire product line across the global market.

Airbus executives launched internal studies to determine which additional aircraft sizes and market segments to pursue. Their objective was to offer airlines what Boeing already could — the ability to obtain most, if not all, of their fleet requirements from a single manufacturer. But developing new airplanes was only part of the challenge. Airbus also had to create the broader ecosystem needed to support them, including a global supply chain, maintenance and service networks, spare-parts availability, and pilot and crew training. Without that infrastructure, Airbus would continue to face a disadvantage during sales campaigns.

Until 1992, Airbus had consistently ranked third in commercial-aircraft market share, behind Boeing and McDonnell Douglas. The growing popularity of the A320 changed that. By the end of 1992, Airbus had overtaken McDonnell Douglas in annual sales and established itself as Boeing's principal emerging competitor.

Airbus had already begun planning its next move. In 1987, only a few years after launching the A320, it initiated development of two new widebody aircraft: the A330 and A340, continuing the naming sequence that had begun with the original A300.

Although they carried different model numbers, the A330 and A340 were closely related aircraft. They shared the same basic fuselage, cockpit architecture, major systems, and much of their wing structure and manufacturing tooling. The high degree of commonality allowed Airbus to develop two aircraft for different missions without bearing the full cost of creating two entirely separate designs.

The principal distinction was propulsion and mission. The A330 was a twin-engine, medium to long-range widebody designed for high-capacity, fuel-efficient service on routes with moderate to heavy demand. The A340, by contrast, used four engines and was intended for long and ultra-long-range operations, particularly on routes where contemporary ETOPS restrictions limited the use of twin-engine aircraft. With four engines, airlines could operate the A340 across remote oceans and other sparsely served regions without being constrained by diversion-time rules governing twins.

Airbus raised eyebrows with advertising that implied four engines were inherently preferable — and safer — than two on long overwater flights. The campaign was a pointed challenge to Boeing, which was placing its future long-range strategy behind the twin-engine 777. The message was controversial because aircraft manufacturers had traditionally avoided using safety as a weapon against competitors. Suggesting that a rival configuration was less safe risked undermining confidence not merely in one aircraft, but in commercial aviation more broadly.

Both the A330 and A340 incorporated the fly-by-wire controls, cockpit automation, and the same design philosophy introduced on the A320. But Airbus was pursuing something more consequential than common technology. It was developing a family of aircraft built around a common cockpit and operating philosophy

— a concept it believed could fundamentally distinguish Airbus from its competitors.

***

Airline pilots are ordinarily assigned to one aircraft type and may remain on it for years before moving to another, often as the result of a promotion, a change of base, or a shift in fleet requirements. One reason is the considerable cost of qualifying a pilot on a new aircraft.

Transition training requires weeks of classroom instruction, self-study, procedural practice, and sessions in a full-flight simulator. These simulators reproduce an aircraft's cockpit, systems, visual environment, and motion with extraordinary fidelity. They are also expensive to purchase, maintain, and operate, with the most advanced units costing many millions of dollars. Large airlines may own several simulators for each major aircraft type, while smaller carriers often lease training time from manufacturers, training centers, or other airlines.

After completing ground and simulator instruction, pilots must undergo supervised operating experience in the aircraft before being released to fly regular schedules without an instructor. The entire process can occupy several weeks of full-time training, during which the pilot is being paid but is not contributing to the airline's normal flight schedule. These costs traditionally encouraged airlines to divide their pilot corps by aircraft type. A pilot qualified on one model could not simply be reassigned to another when schedules changed. That restriction reduced operational flexibility and required airlines to maintain separate pools of pilots for each fleet.

Boeing addressed part of this problem when it developed the 757 and 767 concurrently. Although one was a narrow-body and the other a widebody, the two aircraft shared similar cockpits, avionics, systems logic, and operating procedures. Their high degree of commonality allowed regulators to approve a common pilot type rating for both aircraft.

A pilot qualified on the 757 could therefore transition to the 767 — or vice versa — with comparatively limited additional training. Provided the pilot completed the required differences instruction and maintained currency on both models, an airline could assign that pilot to either aircraft. A crew member might fly a 757 from New York to Miami during one trip and operate a 767 from New York to Europe on the next. This flexibility reduced training costs and allowed airlines to use their crews more efficiently. It became one of the most important selling points of the jointly developed 757 and 767 families.

Airbus adopted the same general principle when it developed the A330 and A340. The two widebodies shared a common cockpit, systems architecture, and pilot type rating despite one having two engines and the other four. A pilot qualified on one could transition to the other with limited differences training.

Airbus then carried the concept considerably further. Because the A320, A330, and A340 employed digital fly-by-wire flight controls, engineers could use software and common control laws to give aircraft of vastly different sizes and performance capabilities more consistent handling characteristics. Airbus did not make the airplanes aerodynamically identical; their weight, inertia, thrust, and performance remained inherently different. But it standardized how they responded to pilot inputs and how pilots interacted with their systems.

The cockpits also followed a common design philosophy. Displays, controls, flight-management systems, checklists, and many routine procedures were deliberately arranged to appear and function in similar ways. An engine start, electrical-system configuration, or flight-management entry on one Airbus model would therefore feel familiar to a pilot on another.

The A320 and A330/A340 did not share a single type rating, but Airbus developed a streamlined cross-crew qualification program that substantially reduced the training needed to move between them. Pilots still had to learn the performance, limitations, systems differences, and emergency procedures of the different aircraft, but they did not have to begin again with an entirely unfamiliar cockpit philosophy.

For airlines operating several Airbus models, the savings could be substantial. Training courses were shorter, simulator time was reduced, and pilots could move among fleets more readily as staffing and network requirements changed. Commonality also simplified the development of training materials, operating procedures, maintenance practices, and spare-parts support.

Airbus was offering flexibility on a much broader scale than Boeing had achieved with the 757 and 767. Boeing's two aircraft shared a common type rating and substantial cockpit commonality, but the principle did not extend across the rest of its product line. A pilot moving from a 737 to a 757, 767, or 747 still faced qualification on a substantially different cockpit and aircraft type.

The 757 and 767 also retained conventional flight-control architectures in which pilot inputs were transmitted through mechanical cables to hydraulically powered control surfaces. Their similar cockpits did not make their handling identical, and their considerable differences in size, weight, and mission remained apparent to pilots. The 777, introduced later, became Boeing's first commercial aircraft with a fully digital fly-by-wire primary flight-control system.

Airbus intended to carry its commonality philosophy into every aircraft family it developed. Its broader sales proposition rested not merely on fuel efficiency and modern technology, but on lower training costs, simplified fleet transitions, standardized procedures, and greater flexibility in deploying crews. Fly-by-wire made it possible to push cockpit and operational commonality farther than Boeing had done.

Boeing remained the dominant commercial-aircraft manufacturer, and its airplanes were far from obsolete. The 737, 757, 767, and 777 remained highly capable aircraft with enormous customer bases and strong economic performance. But billions of dollars had already been invested in those programs, making it impractical for Boeing to redesign its entire product line around a common digital cockpit and fly-by-wire architecture.

Airbus had created an advantage that Boeing could not quickly duplicate. With each successful sales campaign, the European manufacturer demonstrated that airlines were buying not merely individual airplanes, but an integrated family of aircraft.

Boeing's senior leadership could not ignore the growing threat. Airbus was no longer simply another competitor offering discounted airplanes from Toulouse. It was presenting airlines with a coherent operating system for an entire fleet — and gaining ground with every new model it introduced.

***

Like an addict returning to an old dealer for a quick fix, Boeing turned once again to Uncle Sam. The narcotic was government money, and Philip Condit seemed unable to resist its promise of reliable revenue.

By the time Condit became chief executive, Boeing had substantially reduced its dependence on government contracts. Commercial aviation had become the company's bread and butter, fulfilling the vision Bill Boeing had pursued after World War I, when the company struggled to survive without military and government work. Nearly eight decades later, Boeing had finally become what its founder had imagined — a company sustained primarily by building airplanes for commercial passengers and cargo. Condit now sought to reverse some of that evolution.

The post–Cold War peace dividend meant that large new defense programs would be launched less frequently, limiting Boeing's opportunities to win major government contracts through traditional competition. Condit therefore looked for a faster route into the defense business — acquiring companies that already possessed valuable military programs and established streams of government revenue.

The opportunities were disappearing rapidly. Consolidation was sweeping through the defense industry, and the number of significant acquisition targets was shrinking. Condit believed Boeing had to move quickly.

One of the first major acquisitions under his leadership was Rockwell International's aerospace and defense businesses. Rockwell had developed and built the B-1 bomber, conceived as a successor to the B-52 but ultimately produced in numbers too small to replace it. The company had also served as the prime contractor for NASA's Space Shuttle orbiters.

The acquisition gave Boeing valuable capabilities in modern military aircraft, space systems, missiles, electronics, and defense-related support services. It also brought established government programs and their associated revenues into Boeing's portfolio, while strengthening the company's credentials for future Department of Defense and NASA competitions.

Rockwell, however, was no longer producing either the B-1 or the Space Shuttle. Much of the value that remained lay in upgrades, maintenance, engineering support, and other continuing work associated with those programs. Although the acquisition broadened Boeing's defense and space capabilities, it did not provide the scale Condit was seeking. He wanted more than a secondary defense business to cushion Boeing during commercial downturns. His ambition was to transform the company into a more evenly balanced aerospace conglomerate, with government work eventually accounting for roughly half of its revenue.

Rockwell was only the beginning.

Storied Pasts Undone by Uncertain Futures

Donald Wills Douglas Sr. was born in Brooklyn, New York, in 1892. Having witnessed the dawn of aviation, he became fascinated with flying machines at an early age. He enrolled at the U.S. Naval Academy, where he spent some of his free time building model airplanes in his dormitory. But the pull of aviation proved too great. In 1912, during his junior year, he resigned from the academy to pursue his passion. Douglas subsequently enrolled at the Massachusetts Institute of Technology, where he earned a Bachelor of Science degree in aeronautical engineering in 1914, becoming the first person at MIT to receive the degree.

He went to work for several aviation companies, including the Glenn L. Martin Company, where, at only 23 years old, he became chief engineer. In 1920, he formed the Davis-Douglas Company with a partner, David Davis. The partnership dissolved after only a year, and Douglas established the Douglas Aircraft Company in 1921.

The company initially focused on both military and commercial aircraft, making its first major mark with the DT-2 torpedo bomber developed for the U.S. Navy. In 1924, Douglas modified five DT-2s into aircraft designated as Douglas World Cruisers. The U.S. Army Air Service used them in the first aerial circumnavigation of the globe.

The aircraft departed Seattle, Washington, on April 6, 1924. Only two — the Chicago and the New Orleans — completed the 27,553-mile journey, returning to Seattle on September 28 after 175 days. The expedition included 371 hours of flying time and stops in 22 countries, securing Douglas a prominent place in aviation history.

Douglas went on to manufacture progressively more complex aircraft. Another early design that made aviation headlines was the M-2, which defeated a competing Boeing aircraft for the airmail contract awarded in 1925. Douglas subsequently designed and built numerous aircraft with varying degrees of commercial success, although nearly all the company's earliest designs were intended for the military. Beyond the DT-2 and M-2, the company developed transports, bombers, trainers, and other specialized aircraft in rapid succession.

Like Boeing, Douglas wanted to bring air travel to the public. But the early development of aviation was driven largely by the federal government, which possessed both the financial resources to support aircraft development and, following World War I, a growing interest in aviation's military potential. It would take Boeing's development of the revolutionary Model 247 to propel Douglas decisively into commercial aviation.

When Boeing promised the first 60 Model 247s exclusively to Boeing Air Transport, a production run that would take nearly two years to fulfill, Transcontinental & Western Air found itself in a dangerous position. Boeing Air Transport intended to place the new airplanes on its coast-to-coast routes. Without a comparable aircraft, TWA feared that it — and perhaps several other carriers — would be unable to compete. After Boeing informed TWA that it could not receive a Model 247 until Boeing Air Transport's order had been completed, the airline turned to Douglas.

Douglas had never built a commercial airliner, but its experience designing large military aircraft made it a logical candidate. The company went to work on an airplane capable of competing with Boeing's revolutionary design. The resulting prototype, designated the DC-1, did more than compete with the Model 247 — it surpassed it.

The initials "DC" stood for "Douglas Commercial," while the "1" identified it as the company's first commercial model. The DC-1 was slightly larger than the Boeing 247 and could carry two additional passengers. Much of its advantage came from its more powerful Curtiss-Wright engines, which also gave it a higher cruising speed and greater range.

Had Douglas placed the DC-1 into production, it likely would have been sufficient to overtake the Boeing 247. But as the prototype began flying, Curtiss-Wright was preparing to introduce an even more powerful engine. Douglas took advantage of the opportunity for using the more powerful engine to enlarge and strengthen the design. The resulting airplane offered greater passenger capacity, speed, and range than the DC-1 and was designated the DC-2.

The DC-2 could carry 14 passengers — approximately 40 percent more than the Boeing 247 — and quickly became the preferred airliner. Within a year of its maiden flight, the Boeing 247, the airplane that had revolutionized commercial aviation, had itself been rendered obsolete. Douglas produced 192 DC-2s before the model was superseded by another airplane that would transform commercial aviation and become one of the most celebrated aircraft in history.

The DC-2 was enjoying a successful production run when, in 1935, American Airlines president C. R. Smith asked Douglas to build a larger airplane. Smith wanted an aircraft that could operate with conventional seating during daytime flights and be converted into sleeping berths for overnight service. The DC-2's fuselage was too narrow to accommodate side-by-side berths while leaving sufficient room for an aisle. Smith therefore asked Douglas to develop an airplane with a wider cabin and promised to purchase at least 20 if the company agreed to build it.

With newer and more powerful engines becoming available, Douglas' engineers effectively took the DC-2 design and put it on steroids. The result was the DC-3, an airplane capable of accommodating 16 sleeping berths for overnight service or as many as 32 passengers in a daytime configuration.

The DC-3 surpassed the DC-2 in performance and established the standard by which commercial aircraft — and the airlines operating them — would be judged. It set new benchmarks for passenger comfort, speed, range, reliability, and operating economy. It was also rugged, durable, and remarkably adaptable.

Its ten-year production run, from 1935 through 1945, produced more than 16,000 aircraft. More than 600 were built as civilian airliners, while over 15,000 military variants were produced during World War II, most notably the C-47 troop and cargo transport. Even decades later, a significant number of DC-3s and their derivatives remained in service, performing sightseeing, cargo, charter, and other specialized missions.

Civilian aircraft production largely ceased during World War II, but Douglas became one of the nation's most important military suppliers. In addition to more than 15,000 C-47s and related variants, the company produced approximately 7,400 A-20 Havoc light bombers; more than 5,900 SBD Dauntless dive bombers, which played a critical role in the Battle of Midway; more than 2,500 A-26 Invader light bombers; and over 1,100 C-54 Skymaster transports. The C-54 entered service during the war but would find new life as a civilian airliner once the fighting ended.

The DC-2 and DC-3 marked the beginning of the rivalry between Douglas and Boeing that would propel both companies forward in commercial and military aviation. When Douglas' designs overwhelmed the Boeing 247 in the marketplace, Boeing attempted to regain the advantage with its large, four-engine, pressurized Model 307 Stratoliner. But the onset of World War II curtailed the 307's commercial prospects, and Boeing was unable to regain a meaningful foothold in the passenger-aircraft market immediately after the war.

Douglas, by contrast, built upon the success of the DC-2 and DC-3 with a succession of increasingly large and capable commercial aircraft. Its postwar airliners evolved into four-engine piston-powered designs, beginning with the DC-4 — the civilian counterpart of the C-54 Skymaster — and continuing with the DC-6 and DC-7. Douglas also produced the DC-5, a twin-engine, high-wing aircraft whose low fuselage made passenger and cargo loading easier. Like Boeing's Model 307, however, its production was cut short by the war, and only 12 were built.

The DC-6 originated with a military contract awarded in 1944, although the military version did not fly until after the war had ended. Douglas modified the design for civilian service and gave it the DC-6 designation. The approach was not unlike Boeing's conversion of technology derived from the B-29 into the Model 377 Stratocruiser. But whereas Boeing produced only 56 Stratocruisers, Douglas built more than 700 DC-6s before production ended in 1958.

From 1945 through the late 1950s, Douglas dominated the commercial airliner market with the DC-4, DC-6, and DC-7. Its piston-powered aircraft set the standard for the industry. Other than Lockheed, no commercial aircraft manufacturer came close to matching Douglas in either performance or popularity.

Boeing, meanwhile, struggled to gain a foothold in the postwar commercial market, although it overwhelmingly dominated the military side of the business with the B-47 and B-52 bombers, the C-97 Stratofreighter, and their derivatives.

Douglas also retained a substantial military presence. Most of its combat aircraft were designed for carrier operations with the U.S. Navy. Among them were the A-1 Skyraider, a single-engine propeller-driven attack aircraft; the F3D Skynight, a twin-engine jet-powered night fighter; the A-3 Skywarrior, a carrier-based strategic bomber; and the A-4 Skyhawk, a compact, single-engine attack jet. Nevertheless, in the years immediately before and after World War II, the Douglas name remained most closely associated with commercial airliners.

When the de Havilland Comet introduced the world to jet-powered passenger travel, Douglas redirected its engineering resources toward developing a commercial jet of its own. The resulting aircraft was designated the DC-8. It made its maiden flight in May 1958, several months after Boeing's competing 707 had first flown, and entered commercial service nearly a year behind it.

Boeing's advantage stemmed largely from the work it had already completed on the Dash 80 prototype and the KC-135 tanker. That experience gave Boeing a considerable head start in developing its commercial jet.

Unfortunately for Douglas, the 707 proved more popular with the world's airlines. The DC-8 initially accumulated respectable sales momentum and offered nominal advantages in certain performance and payload measures. But Boeing continually modified the 707, introducing new versions in rapid succession. These variants not only improved the aircraft's performance but also gave airlines the flexibility to serve routes of different lengths and traffic volumes.

Douglas, by comparison, remained committed for too long to the basic DC-8 airframe. Most of its early improvements centered on the installation of newer engines rather than the introduction of substantially different versions. As Boeing expanded the 707 family, airlines operating DC-8s increasingly found themselves turning to Boeing when they wanted aircraft suited to new markets. By the end of 1965, orders for the 707 family were approaching twice those of the DC-8.

Douglas moved more slowly in part because it was smaller than Boeing and operated under tighter structural and financial constraints. Boeing was aggressive, better capitalized, and willing to redesign the 707 quickly. Its enormous government contracts for the C-97, B-47, B-52, and KC-135 had provided financial resources and engineering capacity that Douglas could not easily match.

Douglas finally began offering more extensively modified DC-8 variants during the second half of the 1960s, but by then much of the damage had been done. Boeing had achieved runaway success with the 707 and its derivatives and would ultimately sell nearly twice as many aircraft as Douglas sold DC-8s.

Douglas enjoyed far greater success with the DC-9, a short- to medium-range twin-engine jet transport. Once again, the aircraft followed a Boeing competitor into service, arriving almost two years after the 727. Yet the DC-9's compact design, operating economy, and adaptability made it extremely popular with airlines.

The original DC-9 and the later aircraft derived from its basic design would become one of the most successful commercial jet families in history, with thousands produced before the lineage finally came to an end.

But eventual commercial success did not eliminate the immediate financial strain of developing the aircraft. During the early 1960s, the DC-9 program consumed enormous amounts of capital, and Douglas initially sold some aircraft at a loss as it attempted to secure orders, increase production, and establish the model in the marketplace.

The company was caught in a familiar aerospace dilemma. It needed high production volumes to lower its unit costs, but achieving those volumes required accepting unfavorable prices and spending heavily before the resulting revenue arrived. Orders could appear impressive on paper while the company's cash reserves steadily dwindled.

At the same time, sales of the DC-8 were slowing. Although the airplane remained respected and would continue to serve airlines for decades, it had lost the commercial battle to the Boeing 707. Douglas' once-commanding position in the airliner market was eroding just as the cost of competing was becoming greater than ever.

Then, in 1966, Douglas launched the DC-10, its entry into the emerging widebody airliner market. The company regarded the program as essential to its future. Airlines were beginning to demand aircraft capable of carrying far more passengers over long distances, and Douglas could not afford to surrender the market to Boeing, Lockheed, or any other manufacturer.

Yet the DC-10 would be larger, more complex, and vastly more expensive to develop than any commercial aircraft Douglas had previously attempted. Bringing it to market would require extraordinary investments in engineering, tooling, testing, production facilities, suppliers, and labor. Those expenses would have to be incurred years before the company could expect to recover them through aircraft deliveries.

Douglas entered the widebody race at precisely the moment when its financial strength was deteriorating. DC-8 sales were declining. The DC-9 had yet to generate the profits its eventual success promised. Production inefficiencies and delivery delays strained relations with customers and disrupted the flow of desperately needed cash. The company possessed an extraordinary engineering legacy and a strong order book, but it increasingly lacked the financial resources necessary to convert those advantages into stability.

By 1967, Douglas Aircraft was in a precarious position. The company that had created the DC-2, DC-3, DC-4, DC-6, DC-7, DC-8, and DC-9 — airplanes that had helped define commercial aviation — now faced an uncertain future.

Its proud past could not finance the airplanes it needed to build next.

***

James Smith McDonnell was born in 1899. He graduated from Princeton University in 1921 and earned a master's degree in aeronautical engineering from the Massachusetts Institute of Technology in 1925. Over the next fourteen years, he worked for several companies in America's burgeoning aviation industry, including the aircraft division of Ford Motor Company, which produced the famous Ford Trimotor, and the Glenn L. Martin Company. Donald Douglas had also worked for Martin, serving as its chief engineer before establishing his own company. McDonnell's career placed him at the center of the rapid advances in aviation occurring between the world wars and gave him invaluable experience as an aeronautical engineer.

In 1939, McDonnell established the McDonnell Aircraft Corporation near St. Louis. The fledgling company initially struggled to secure contracts of substantial value, but the outbreak of World War II brought a flood of government work. Much of it involved manufacturing components and subassemblies for larger aircraft companies rather than producing complete McDonnell designs. Nevertheless, the work gave the company broad manufacturing experience and helped it develop a skilled workforce.

McDonnell soon became confident enough to offer the government an advanced fighter of its own. Designated the XP-67 and nicknamed the "Bat," the aircraft featured an unusually streamlined configuration in which the fuselage, engine nacelles, and wings blended together. Its ambitious design promised exceptional performance, but engine failures and other developmental problems plagued the program. After the sole prototype was badly damaged by an engine fire in 1944, the Army Air Forces canceled the project. Although the XP-67 never entered production, its development demonstrated that McDonnell possessed both the engineering talent and the ambition to compete as a designer of complete aircraft.

After the war, McDonnell developed several successful jet fighters for the American military. These included the F2H Banshee, a carrier-based Navy fighter used extensively during the Korean War; the F3H Demon, another Navy carrier fighter; and the F-101 Voodoo, a supersonic interceptor and fighter-bomber operated primarily by the Air Force.

The aircraft that transformed McDonnell into a pillar of the American defense industry, however, was the F-4 Phantom II. Originally developed for the Navy, the Phantom became the premier American fighter of its era and was eventually adopted by the Navy, Air Force, and Marine Corps. It also served with numerous allied nations, including Australia, Egypt, West Germany, Greece, Iran,

Israel, Japan, South Korea, Spain, Turkey, and the United Kingdom. A total of 5,195 Phantoms were built, making the F-4 the most widely produced Western supersonic military aircraft. Although the design first flew in 1958, several countries continued operating limited numbers well into the 21st century — an extraordinary record for an aircraft conceived during the opening years of the jet age.

The F-4 solidified McDonnell's standing within the American military-industrial establishment and strengthened its reputation for producing technologically advanced, high-quality aircraft. The program also helped broaden the company's expertise in avionics, missiles, radar, and weapons-integration systems. That growing body of knowledge opened additional doors when the Space Race accelerated during the early 1960s.

McDonnell became a central participant in America's first crewed space programs. It served as the prime contractor for the one-man Mercury spacecraft that carried the first American astronauts into space. NASA then selected the company to build the two-man Gemini spacecraft. Through Gemini, astronauts and engineers developed and demonstrated many of the capabilities required for a lunar mission, including prolonged orbital flight, spacewalking, rendezvous, docking, and precision maneuvering. Those techniques would subsequently prove indispensable during the Apollo program.

By the middle of the 1960s, McDonnell had become one of the country's most important military and space contractors. Its dependence upon government programs nevertheless presented a long-term strategic vulnerability. Defense and space contracts could be enormous, but they also involved risks — shifting military requirements, changing presidential administrations, congressional appropriations, and the unpredictable cycles of government procurement. McDonnell had little presence in the civilian aircraft market that could help offset a future downturn in defense spending on contracts it had been awarded or other government spending.

The company was not facing the immediate financial crisis confronting Douglas Aircraft. On the contrary, McDonnell's successful military and space programs placed it in the stronger position. Its problem was one of strategic concentration rather than imminent insolvency because too much of its business depended upon a single customer — the United States government.

Douglas offered McDonnell an opportunity to diversify. It possessed one of the most respected names in commercial aviation, a worldwide airline customer base, and two modern jetliners — the DC-8 and DC-9. What Douglas lacked was cash. Production delays, cost overruns, and the enormous working-capital demands created by its commercial backlog had pushed the company into a severe liquidity crisis.

The two companies therefore appeared to possess complementary strengths and weaknesses. McDonnell had financial resources and a thriving military and space business but almost no commercial presence. Douglas had commercial aircraft, airline customers, and an enormous order book but lacked the capital needed to build and deliver those airplanes efficiently. For McDonnell, acquiring Douglas offered a path into the civilian market and protection against future fluctuations in military procurement. For Douglas, a combination with McDonnell offered something more urgent: survival.

***

By late 1966, Douglas Aircraft was becoming financially desperate, while McDonnell Aircraft was searching for a way to diversify beyond its heavy dependence on government contracts. Their circumstances were not equally dire. Douglas faced an immediate threat to its survival, whereas McDonnell remained profitable but vulnerable to the shifting cycles of military and space procurement. Each nevertheless possessed what the other lacked. Douglas had a storied name in commercial aviation, a worldwide airline customer base, and two modern jetliners. McDonnell had financial strength, disciplined management, and a robust stream of defense and space contracts. A combination offered an obvious logic.

On April 28, 1967, McDonnell Aircraft and Douglas Aircraft formally combined to create the McDonnell Douglas Corporation. Although described as a merger, the balance of power left little doubt that McDonnell had acquired Douglas. McDonnell was the financially stronger and more valuable company, and its executives assumed control of the new corporation. James McDonnell became chairman and chief executive officer, while McDonnell president David Lewis became president. Donald Douglas was named honorary chairman, a ceremonial position that gave the aviation pioneer little influence over the company he had founded more than four decades earlier.

The new corporation would bear both names, but its leadership, financial philosophy, and organizational discipline came primarily from St. Louis. McDonnell's influence would increasingly permeate the combined company's management culture.

James McDonnell and David Lewis began consolidating the two organizations, imposing tighter financial controls, eliminating duplication, and restoring stability to Douglas' troubled commercial-aircraft operations. Financial discipline became an overriding priority. The company benefited from a strong flow of government business, led by production of the F-4 Phantom II and its work on other defense and space programs. Those contracts provided the stability needed to reorganize the commercial side of the company.

The DC-9 appeared to have the brighter future. Airlines had responded favorably to the twinjet, and the company concentrated on increasing production and making the program consistently profitable. The aircraft also lent itself readily to further development. The DC-9-30, which had been developed before the merger, offered greater capacity than the original Series 10. It was followed by the Series 20 and Series 40, which provided different combinations of capacity, field performance, and operating capability. Later versions would continue expanding the family.

The DC-8 presented a more difficult challenge. Its sales had fallen well behind those of Boeing's 707, and the program had struggled to recover its enormous development costs. Douglas had already begun developing the Super 60 series before the merger, but McDonnell Douglas pushed the improved models into the market. The DC-8-61 incorporated a dramatically lengthened fuselage for high-capacity routes. The DC-8-62 combined a more modest fuselage extension with aerodynamic improvements and additional fuel capacity to provide greater range. The DC-8-63 joined the long fuselage of the -61 with the improved wing and greater operating weights of the -62.

In high-density configuration, the largest Super 60 models could carry as many as 259 passengers, making the DC-8 the highest-capacity commercial jetliner then in service. It would retain that distinction only until the Boeing 747 entered airline service in 1970.

The Super 60 series gave the DC-8 a valuable second life, but it could not alter the direction in which the commercial aircraft market was moving. Boeing's 707 had already established itself as the dominant long-range jetliner, and Boeing's recently launched 747 promised to transform high-capacity international travel. The era in which a stretched single-aisle narrow-body could remain the largest aircraft in an airline fleet was drawing to a close.

McDonnell Douglas therefore placed much of its commercial future on an entirely new aircraft: the DC-10. The widebody trijet was intended to succeed the DC-8 and compete for the growing market between conventional narrow-body jets and the enormous Boeing 747. Its development moved forward while the company was still consolidating two organizations and imposing strict controls on spending.

The stakes rose considerably when Lockheed entered the same market. In April 1966, American Airlines had circulated a requirement for a twin-aisle aircraft smaller than the 747 and capable of operating from existing airport runways. Both Douglas and Lockheed responded. Available engines and the performance requirements eventually pushed both manufacturers toward broadly similar three-engine configurations — one engine under each wing and a third in the tail.

Lockheed's proposed aircraft, eventually designated the L-1011 TriStar, would compete directly with the DC-10 for many of the same airline customers. Its entry alarmed McDonnell Douglas executives. The anticipated market was substantial, but many within the industry doubted that it could profitably support two nearly comparable American widebody trijets — particularly with Boeing's 747 occupying the upper end of the market.

American Airlines had helped define the requirement, but it had not committed itself to McDonnell Douglas. Other major carriers were also evaluating both designs, and their early decisions would determine whether either program accumulated enough orders to justify production. What had initially appeared to be Douglas' opportunity to secure its future had become a race against one of the most technically accomplished aerospace companies in the world.

For McDonnell Douglas, losing that race could mean more than the failure of a single aircraft. The DC-8 was nearing the end of its commercial life, and the DC-9 alone could not sustain the company's ambitions as a full-line commercial-aircraft manufacturer. The DC-10 had to attract enough orders, enter service quickly enough, and be built economically enough to secure the company's place in the market.

The competition would place extraordinary pressure on an organization already preoccupied with consolidation, cost control, and financial stability. Under that pressure, McDonnell Douglas would make decisions that had consequences far beyond the profitability of the DC-10. They would expose weaknesses in the company's engineering and management processes, damage its reputation, and help establish patterns that would remain embedded in its culture for decades.

***

McDonnell Douglas ultimately won the American Airlines competition with an order for 25 DC-10s, but the victory came at considerable financial cost. American used the threat of selecting Lockheed's L-1011 to extract substantial pricing concessions. United Airlines followed with an order for 25 aircraft and options for 30 more. Continental, Western Airlines, KLM, and other carriers soon placed orders as well. Compared with the L-1011, the DC-10 appeared to be a runaway commercial success.

One reason was that the DC-10 gave airlines a choice of engines from two major manufacturers — General Electric and Pratt & Whitney. That flexibility allowed carriers to negotiate between competing suppliers and, in some cases, select an engine family already supported by their maintenance operations.

The DC-10 possessed another advantage that airlines could not ignore: price. In 1970, depending on configuration, order size, and the bargaining strength of the customer, a DC-10 generally sold for approximately $15 million to $20 million. A similarly configured L-1011 could cost roughly $18 million to $25 million. At the upper end, the difference could approach 20 percent. The price disparity reflected, in part, the fundamentally different financial environments in which the two aircraft were developed.

After McDonnell acquired Douglas in 1967, it imposed strict financial discipline throughout the company. Management proceeded with the DC-10 because the aircraft represented the future of the commercial division and, perhaps, the long-term viability of the merged corporation. But the program was not exempt from the new emphasis on cost control. It was assigned a firm development budget intended to prevent the overruns that had plagued earlier Douglas programs, including the DC-9. McDonnell Douglas leadership was determined that the DC-10 would be profitable from the outset.

As a result, the DC-10 was designed as a comparatively straightforward aircraft. Wherever practical, Douglas engineers adapted established design practices and technologies from the DC-8 rather than developing entirely new systems. This reduced development time and manufacturing costs, but it also left the aircraft with less redundancy and fewer safeguards in certain areas than its principal competitor.

That does not mean McDonnell Douglas deliberately ignored safety. Rather, when several technically acceptable solutions were available, management pressure favored the least expensive one. In some cases, those choices provided less protection against an unusual failure or an unforeseen combination of events.

One example involved the aircraft's hydraulic systems. Like other large transports, the DC-10 used multiple hydraulic circuits to operate its flight controls, landing gear, brakes, and other systems. Its three independent hydraulic systems met regulatory requirements and were intended to provide redundancy if one system failed.

The Boeing 747 and Lockheed L-1011, however, had four hydraulic systems. More important than the number alone was how the lines were routed. On competing aircraft, hydraulic lines were separated wherever practical so that damage in one area would be less likely to disable every system. In the wings, lines could be positioned farther aft, where the wing structure offered greater protection from damage near the leading edge. Even though Airbus' A300 had only three hydraulic circuits as well, those circuits were routed separately, and inside the wing they were embedded farther aft from the leading edge.

On the DC-10, portions of all three hydraulic systems passed close together, including areas near the wing's main structural members. Grouping the lines simplified installation and kept them near the engines and engine-driven hydraulic pumps. It also created the possibility that a single structural failure could damage all three systems at once — a vulnerability that would later have catastrophic consequences.

A second example was the rear cargo door. Because an inward-opening door would occupy valuable space inside the cargo compartment, Douglas designed the door to open outward. This maximized usable volume and allowed baggage or freight to be loaded directly behind it, an attractive feature for airline customers. However, an outward-opening door on a pressurized aircraft requires a particularly reliable locking mechanism because cabin pressure continually pushes the door away from the fuselage. The DC-10's rear cargo door used an electrically driven actuator and a system of linkages and locking sectors. The mechanism was comparatively simple and inexpensive, but it could be vulnerable to mis-rigging, deformation, or incomplete engagement. Even more troubling, the external handle could sometimes be forced into its closed position despite the locking system not being fully engaged.

A third example involved the aircraft's stall-warning system. McDonnell Douglas provided a stick shaker on the captain's control column as standard equipment, but a separate shaker for the first officer was optional and cost extra for an airline to have installed. A stick shaker vibrates the control column vigorously when the aircraft approaches an aerodynamic stall, providing an unmistakable warning that immediate corrective action may be required. Because either pilot might be flying the aircraft, competing widebody designs generally ensured that both pilots received the warning. On the DC-10, an airline that wanted equivalent warning capability for both pilots had to purchase the additional equipment.

A fourth example concerned the engine pylons — the structures connecting the wing-mounted engines to the aircraft. McDonnell Douglas designed the pylon attachments with tight clearances and highly efficient load paths. This reduced structural weight and avoided the cost of larger, heavier components. However, this arrangement complicated engine maintenance. To avoid any damage to surrounding components, McDonnell Douglas stipulated in its maintenance procedures manual that the engine was to be separated from the pylon before the pylon itself was removed from the wing. Other widebody aircraft used more robust attachment arrangements that made it practical to remove the engine and pylon together as a single assembly.

This particular engine removal procedure on the DC-10 required more time and labor. Some airlines therefore developed faster methods of removing the engine and pylon together, placing loads on the attachment structure that it had not been designed to withstand. What began as a decision intended to reduce weight and manufacturing expense would later interact disastrously with an airline's effort to reduce maintenance time and cost.

***

Lockheed's L-1011, the direct competitor to the DC-10, developed a reputation as one of the safest and most technologically advanced commercial aircraft of its era. It incorporated numerous features intended to reduce pilot workload and provide additional safeguards against human error. Although such capabilities are common aboard modern airliners, the L-1011 was among the first commercial aircraft designed from the beginning around a highly integrated automatic-flight system, and it became the first widebody certified for Category III automatic landings. Under the appropriate conditions, its systems could guide the aircraft down the instrument landing system, control its speed and descent, flare it above the runway, and complete the touchdown without the pilots touching the controls or throttles.

Lockheed intended this technological sophistication to be one of the airplane's greatest selling points. Although the L-1011 and DC-10 offered broadly comparable capacity, range, and performance, the TriStar was generally regarded within aviation circles as the more advanced aircraft. Its automatic-flight system, direct-lift-control system, redundant systems, and carefully engineered handling characteristics reflected Lockheed's determination to produce an airliner that would establish a new standard for safety and refinement. That engineering ambition, however, brought problems of its own.

When Lockheed formally launched the L-1011 in 1968, Rolls-Royce was developing a new generation of high-bypass turbofan engine known as the RB211. Its innovative three-spool architecture promised better fuel efficiency, lower noise, and an attractive power-to-weight ratio. It was also more compact than other competing engine designs from GE and Pratt & Whitney, reducing the cross-sectional area exposed to the airflow and potentially lowering aerodynamic drag.

Convinced that the RB211 offered the best combination of performance and technology, Lockheed elected to offer it as the L-1011's only engine. The airplane — particularly the center-engine installation and its curved intake duct — was designed around the RB211's dimensions. Unlike the DC-10, which could be ordered with engines from General Electric or Pratt & Whitney, the TriStar was dependent upon a single supplier and a single engine program.

That decision proved disastrous when the enormous cost of developing the RB211 drove Rolls-Royce into receivership in February 1971. Work on the engine was thrown into doubt, disrupting L-1011 production and leaving Lockheed with an almost completed airplane but no assured source for its engines.

Substituting another manufacturer's engine would have required extensive and prohibitively expensive redesign work.

The British government quickly intervened, nationalizing Rolls-Royce's principal operations and creating a reorganized company to continue the RB211 program. The rescue also required a renegotiated engine contract and American government guarantees for loans needed by Lockheed to complete the TriStar. Although the engine program survived, the crisis cost Lockheed valuable time and further strained a company already under considerable financial pressure.

McDonnell Douglas took advantage. The DC-10 entered commercial service with American Airlines in August 1971, while Lockheed did not deliver its first L-1011 to Eastern Air Lines until April 1972. The difference amounted to less than a full year, but in the fiercely competitive commercial-aircraft business, those months mattered. McDonnell Douglas was delivering airplanes and collecting revenue while Lockheed was still struggling to overcome doubts about the security of its engine supply.

The Rolls-Royce crisis also gave the L-1011 an early black eye. Airlines questioned whether Lockheed had produced an aircraft that was too technologically ambitious and too dependent upon unproven systems. When the first TriStars entered service with Eastern and TWA, the carriers encountered the inevitable technical problems associated with introducing an unusually sophisticated new airplane. Those early difficulties reinforced the perception that the L-1011, however capable, was complicated and expensive to support.

But did the DC-10's simpler and less sophisticated design — shaped by McDonnell Douglas' almost obsessive adherence to cost containment — make it the better airplane for an airline to operate?

McDonnell Douglas intended to convert simplicity into two commercial advantages — a lower purchase price and earlier availability. Even before the Rolls-Royce crisis disrupted the L-1011 program, the company was determined to beat Lockheed into service. The competitive objective was sometimes summarized within the program by the phrase "Fly before they roll." The meaning was unmistakable: McDonnell Douglas wanted the DC-10 flying before the first L-1011 had even rolled out of Lockheed's assembly plant.

By 1971, that strategy appeared to have succeeded. The DC-10 had entered airline service ahead of the L-1011, its order book was growing, and initial operating reports were encouraging. For James McDonnell, the moment appeared favorable for beginning a transfer of executive authority after more than three decades at the head of the company he had founded. His nephew, Sanford McDonnell, had joined McDonnell Aircraft in 1948 as an engineer and had advanced through a succession of engineering and management positions. He became president of McDonnell Douglas in 1971 and succeeded his uncle as chief executive officer the following year. James McDonnell did not withdraw completely, however. He remained chairman of the corporation and continued to exert influence over its direction until his death in 1980.

James McDonnell's son, John, was sixteen years younger than Sanford and was also rising through the company's management ranks. He would eventually become president in 1980 and succeed Sanford as chairman and chief executive officer in 1988. For more than two decades after the merger, authority over McDonnell Douglas would therefore remain concentrated within the McDonnell family.

Both Sanford and John had been deeply influenced by and immersed within the management culture James McDonnell had created — a culture characterized by organizational discipline, conservative financial management, and relentless attention to cost. Those principles had helped transform McDonnell Aircraft into one of the country's most successful defense contractors. Applied to the commercial aircraft business, however, the same principles would sometimes produce very different results.

When Flaws Bred Calamity

On June 12, 1972, American Airlines Flight 96, a DC-10 that had been in service for less than a year, departed Detroit Metropolitan Airport for Buffalo, New York. The aircraft carried 56 passengers and 11 crew members on what should have been a routine 90-minute flight.

As the DC-10 climbed over Windsor, Ontario, and passed through 11,750 feet, the crew heard a tremendous thud. Dust and debris erupted through the cockpit, the rudder pedals suddenly moved to the full-left position, the three thrust levers snapped back toward idle, and the airplane yawed violently to the right. Unbeknownst to anyone aboard, the left rear cargo door had blown open and separated from the aircraft.

Because the cargo door was mounted beneath the passenger cabin near the rear of the fuselage, its sudden opening caused the pressurized air inside the cargo compartment to escape almost instantaneously, but the passenger cabin above it remained pressurized for a fraction of a second longer. That momentary imbalance placed an enormous downward load on the cabin floor — one it had never been designed to withstand — causing the floor to partially collapse into the cargo compartment below.

Running through the passenger cabin floor beams were the cables that connected the cockpit controls to the rudder, elevators, horizontal stabilizer, and the No. 2 engine mounted in the tail. When the floor buckled, some of those cables were stretched, jammed, or damaged. The pilots initially lost rudder authority, the No. 2 engine's thrust lever became immovable near idle, and elevator response became increasingly sluggish. The two wing-mounted engines remained controllable, however, giving the pilots at least one means of influencing the aircraft's attitude and direction.

The crew declared an emergency and began the difficult task of maneuvering the crippled DC-10 back toward Detroit. By carefully varying the thrust of the two wing engines and applying sustained aileron input, they managed to line up with the runway and land safely. All 67 people aboard survived. Two flight attendants suffered minor injuries when the floor beneath them partially collapsed during the cabin decompression, but no passengers were seriously hurt.

The passengers and crew aboard Turkish Airlines Flight 981 would not be so fortunate.

Turkish Airlines Flight 981 was a scheduled service from Istanbul to London with an intermediate stop at Paris' Orly Airport. On March 3, 1974, the flight was operated by a DC-10 that had been delivered to Turkish Airlines approximately 15 months earlier. Because of a strike affecting British Airways, the Paris-to-London segment was unusually at near capacity. When Flight 981 departed Orly at approximately 12:30 p.m. local time, it carried 335 passengers and 11 crew members.

Approximately ten minutes after takeoff, as the aircraft climbed through roughly 11,000 feet, its left rear cargo door blew open and tore away from the fuselage. The cargo compartment depressurized almost instantaneously, once again creating a tremendous pressure differential between the cargo hold and the passenger cabin above it. This time, however, the cabin floor did not merely buckle. A substantial section collapsed completely.

Two rows of seats — with six passengers still strapped into them — went out through the opening created by the blown cargo door and were completely ejected from the aircraft. Their bodies, portions of the cabin floor, and the separated cargo door were later discovered in a turnip field.

The collapse also severed every cable controlling the elevators and horizontal stabilizer and disabled control of the tail-mounted No. 2 engine. The DC-10 immediately pitched downward and began accelerating rapidly. Unlike the crew of Flight 96, the pilots of Flight 981 no longer possessed enough control authority to recover their aircraft.

For the next 77 seconds, the flight crew fought desperately to save the airplane. Inside the passenger cabin, the explosive decompression had produced chaos. Sections of the rear floor had collapsed, ceiling panels and debris had fallen into the cabin, and a dense fog briefly formed as moisture in the pressurized air condensed during the sudden drop in temperature and pressure.

The aircraft's nose initially dropped to approximately 20 degrees below the horizon. As its speed increased, aerodynamic forces gradually reduced the descent angle. The pilots attempted to use the two functioning wing engines to influence the airplane's attitude, but with the elevator and stabilizer controls destroyed, they could neither arrest the descent nor bring the aircraft level.

Seventy-seven seconds after the cargo door separated, the DC-10 struck the Ermenonville Forest approximately 23 miles northeast of Paris. It was traveling at roughly 430 knots — nearly 500 miles per hour — and was still descending in a shallow nose-down attitude and a left bank. The aircraft disintegrated on impact, pulverizing the DC-10 and its passengers and scattering debris across the forest, instantly killing all 346 people aboard.

The destruction was so complete that identifying the sequence of events might have proved extraordinarily difficult had the cargo door, the displaced section of cabin floor, and the six ejected passengers not been found miles from the main wreckage. Together, they provided investigators with unmistakable evidence that the accident had begun with the failure of the rear cargo door.

Within 96 hours of the crash, the Federal Aviation Administration issued an emergency airworthiness directive requiring DC-10 operators to incorporate cargo-door modifications that McDonnell Douglas had developed following the Flight 96 incident nearly two years earlier. Additional directives soon required a fundamentally improved, closed-loop locking system and changes intended to prevent cabin-floor collapse during a rapid decompression if one started in the cargo hold.

Turkish Airlines Flight 981 was, at the time, the deadliest aviation accident in history. It remains one of the worst disasters in commercial aviation and the deadliest single-aircraft accident in which no one survived.

But the cargo doors that failed on Flights 96 and 981 should not have come as a surprise to McDonnell Douglas. The danger had been identified before the plane ever carried a paying passenger.

***

The warning had come even before the DC-10 entered commercial service. In 1970, while the aircraft was undergoing certification testing, McDonnell Douglas conducted a static ground-pressurization test on one of the test airplanes. Because electrical power was not being supplied to the aircraft while it was being buttoned-up, members of the test team manually closed the left rear cargo door with a crank and then moved the external handle into what they believed was the locked position. They were mistaken.

When the pressure inside the fuselage rose to approximately three pounds per square inch higher than the outside air pressure, the cargo door suddenly blew open with tremendous force. The aircraft was stationary inside a hangar and its cabin was empty, so the consequences were far less severe than they would later be aboard American Airlines Flight 96 and Turkish Airlines Flight 981. Nevertheless, the rapid depressurization of the cargo compartment created a momentary pressure differential between the cargo hold and the passenger cabin above it. Part of the cabin floor collapsed.

The incident exposed two potentially catastrophic weaknesses. First, the outward-opening cargo door could appear to be closed and locked even though its latching mechanism had not fully engaged. Second, the cabin floor could collapse during a sudden decompression and damage the flight-control cables routed through it.

The event should have led McDonnell Douglas to undertake a fundamental review of the door's design and the vulnerability of the floor structure. The people who had mistakenly secured the door were not hurried baggage handlers working on a crowded airport ramp. They were members of the DC-10 test organization who were familiar with the aircraft and had direct access to the engineers who had designed it and the specification manuals associated with the door. If they could believe that the door was locked when it was not, then ordinary cargo handlers around the world could certainly make the same mistake — and probably with greater frequency.

Instead, the incident was largely attributed to the manner in which the test crew had operated the door. Because the aircraft had no electrical power, the door had been closed manually rather than by its normal powered system. The failure was treated as an operational error committed under unusual circumstances rather than as evidence of an inherently deficient design.

That distinction allowed the DC-10 program to continue without a major redesign of either the cargo-door mechanism or the floor structure.

McDonnell Douglas was under intense pressure to certify the airplane and place it into commercial service before Lockheed could establish the competing L-1011 as a formidable competitor. Every delay threatened additional development costs, lost orders, and the possibility that Lockheed might capture a larger share of a market that many industry observers believed was not big enough to support two competing three-engine widebody aircraft.

The financial stakes were particularly high for McDonnell Douglas. The merged company was still attempting to stabilize itself after the combination of McDonnell Aircraft Corporation and Douglas Aircraft Company. Its commercial aircraft division was relying on the DC-10 to become both a technical and financial success. The airplane was not merely another product. It was central to the company's effort to reestablish itself as a major force in commercial aviation.

Although the DC-10 had originated within Douglas, James McDonnell now controlled the merged company. He understood the importance of the aircraft, but it was not his creation in the way that the F-4 Phantom II, the Mercury spacecraft, and the Gemini spacecraft had been products of the company he had built. His strongest connection to the DC-10 was therefore not necessarily emotional or technological. It was financial and strategic.

McDonnell knew that if Lockheed's L-1011 entered service first and gained any kind of a lead, the consequences for the newly merged company could be severe. The directive flowing from the executive offices in St. Louis was therefore straightforward — keep the DC-10's design as simple and economical as possible, control development costs, and avoid delays that might give Lockheed an advantage. That mandate influenced engineering decisions throughout the aircraft.

The DC-10's primary flight-control cables ran together through the floor structure above the rear cargo compartment. Those cables connected the cockpit controls to the hydraulic systems operating the elevators, horizontal stabilizer, rudder, and tail-mounted No. 2 engine. Routing them together through the floor was direct, relatively simple, and economical. It was not, however, the safest possible arrangement.

Although the airplane had multiple control cables, many of them for redundancy purposes in case one failed, their close physical proximity meant that a single structural failure could damage or disable several critical systems at once. The controls were redundant in number, but they were not sufficiently separated from one another.

The danger became evident during the 1970 pressurization test. It became unmistakable aboard American Airlines Flight 96, when the cabin floor partially collapsed and stretched, jammed, or restricted the control cables. The crew retained just enough control to save the aircraft. On Turkish Airlines Flight 981, the floor collapsed more completely, severing the cables controlling the elevators and horizontal stabilizer and making recovery impossible.

Lockheed took a different approach with the L-1011. Its critical control systems were routed with greater physical separation so that damage in one area of the aircraft was less likely to disable every means of controlling it. The design was more complex and more expensive, but it reflected Lockheed's determination to present the L-1011 as the most technologically advanced and safety-conscious widebody aircraft of its generation.

Lockheed had compelling reasons to be able to make that claim. It was returning to the commercial airliner market after an absence of nearly a decade, and its reputation had been damaged by structural problems and enormous cost overruns associated with the C-5 Galaxy military transport. The company intended the L-1011 to demonstrate that it remained capable of producing an airplane of exceptional sophistication.

The contrast between the two companies appeared stark. McDonnell Douglas emphasized simplicity, cost control, and speed to market. Lockheed emphasized technological sophistication, redundancy, and engineering refinement.

But Lockheed's apparent devotion to engineering excellence did not mean that it was an honorable or trustworthy competitor. That fact would become evident through the history of the very DC-10 that was later destroyed as Turkish Airlines Flight 981.

The aircraft had originally been built for All Nippon Airways, commonly known as ANA. The Japanese airline had placed an order for six DC-10s, and production had already begun when ANA canceled the order and selected Lockheed's L-1011 instead.

At the time, the decision appeared to be a major commercial victory for Lockheed and a damaging setback for McDonnell Douglas. Several years later, however, investigators discovered that Lockheed had secretly paid large sums of money to political figures, intermediaries, and other influential individuals in Japan to help secure the ANA order.

The scandal reached the highest levels of the Japanese government. Former Prime Minister Kakuei Tanaka was arrested and later convicted of accepting payments connected to the L-1011 sale. The disclosures became part of a much broader investigation into Lockheed's use of secret commissions and bribes in several foreign countries. The Lockheed scandal became one of the largest international corporate-bribery controversies of its time. Together with revelations involving other American companies, it helped generate the political pressure that led Congress to enact the Foreign Corrupt Practices Act in 1977.

Lockheed may have designed the more sophisticated airplane, but it had not relied solely on engineering merit to sell it. The company that portrayed the L-1011 as a triumph of technological integrity had used bribery and political influence to take business away from its competitor.

Regardless, the canceled ANA order left McDonnell Douglas with partially completed DC-10s that needed new customers. Finding buyers was not easy. Most major North American airlines planning to operate widebody aircraft had already committed to the Boeing 747, the DC-10, or the L-1011. Many of the largest foreign carriers had also placed their orders with one of the three American manufacturers or with the emerging Airbus consortium.

McDonnell Douglas began looking beyond the largest and most established airlines for customers willing to take the orphaned ANA aircraft. One of those customers was Turkish Airlines.

At the time, Turkish Airlines was expanding, but it remained a relatively small and comparatively inexperienced international carrier. Its jet fleet consisted largely of narrow-body aircraft, and operating a widebody aircraft would present significant challenges involving pilot training, maintenance, ground handling, route planning, and passenger demand. The question was not merely whether Turkish Airlines could operate the DC-10 safely. It was also whether the airline could consistently fill an aircraft of that size.

Yet the DC-10 offered an appeal that extended beyond ordinary commercial considerations. The Turkish Air Force reportedly saw value in having access to aircraft capable of transporting approximately 345 troops at a time if war broke out with Greece over Cyprus. Tensions between the two countries were already high, and armed conflict on the island was a genuine possibility. That conflict would, in fact, erupt during the summer of 1974, only months after the crash of Flight 981.

There was also a more immediate inducement — price. McDonnell Douglas offered Turkish Airlines terms and discounts normally associated with a large, prestigious international carrier placing a substantial order. The company needed to find homes for the former ANA aircraft, and Turkish Airlines was willing to accept three of them.

The arrangement served both parties. McDonnell Douglas moved unwanted airplanes that had already entered production, while Turkish Airlines acquired widebody aircraft at a price it might otherwise never have been offered. But the airplanes also carried with them a known design vulnerability that had not been adequately resolved.

Following the near catastrophe involving American Airlines Flight 96, the National Transportation Safety Board recommended that the Federal Aviation Administration require modifications to the DC-10's cargo-door locking mechanism. The NTSB also recommended changes intended to prevent the cabin floor from collapsing during a sudden decompression of the cargo compartment.

The FAA considered issuing an airworthiness directive, which would have made compliance mandatory. McDonnell Douglas argued that the problem could instead be addressed through manufacturer-issued service bulletins and a company-managed modification program that would see the modifications implemented over a period of time. The FAA accepted that approach.

The distinction was critical. An airworthiness directive carried the force of law. A service bulletin issued by a manufacturer was generally a recommendation unless a regulatory authority specifically mandated compliance.

McDonnell Douglas subsequently issued service bulletins calling for changes to the cargo-door mechanism. One of the most important modifications involved the installation of a support plate intended to prevent the external handle and its small vent door from being closed unless the latches had moved fully into the locked position.

When Turkish Airlines agreed to purchase the three DC-10s, it required that the applicable service bulletins be incorporated into the aircraft. McDonnell Douglas' records indicated that the cargo-door modifications had been completed. They had not.

When investigators recovered the cargo door from the aircraft destroyed as Flight 981, they discovered that the required support plate had never been installed. Work associated with the modification may have been started, but it had not been completed. Nevertheless, the paperwork accompanying the aircraft to its new owner-airline indicated that the service bulletin had been complied with.

Whether the inaccurate sign-off resulted from deliberate misrepresentation, careless record keeping, inadequate inspection, or a breakdown in communication could not be conclusively established. But the result was the same — a critical safety modification was represented as complete when it in fact was not.

Other deficiencies compounded the failure. The locking pins were improperly adjusted. The warning switch for the locking pins was incorrectly rigged. Therefore, the mechanism could extinguish a warning light in the cockpit, which would indicate to the pilots that the door was secure even when the latches had not fully engaged. The viewing port through which ground personnel were supposed to confirm the position of the locking pins was also difficult to use.

Together, those conditions allowed the cargo handler at Orly Airport to lower the external handle and close the vent door without realizing that the main latches had not moved fully into position. From the outside, the door appeared to be locked. In the cockpit, the warning light went out. The aircraft departed with the door unsecured.

As Flight 981 climbed and the pressure inside the fuselage increased, the cargo-door mechanism was subjected to forces it could not withstand. The attachment hardware failed, the latches released, and the door blew open and ripped away from the aircraft with tremendous force.

The catastrophe was not the result of a single unforeseeable defect. The danger had first revealed itself during the 1970 ground test. It had appeared again aboard American Airlines Flight 96. The failure to mandate a comprehensive redesign, combined with the incomplete modification of the Turkish aircraft and inaccurate documentation that indicated otherwise, allowed the same sequence to occur a third time. This time, 346 people died. It would be a death toll that would eerily appear again 45 years later in a similar set of circumstances related to financial pressures and how the warning signs were ignored.

The story of Turkish Airlines Flight 981 involved more than the failure of a cargo door. It was also the story of two manufacturers struggling for survival and market dominance.

McDonnell Douglas allowed competitive urgency, financial pressure, and an emphasis on simplicity to influence decisions affecting redundancy and safety. Lockheed built the more technically sophisticated aircraft but used bribery and political influence to win orders that might not otherwise have gone its way. Neither company emerged from the competition with clean hands.

For McDonnell Douglas, the consequences were immediate and devastating. Flight 981 revealed what could happen when a known design vulnerability was treated as an operational problem, when voluntary corrective measures substituted for mandatory action, and when paperwork was accepted at face value as proof that critical work had been completed.

The DC-10 had been designed to be economical, competitive, and quick to market. After Flight 981, the public increasingly suspected that McDonnell Douglas had achieved those objectives by accepting risks that should never have been accepted.

That suspicion would deepen only five years later, when another DC-10 fell from the sky and once again placed the aircraft — and the company that built it — at the center of one of aviation's worst disasters.

***

On May 25, 1979, American Airlines Flight 191, a McDonnell Douglas DC-10, taxied toward Runway 32R at Chicago's O'Hare International Airport for its scheduled flight to Los Angeles. At the controls were three experienced crew members. Behind them were ten flight attendants and 258 passengers.

Passengers aboard the aircraft could watch the takeoff through a relatively new feature American Airlines had installed on its DC-10 fleet only months earlier. A camera mounted in the cockpit, behind the captain's right shoulder, transmitted images of the pilots and the view through the windshield to large screens positioned throughout the passenger cabin. What had been intended as entertainment would take on a horrifying significance during the flight's final seconds.

Flight 191 began its takeoff roll at 3:02:45 p.m. Central Daylight Time. Approximately 46 seconds later, as the aircraft accelerated through roughly 145 knots, the captain called, "Rotate." First Officer Dillard, who was flying the aircraft, pulled back on the control column and raised the nose.

At almost the same moment, the No. 1 engine beneath the left wing, together with the pylon attaching it to the aircraft, separated completely. The enormous engine-and-pylon assembly pitched upward, passed over the top of the wing, tore away approximately three feet of its leading edge, and crashed onto the runway behind the accelerating DC-10. The plane's main landing gear lifted off the pavement a moment later.

Witnesses saw a stream of white vapor pouring from the damaged left wing. It consisted largely of fuel and hydraulic fluid escaping from lines that had been torn open when the engine and pylon ripped away. From the cockpit, however, the pilots could not see what had happened. They knew they had lost thrust from the left engine, but they had no way of knowing that the engine itself — and the structure that had supported it — was no longer attached to the airplane.

Flight 191 initially continued climbing. The crew responded as they had been trained to respond to an engine failure during takeoff. The aircraft reached approximately 325 feet above the ground and remained airborne for 31 seconds. Then the left wing stalled.

The DC-10 began rolling sharply to the left. Its bank angle increased beyond 90 degrees, its nose dropped, and the aircraft descended almost inverted into an open field and trailer park approximately 4,600 feet beyond the end of the runway. The impact produced an enormous fireball and scattered wreckage across a wide area.

All 271 people aboard were killed, as were two people on the ground. Two others on the ground were seriously injured. With 273 fatalities, Flight 191 remains the deadliest aviation accident in the United States.

It is impossible to know how many passengers were looking at the cabin screens during those final seconds. Those who were may have watched the left wing drop and the ground rise toward them. Passengers seated along the left side of the aircraft may have seen the massive engine and pylon separate during the takeoff roll and pass over the wing.

The wreckage of the engine and pylon lying on the runway immediately told investigators where to begin. The No. 1 engine had not failed internally and fallen away on its own. The structure attaching it to the wing had fractured, allowing the engine and pylon to separate as one unit.

Initial inspections revealed cracking in the pylon's rear attachment structure. The FAA ordered operators to inspect the comparable attachment areas on other DC-10s. Those inspections uncovered similar cracks on additional aircraft operated by American Airlines and Continental Airlines. Ultimately, investigators found nine other cracked pylon mounts.

The discovery raised the terrifying possibility that the problem might be inherent in the DC-10's design. If so, any aircraft in the fleet might be vulnerable to losing an engine and pylon in flight.

On June 6, 1979, just 12 days after the crash, the FAA took the extraordinary step of suspending the DC-10's type certificate. Every DC-10 registered in the United States was grounded, and foreign-registered DC-10s were prohibited from operating in American airspace. Many foreign regulatory authorities followed the FAA's lead, effectively grounding the worldwide fleet.

Nearly 250 DC-10s had been delivered by then. Their sudden removal disrupted airline schedules around the world as carriers reassigned other aircraft, canceled flights, consolidated services, and attempted to accommodate passengers on whatever capacity remained available.

The suspension was far more serious than an ordinary inspection order. Without a valid type certificate, the DC-10 could not legally carry passengers. Aircraft could fly only when specifically authorized for testing, repositioning, or other limited purposes. A pall settled over McDonnell Douglas.

The company's engineers had to determine whether the engine-pylon attachment contained a fundamental structural weakness, how extensive the problem might be, and what modifications would be required before the fleet could fly again. Management also confronted the possibility of enormous legal liability, lost sales, compensation claims from airlines, and lasting damage to the reputation of both the DC-10 and its manufacturer.

The investigation ultimately determined that the initial damage had not resulted from an inherent failure of the pylon during normal operation. It had been inflicted during maintenance work performed by American Airlines approximately eight weeks before the crash.

The work had been undertaken to comply with McDonnell Douglas service bulletins calling for inspection and replacement of bearings in the engine-pylon attachment system. The manufacturer's maintenance procedures contemplated removing the engine from the pylon first and then removing the pylon from the wing as a separate operation. That procedure was laborious. Removing the engine and pylon separately required more time, more labor, and the disconnection and reconnection of additional fuel, hydraulic, electrical, and control lines. Each additional connection created another opportunity for a leak, improper fitting, or maintenance error. It also kept the aircraft out of revenue service longer.

American Airlines had devised its own procedure to save time and money. Instead of removing the engine and pylon separately, maintenance personnel would detach the pylon from the wing and lower the entire engine-and-pylon assembly as one unit. Continental Airlines independently developed a similar procedure.

Removing the assembly as a unit promised substantial savings in labor and aircraft downtime. American used a forklift fitted with an engine-support fixture to hold and maneuver the combined assembly as mechanics disconnected it from the wing.

The procedure was not contained in the McDonnell Douglas maintenance manual, and American adopted it without obtaining the manufacturer's concurrence. Although the airline evaluated portions of the procedure, it did not adequately account for the extraordinary precision required to maneuver the assembly without damaging the tightly fitted pylon attachment structure. The combined engine and pylon weighed thousands of pounds. Even a small unintended movement of the forklift could force the pylon's rear attachment flange against the corresponding structure in the wing. And that was precisely what happened to the aircraft that became Flight 191.

During reinstallation, the engine-and-pylon assembly shifted. The rear attachment structure contacted the wing fitting with sufficient force to crack the upper flange of the pylon's aft bulkhead. The damage was in a confined area that was difficult to inspect once the assembly had been reinstalled, and it went undetected. The aircraft returned to service.

During the following eight weeks, the crack grew under the repeated stresses imposed during takeoffs, landings, and normal flight. On May 25, as Flight 191 reached rotation speed and the wing began producing lift, the weakened rear attachment failed. The remaining attachment points could not carry the resulting load, and the entire engine-and-pylon assembly tore away. The maintenance damage began the accident sequence, but it did not, by itself, make the crash inevitable.

Commercial-aircraft disasters rarely result from one isolated failure. They usually occur when multiple weaknesses, errors, and circumstances align. Aviation safety experts often describe this as the Swiss-cheese model. Each layer of protection contains imperfections, like holes in slices of Swiss cheese. Ordinarily, another layer blocks a hazard from progressing. Disaster occurs when the holes temporarily align, allowing the hazard to pass through every defense.

On Flight 191, the first hole was the decision to use an unapproved maintenance method. The next was the inability of the forklift to position the massive assembly with sufficient precision. Another was the failure to detect the damage before returning the aircraft to service. But the engine separation still did not have to bring the airplane down.

The DC-10 had been designed to continue flying after losing one engine. Flight 191 did not crash simply because its left engine was gone. It crashed because the departing engine and pylon caused additional damage that changed the aerodynamic configuration of the left wing while simultaneously depriving the pilots of the warnings that might have revealed what was happening.

When the engine-and-pylon assembly passed over the wing, it tore through hydraulic lines running through the wing's leading edge. Those lines supplied pressure to the actuators that controlled the leading-edge slats outboard of the No. 1 engine. Slats extend from the front of a wing during takeoff and landing, changing its shape and allowing it to produce sufficient lift at lower speeds. On the DC-10, hydraulic pressure held the slats in their commanded position. The system did not include a mechanical lock that would keep them extended if that pressure disappeared.

When the hydraulic lines were severed, pressure drained away. Aerodynamic forces then pushed the outboard slats on the left wing back into their retracted position. The slats on the right wing remained extended. The aircraft now had two wings configured differently from one another. The left wing required a significantly higher airspeed to remain flying than the right. The pilots did not know this.

The separation of the No. 1 engine also removed its electrical generator. That loss deprived several cockpit instruments and warning systems of power, including the captain's flight instruments, the left stall-warning computer, the slat-disagreement warning system, and the stick shaker attached to the captain's control column.

The stick shaker was designed to vibrate the control column forcefully as the aircraft approached a stall. Only the captain's control column had one. A second stick shaker for the first officer's side was available as an option but was not installed on the aircraft. The crew therefore received no tactile warning that the left wing was approaching a stall. They also received no slat-disagreement indication telling them that the slats on the two wings were in different positions.

Believing they had experienced a conventional engine failure, the pilots followed American Airlines' prescribed procedure. After the airplane initially accelerated to approximately 172 knots, they allowed it to decelerate toward V2, the designated engine-out climb speed. Under ordinary circumstances, V2 provided an adequate margin above stall speed with one engine inoperative. But those calculations assumed that the airplane still had both wings intact and symmetrically configured. They did not account for the loss of part of the wing's leading edge or the retraction of the slats on only one side.

As the aircraft slowed to approximately 159 knots, the left wing stalled. The right wing continued producing lift, creating a powerful rolling moment that the pilots could not overcome. Flight 191 rolled left, pitched downward, and struck the ground.

The cockpit crew had done what their procedures and training told them to do. They had no way of knowing that the airplane they were flying was no longer aerodynamically equivalent to one that had merely lost an engine. The warnings that could have alerted them to the asymmetrical slat configuration and impending stall had disappeared with the same engine whose loss created the emergency. The causal chain was therefore much longer than a single maintenance mistake.

American Airlines used an unauthorized procedure to remove and reinstall the engine and pylon as one assembly. The forklift damaged the pylon's rear attachment structure. The damage was not detected. A fatigue crack grew during subsequent flights. The attachment failed during takeoff. The engine and pylon separated and tore through the wing's leading edge. Hydraulic pressure was lost, allowing the left outboard slats to retract. Electrical power was lost to the stall-warning and slat-disagreement systems. The pilots followed an engine-out procedure that slowed the aircraft below the stall speed of the damaged left wing. The wing stalled, and the airplane rolled beyond recovery. Breaking any one of several links in that chain might have prevented the crash.

The accident's initiating damage was caused by American Airlines, not by McDonnell Douglas. The NTSB concluded that American's maintenance procedure had not been properly evaluated and had directly damaged the pylon structure. But the investigation did not leave McDonnell Douglas entirely untouched.

The DC-10's slats depended on hydraulic pressure to remain extended and had no alternate mechanical locking system. The design allowed the loss of hydraulic pressure in one area to produce an asymmetrical wing configuration. The loss of a single engine-driven generator disabled critical stall-warning and slat-disagreement indications. The stick-shaker installation lacked sufficient redundancy. Most importantly, the aircraft gave its pilots no means of recognizing that one wing had entered a stall.

The investigation also exposed weaknesses in communication across the industry. Continental Airlines had previously damaged the same pylon structure on two DC-10s while using a similar forklift procedure. In both cases, maintenance personnel heard the structure fracture and repaired the damage. Continental treated the events as internal maintenance errors and did not report them to McDonnell Douglas, the FAA, or other operators. Had that information been shared, American might have recognized the danger before Flight 191.

The FAA reinstated the DC-10's type certificate on July 13, 1979, after a 37-day suspension. Investigators had established that the pylon cracking was associated with the airlines' maintenance methods rather than a fleet-wide structural defect developing under normal operation. New directives prohibited removal of the engine and pylon as one assembly and required inspections and revised maintenance procedures. The DC-10 was legally cleared to fly again, but the greater battle had only begun.

To much of the public, the technical distinction between a design failure and maintenance-induced damage was almost meaningless. They had already seen photographs of Flight 191 falling from the sky with one engine missing. Flight 191 followed the cargo-door failure that had nearly destroyed American Airlines Flight 96 and had killed all 346 people aboard Turkish Airlines Flight 981. Another DC-10 catastrophe made the aircraft appear fundamentally unsafe, regardless of what investigators ultimately concluded.

The cockpit camera intensified the horror surrounding Flight 191. Stories circulated that passengers had watched their own aircraft roll toward the ground on the cabin screens. Whether every screen remained operating or how many passengers were watching could never be known, but the image became inseparable from the disaster. The name DC-10 itself became toxic. A grim joke circulated that the letters "DC" stood for "Death Cruiser."

American Airlines had promoted its aircraft as the "DC-10 LuxuryLiner," which was emblazoned on the fuselages of its DC-10s. After the crash, it removed the model designation from the fuselages and repainted them simply as "American Airlines LuxuryLiner." The change was an unmistakable attempt to separate the airline's brand from the aircraft's damaged reputation.

McDonnell Douglas and the airlines operating the DC-10 mounted an extensive campaign to restore public confidence. Print and television advertisements emphasized the inspections conducted during the grounding and portrayed the DC-10 as one of the most thoroughly examined aircraft ever returned to service. Airlines stressed the airplane's capabilities, reliability, and millions of accumulated flight hours.

The campaign faced a difficult reality because the public had lost trust in the airplane. That distrust threatened not only

McDonnell Douglas but also the airlines whose long-haul route systems depended on the DC-10. American, United, Continental, and other major carriers had invested heavily in the aircraft, its crews, its maintenance infrastructure, and the schedules built around it. They could not simply remove it from service without enormous financial consequences.

For McDonnell Douglas, Flight 191 arrived at an especially damaging moment. The DC-10 program had already been weakened by its costly competition with Lockheed's L-1011. The market for widebody trijets had never become large enough to provide both manufacturers with the sales volumes they had originally anticipated.

Now McDonnell Douglas faced the possibility that airlines would avoid future purchases, the reality that existing customers would demand compensation for the grounding, and that lawsuits arising from the deaths of 273 people would impose substantial additional costs. Even though the investigation placed primary responsibility for the initiating damage on American Airlines' maintenance procedure, the DC-10's name — and McDonnell Douglas' name — remained attached to the catastrophe.

Orders slowed dramatically. Fewer than 30 additional commercial DC-10s were ordered between the crash in May 1979 and McDonnell Douglas' announcement in August 1983 that production would be phased out. The program received an important lifeline from the United States Air Force, which ordered 60 military tanker-and-transport versions designated the KC-10 Extender. The order kept the production line active longer than commercial demand alone might have justified, but it could not restore the DC-10's former position in the airline market.

When the aircraft entered service, McDonnell Douglas reportedly estimated that it needed to sell approximately 438 units to recover the program's development and production costs. By the time manufacturing ended, 446 DC-10s and KC-10s had been built

— 386 commercial aircraft and 60 military tankers. On paper, the program had barely exceeded its original break-even quantity.

But the raw production total concealed the program's true economics. McDonnell Douglas had offered substantial discounts in its competition with Lockheed, paid for engineering changes and fleet modifications, defended lawsuits, negotiated settlements, supported operators through the grounding, and absorbed the reputational cost of accidents associated with the aircraft. Whatever the original calculation had been, simply delivering 446 airplanes under these circumstances meant the program absorbed massive losses. The DC-10 had outsold the L-1011, but the victory was hollow.

In December 1981, Lockheed announced that it would leave the commercial-airliner business after completing its remaining L-1011 orders. The final aircraft was delivered in 1984. Under other circumstances, Lockheed's withdrawal might have vindicated McDonnell Douglas and established it as the winner of the trijet war. Instead, the contest had become a pyrrhic victory. McDonnell Douglas had survived its competitor, but the market both companies had fought to control had been badly damaged. The enormous development costs, aggressive discounting, technical failures, accidents, lawsuits, and destruction of public confidence had consumed much of whatever reward victory might have offered.

But for McDonnell Douglas, it became another unmistakable warning. Cost, simplicity, maintainability, regulatory oversight, airline practices, and engineering redundancy could never be treated as separate concerns. Decisions made in one area could create vulnerabilities in another, and those vulnerabilities could remain hidden until every hole in the system aligned.

The DC-10 had been created to secure McDonnell Douglas' future in commercial aviation. Instead, its troubled history helped establish a pattern that would shape nearly every commercial aircraft program the company undertook thereafter. McDonnell Douglas had won the trijet war but in the process, it had destroyed the spoils.

A Turnaround Without a Future

The 1980s witnessed Airbus emerge as a credible challenger in a commercial aircraft market still dominated by Boeing and McDonnell Douglas. Rapid advances in jet-engine technology and avionics, coupled with explosive growth in air travel, were ushering in a new generation of airliners increasingly built around twin-engine designs. Engines were becoming more powerful, fuel-efficient, and reliable, creating a potentially lucrative opportunity for manufacturers with the financial resources to develop aircraft with capabilities far beyond those of the previous generation.

Boeing and Airbus were eager to exploit that opportunity. McDonnell Douglas, however, had become increasingly reluctant to undertake another major commercial-aircraft program — and with good reason. The financial damage inflicted by the DC-10 had followed the mediocre commercial performance of the DC-8 and years of disappointing returns from the DC-9. By the 1980s, the company was not merely gun-shy. Its weakened financial position left it increasingly incapable of competing in the costly race to develop the next generation of airliners.

When Douglas had begun developing the DC-10, it considered a twin-engine configuration but ultimately rejected it for two principal reasons. First, the engines then available could not provide the performance required at hot-and-high airports. That was particularly important because Douglas was courting United Airlines as a launch customer, and United maintained a major operation at Denver, more than 5,000 feet above sea level. At the payloads and ranges envisioned for the DC-10, a twin-engine design would have faced significant operating restrictions there and at similarly demanding airports.

The second obstacle was the restrictive overwater operating rules then applied to twin-engine aircraft. A twin was generally confined to routes that remained within 60 minutes of a suitable diversion airport following an engine failure — and even that authority typically required a period of proven service. A third engine addressed both concerns. It provided greater takeoff performance and permitted more direct long-range routes over oceans and other remote regions. Douglas therefore adopted for the trijet configuration for the DC-10. Lockheed reached much the same conclusion while developing the L-1011.

By the 1980s, however, the technological and regulatory environment was changing rapidly. Airlines wanted newer aircraft that were more efficient than anything then flying, and both Boeing and Airbus began preparing to meet that demand.

Airbus pursued a dual-aircraft strategy with the A330 and A340, which were developed around a common airframe. The twin-engine A330 was intended primarily for high-density medium and long-haul routes, while the four-engine A340 was designed for long and ultra-long-range missions. Airbus formally launched the A330/A340 family in 1987.

The A340 demonstrated just how far engine and airframe technology had advanced since the development of the DC-10. Despite having four engines rather than three, it could operate more efficiently than the older trijets while carrying a larger payload over greater distances. Airbus believed the commonality between the A330 and A340 would also appeal to airlines because the two aircraft shared many components, systems, and cockpit features.

Boeing chose a different path. Rather than develop both a twin-engine and four-engine aircraft, it placed its bet on a large, long-range twin. General Electric, Pratt & Whitney, and Rolls-Royce were promising a new generation of engines powerful enough to allow a two-engine aircraft to perform missions that had previously required three or four. Boeing waited for that technology to mature and then committed itself to the aircraft that would become the 777.

The 777 would eventually be capable of flying many of the same routes as a trijet or quad-jet while offering the economic advantages of having only two engines. Boeing was effectively betting that advances in engine reliability, performance, and efficiency — combined with the gradual relaxation of restrictions on long-range twin-engine operations — would redefine the widebody market.

By the mid-1980s, with Airbus and Boeing making their plans for a new generation of widebody aircraft increasingly clear, McDonnell Douglas was forced to consider the future of its own commercial-airliner business.

Before the crash of American Airlines Flight 191, the company had studied several modifications to the DC-10 that would have provided airlines with greater capacity and range. After the disaster, however, DC-10 orders quickly dried up. McDonnell Douglas turned its attention toward adapting the aircraft to take advantage of newer developments in propulsion, aerodynamics, and avionics.

The company's financial condition made the decision especially difficult. McDonnell Douglas remained a major force in the defense industry. It had developed the F-15 Eagle for the United States Air Force, the F/A-18 Hornet for the United States Navy, and the AV-8B Harrier II for the Marine Corps. These aircraft served as frontline combat platforms for their respective services, and they had also attracted substantial orders from foreign governments. The company was also beginning development of the C-17 military transport for the Air Force.

Those government contracts provided McDonnell Douglas with a dependable stream of revenue and helped stabilize its stock price. Yet the company had drifted back into much the same predicament that had led to the 1967 merger of McDonnell Aircraft and Douglas Aircraft. Its profit margins remained exceedingly thin, generally hovering between 2 and 3 percent. Research-and-development costs were high, as were the capital investments required for tooling and production facilities. The bulk of the company's revenues came from government contracts, but those contracts generated only modest returns given the high costs the company was incurring.

The commercial-aircraft division was in even greater difficulty. Although the MD-80 series of aircraft continued generating revenue, the business was being weighed down by the commercial failure, development costs, and lingering liabilities associated with the DC-10. The MD-80 was also coming under intense pricing and technological pressure from Boeing and the rapidly growing Airbus consortium. If McDonnell Douglas did not bring forward a new aircraft capable of competing against the next generation of Boeing and Airbus widebodies, it risked being effectively shut out of the commercial-airliner market.

Developing an entirely new aircraft, however, was enormously expensive. By the middle to late 1980s, designing, certifying, manufacturing, and bringing a clean-sheet widebody to market could require an investment approaching $5 billion — roughly $16 billion in 2026 dollars. Such a commitment represented a massive financial risk for a company that was barely maintaining profitability.

Sanford McDonnell therefore decided to follow the same strategy the company had used with the DC-9 and MD-80. Although the DC-9's basic design dated to the early 1960s, McDonnell Douglas had kept the aircraft commercially relevant through a series of incremental improvements. During the early 1980s, the company renamed the latest versions the MD-80 series, helping distance them from the aging DC-9 designation. In 1989, the company began yet another program to upgrade the airframe by stretching the fuselage even further and adding, finally, high-bypass engines and an all-glass cockpit to better compete in the market place. This new model would be named the MD-90.

Yet the MD-80 and MD-90 remained, in essence, an updated DC-9 equipped with newer engines, modernized cockpit systems, a lengthened fuselage, and several performance refinements. McDonnell Douglas now attempted to apply that same evolutionary approach to the DC-10.

In the early 1980s, Sanford McDonnell directed the company's engineers to revisit the DC-10 design and determine how it could be improved using the latest engines, avionics, materials, and aerodynamic refinements. By fitting newer powerplants from General Electric or Pratt & Whitney, modifying the wings, lengthening the fuselage, and reducing unnecessary weight, the company believed it could substantially increase the aircraft's payload and range.

Advances in avionics offered another important improvement. The DC-10 had been designed around a three-person cockpit consisting of two pilots and a flight engineer. New digital systems and automated controls would allow the redesigned aircraft to use a two-pilot, glass cockpit, eliminating the flight engineer's position and reducing airline crew costs.

McDonnell Douglas believed the resulting aircraft could compete directly with the new widebodies being planned by Boeing and Airbus. More importantly, because it was derived from an existing design, the company believed it could bring the airplane to market several years ahead of its competitors. That head start might allow McDonnell Douglas to capture orders from airlines that were already considering their next generation of medium and long-range widebody aircraft.

Although the new airplane was an evolution of the DC-10, the company understood that it needed to distance the program from the older aircraft's damaged public image. Therefore, it designated the new model the MD-11.

The MD-11's lineage nevertheless offered a potential advantage. Just as the MD-80 operated under the same basic type certification lineage as the DC-9, the MD-11 would retain substantial commonality with the DC-10. McDonnell Douglas believed airlines already operating DC-10s would be particularly interested in acquiring a modernized version that promised better performance and economics while remaining relatively easy to integrate into their existing fleets.

In December 1986, McDonnell Douglas formally launched the MD-11 with commitments from ten airlines for more than 50 aircraft and options for approximately 40 additional airplanes. The initial response appeared to validate the company's strategy.

Once the development program was underway, Sanford McDonnell prepared to step down as chief executive. In 1988, control of the company passed to John McDonnell. The leadership transition was expected to mark the beginning of a promising new chapter for McDonnell Douglas. Unfortunately, it would not turn out that way.

The prototype MD-11 was originally scheduled to make its first flight in early 1989. Manufacturing difficulties and delays in obtaining parts from suppliers pushed that milestone into January 1990. Those delays proved to be only the beginning of the program's problems.

Almost as soon as the first aircraft entered service, serious performance shortcomings became apparent. The MD-11 was unable to meet several of its promised payload and range guarantees. American Airlines took delivery of 19 MD-11s but discovered that the aircraft could not efficiently fly some of the long-range routes for which the airline had acquired them. Singapore Airlines, having learned of the performance problems before accepting its aircraft, canceled its order for 20 MD-11s and selected the Airbus A340 instead.

McDonnell Douglas responded by working with General Electric, Pratt & Whitney, and other suppliers to develop a performance-improvement program, commonly known as a PIP. Engineers sought to eliminate unnecessary airframe weight, reduce aerodynamic drag, refine the engine installations, and improve the aircraft's operating performance. The modifications eventually allowed the MD-11 to meet much of its originally promised performance goals. But by the time the improvement program achieved its intended results in 1995, the damage had already been done.

American Airlines disposed of its MD-11 fleet, selling the aircraft to FedEx, one of the program's original launch customers. Delta Air Lines retained its MD-11s for a time but canceled options for additional aircraft. Other airlines similarly reduced or abandoned their plans for the model.

The MD-11 found a more durable role among cargo operators. Freight carriers such as FedEx generally placed greater emphasis on payload capacity than on the extreme range demanded by some passenger airlines. In that role, the MD-11 offered excellent operating economics. Its large fuselage, heavy-payload capability, and ability to accommodate substantial freight volume made it particularly well suited to cargo operations. As a passenger aircraft, however, the MD-11 was caught between generations.

McDonnell Douglas had concentrated on avoiding the enormous cost of developing a clean-sheet aircraft and had therefore remained committed to a basic airframe conceived during the 1960s. The DC-10 had been created as a trijet because the engine technology and regulatory environment of that period made a large, long-range twin impractical. By the late 1980s and early 1990s, both of those conditions were rapidly disappearing.

Twin-engine widebodies could increasingly perform the missions for which the MD-11 had been designed. Engine reliability had improved dramatically, and regulations governing extended overwater operations were being relaxed as manufacturers and airlines accumulated experience with modern twin-engine aircraft. A new generation of powerful turbofans made it possible for aircraft such as the Airbus A330 and Boeing 777 to carry heavy payloads across intercontinental distances using only two engines.

The economic consequences were decisive. Two engines were substantially less expensive to purchase, fuel, inspect, overhaul, and maintain than three or four. The very justification for placing a third engine in the DC-10's tail was evaporating during the years in which the MD-11 was being developed and introduced.

McDonnell Douglas had produced an improved version of an aircraft optimized for a regulatory and technological environment that no longer existed. The MD-11 also proved expensive to manufacture and purchase. Its sales price compared unfavorably with the Airbus A330 and A340, while high production costs forced McDonnell Douglas to sell some of the earliest aircraft at a loss.

Once again, the fundamental weakness that had plagued the company since the 1967 merger had resurfaced. McDonnell Douglas generated enormous revenues through its military programs, but it struggled to earn adequate profits from them. Its commercial-aircraft division offered few decisive technological or economic advantages over its competitors and, even worse, was losing money.

The company had expected to sell approximately 300 MD-11s — enough, it believed, to preserve its place as a major commercial-aircraft manufacturer. Instead, the program's early performance failures, changing market conditions, and the superiority of emerging twin-engine competitors limited total production to just 200 aircraft.

When the final MD-11 left the production line in 2001, most of the surviving airplanes were serving as freighters rather than passenger airliners. The aircraft had ultimately found a productive role, but not the one McDonnell Douglas had envisioned when the program was launched. The MD-11 had been intended to secure the company's future in commercial aviation. Instead, it demonstrated how little room remained for McDonnell Douglas in a market increasingly dominated by Boeing and Airbus.

***

In 1994, after only six years at the helm, John McDonnell decided to step down as chief executive of McDonnell Douglas. The company had become trapped in a quagmire from which there appeared to be no easy escape. Its commercial aircraft business was losing ground, its profit margins remained thin, and the cost of remaining competitive against Boeing and Airbus was rapidly moving beyond the company's financial reach.

John McDonnell concluded that the company needed a fresh face — someone from outside the McDonnell family who could confront its strategic and operational problems without being constrained by tradition or sentiment. More specifically, it needed an executive with a demonstrated ability to restore discipline and profitability to a struggling aerospace company. In September 1994, the McDonnell Douglas board approved the appointment of Harry Stonecipher as chief executive officer.

Harry C. Stonecipher was born in 1936 and earned a bachelor's degree in physics from Tennessee Technological University in 1960. He began his career at General Motors' Allison Division, where he worked briefly on jet engines before moving to General Electric and joining its aircraft-engine business.

Stonecipher spent more than a decade at GE, gaining extensive experience in aerospace engineering, customer relations, and corporate management. His work brought him into contact with major aircraft manufacturers, including Boeing and Douglas, while his rise through GE's managerial ranks exposed him to the emerging leadership philosophy of Jack Welch.

Welch was then building a reputation within GE for his aggressive approach to performance, cost control, and corporate restructuring. His philosophy was straightforward — a GE business should be first or second in its market, or the company should fix it, sell it, or close it. Businesses, functions, and employees were expected to justify their continued place within the corporation through measurable performance. Stonecipher absorbed much of that philosophy. He came to believe that an organization should retain only those activities it could perform exceptionally well and at a competitive cost. Work that could be performed more efficiently by outside suppliers should be outsourced, allowing the company to concentrate its resources on the activities it considered most strategically valuable.

His management style became defined by an intense focus on operational efficiency, financial accountability, and cost control. Like Welch, Stonecipher managed with a blunt, unsentimental, and results-oriented discipline. Programs, facilities, suppliers, and employees were judged by measurable performance. Anything that failed to meet his standards became a candidate for restructuring, outsourcing, sale, or elimination.

During the early 1970s, Stonecipher left GE for Sundstrand Corporation, a major producer of aerospace and industrial systems and components for commercial and government customers. He steadily advanced through the company's executive ranks and, in 1987, became its chief executive officer. Stonecipher assumed control of Sundstrand at a difficult moment. The company was facing serious financial and reputational problems, including the consequences of a government-contracting fraud scandal that resulted in a then record-setting settlement of approximately $115 million with the United States government. He responded by imposing strict financial discipline, strengthening internal oversight, and demanding greater accountability throughout the organization. Under his leadership, Sundstrand stabilized its finances and restored its profitability.

A central element of Stonecipher's strategy was to reduce the number of activities performed internally. Sundstrand began transferring the production of certain subcomponents — including electronics and machined parts — to outside suppliers. Stonecipher viewed many of those manufacturing activities as relatively low-value work that consumed capital, management attention, and factory resources while exposing the company to additional cost and quality risks.

Rather than manufacture every component itself, he wanted Sundstrand to concentrate on system design, integration, final assembly, sales, and post-delivery support. Outside suppliers would manufacture many of the individual parts, while Sundstrand would assemble them into complete systems, deliver them to customers, and retain responsibility for the customer relationship.

The strategy reduced Sundstrand's capital requirements and allowed the company to focus on the activities Stonecipher believed produced the greatest value. It also reflected a broader transformation taking place across American industry, as corporations increasingly abandoned the vertically integrated model in favor of outsourcing, supplier consolidation, and leaner internal operations.

Stonecipher's philosophy was, in many respects, an extension — and perhaps an acceleration — of the management system Jack Welch had introduced at General Electric. Financial performance would take precedence over institutional tradition. Internal capabilities would be retained only when they could be justified economically. Suppliers would assume a greater portion of the manufacturing burden, while the corporation itself would increasingly act as a designer, integrator, assembler, and marketer of complex products.

That philosophy had helped Stonecipher rehabilitate Sundstrand. He would now carry it with him into the greatest professional undertaking of his career — the attempted rescue of McDonnell Douglas.

***

The end of the Cold War in 1991 triggered a sweeping consolidation of the American defense industry. Although McDonnell Douglas remained one of the nation's largest defense contractors and was expected to retain an important role, the industry entered a period of rapid contraction as Washington sought to capture the so-called peace dividend made possible by the collapse of the Soviet Union.

McDonnell Douglas was still producing two of the most prominent combat aircraft in the American arsenal. The F-15 Eagle remained in production for the United States Air Force and several foreign air forces, while the F/A-18 Hornet continued to be manufactured for the United States Navy and international customers. The company had also begun producing the C-17 Globemaster III transport under an initial Air Force requirement for 120 aircraft.

Those programs provided McDonnell Douglas with substantial revenues, but the seemingly bottomless well of Department of Defense spending that had sustained the company for decades could no longer be taken for granted. Weapons programs were being reduced, delayed, consolidated, or canceled as military requirements were reassessed for a post-Cold War world.

McDonnell Douglas had been counting heavily on winning and retaining the contract for the A-12 Avenger II, an advanced carrier-based stealth aircraft being developed for the Navy. The program was intended to provide the company with a major new source of military revenue and help carry its combat aircraft business into the next century. Instead, the A-12 became engulfed in technical difficulties, schedule delays, and enormous cost overruns. In January 1991, with global tensions receding and the program's problems mounting, the Department of Defense canceled the program.

The cancellation dealt McDonnell Douglas a severe financial and strategic blow. It eliminated a program on which the company had placed considerable hopes and exposed it to years of litigation over responsibility for the failed contract. More broadly, it demonstrated that even the company's traditionally dependable defense business was no longer immune from the forces reshaping the aerospace industry.

Thus, when John McDonnell decided to step down as chief executive, the company was in a precarious position. Its bread-and-butter stream of government revenue had become less secure, while profit margins on many of those contracts remained exceedingly thin because of high production costs, organizational inefficiencies, and the immense expense of developing and manufacturing advanced military systems.

The commercial aircraft division was in even worse condition. The MD-11 had been outmaneuvered by increasingly capable aircraft from Airbus and Boeing, while the MD-80 and the new MD-90 offered too little technological advantage to secure the company's long-term future. McDonnell Douglas' commercial-airliner business was not merely struggling. It was circling the drain.

The company's predicament bore an uncomfortable resemblance to the circumstances that had produced the 1967 merger of McDonnell Aircraft and Douglas Aircraft. Twenty-seven years later, remarkably little seemed to have changed.

McDonnell Douglas still possessed a substantial but only modestly profitable defense business, while its commercial aircraft division continued to deteriorate. The crucial difference was that, in 1967, Douglas at least had the proposed DC-10 — a new aircraft that appeared capable of opening a path toward future growth. By 1994, no comparable program existed.

There was no clean-sheet commercial aircraft under serious development, no revolutionary new product waiting in the wings, and no credible plan for challenging the next generation of Boeing and Airbus airliners. McDonnell Douglas' commercial aircraft business was dying, and nearly everyone in the industry knew it. Harry Stonecipher knew it as well when John McDonnell approached him about occupying the chief executive's office in St. Louis.

Through his years at General Electric and Sundstrand, Stonecipher had been deeply involved in major aerospace programs and had developed extensive personal and professional relationships with senior executives throughout the industry. He was widely known and enjoyed a strong reputation as a disciplined, demanding, and results-driven executive.

His turnaround of Sundstrand made him especially attractive to John McDonnell. Stonecipher had taken control of a company facing financial and reputational peril and restored its stability through rigorous oversight, aggressive cost control, operational restructuring, and a relentless focus on measurable performance. John McDonnell came to view him as the safest possible choice — the executive most capable of turning McDonnell Douglas around, preserving its independence, and protecting the McDonnell family's industrial legacy.

Yet the qualities that made Stonecipher appear to be the ideal candidate were also closely aligned with the management philosophy that had helped create the company's predicament. Beginning with James McDonnell in the late 1960s, cost containment had become an institutional obsession. Faced with the enormous financial demands of the DC-10 program and the narrow margins of the defense business, McDonnell imposed increasingly elaborate systems for reviewing expenditures, approving investments, measuring performance, and controlling risk.

That emphasis continued under the family members and longtime insiders who succeeded him. Over time, the company's organizational DNA changed. McDonnell Douglas had once been defined principally by engineering ambition and technical achievement. It increasingly became a corporation governed by cost analyses, financial thresholds, approval templates, and layers of managerial review. Every investment had to be justified. Every department had to defend its expenses. Every program was scrutinized for opportunities to reduce staffing, shift work to suppliers, defer capital spending, or eliminate functions that could not demonstrate an immediate financial return.

Financial discipline was necessary for any manufacturer competing in the aerospace industry. At McDonnell Douglas, however, it gradually became more than a management tool. It became the dominant philosophy through which nearly every decision was evaluated. Stonecipher did not challenge that philosophy. He intensified it.

His reputation had been built on cost reduction, outsourcing, organizational streamlining, and uncompromising financial accountability. Those attributes were a major reason John McDonnell believed he was the perfect successor. Stonecipher seemed ideally suited to impose discipline on a company whose operations had become inefficient and whose margins were dangerously thin. But in selecting him, John McDonnell and the board were effectively doubling down on the same formula that had contributed to the company's decline.

The heavy-handed emphasis on containing costs had shaped the DC-10 program, constrained investment in new commercial aircraft, and gradually weakened the company's engineering foundation. Rather than restoring the balance between financial discipline and technical ambition, Stonecipher was poised to push the company even farther toward the former.

To many employees, his arrival promised not a fundamental change in direction but a harsher version of what they had already endured. It would be more of the same — only more so. Within the halls and factories of McDonnell Douglas, employees had long joked that the company employed more bean counters than engineers. Stonecipher's appointment seemed to confirm their worst suspicions. He was regarded as a lion tamer brought in to discipline a restless organization. But the lions at McDonnell Douglas did not wield sharp teeth. They wielded sharpened pencils.

***

When Harry Curtis Stonecipher took the helm of McDonnell Douglas, one of his first priorities was to reassure investors that the company's commercial aircraft business would remain central to its future. He publicly committed himself to reviving Douglas Aircraft, the division responsible for producing the company's commercial airliners.

Stonecipher's early statements suggested that Douglas had fallen into a distant third place behind Boeing and Airbus not because its products were fundamentally uncompetitive, but because the company had failed to market them aggressively enough. He emphasized that the DC-9 and MD-80 families remained commercially viable and pointed to the forthcoming MD-90 as evidence that McDonnell Douglas still possessed a formidable product line for the short and medium-haul market.

The MD-90 was the latest evolution of the DC-9 platform, incorporating newer engines, updated avionics, and other improvements intended to make it more economical and competitive. Initial deliveries were expected within the following year. Stonecipher pledged to strengthen the division's marketing organization, deepen its relationships with existing airline customers, and use those relationships to jump-start the commercial side of the company.

He also announced plans to reinforce McDonnell Douglas' position in the one segment of the commercial market where it still retained a meaningful competitive presence. The company would develop another derivative of the DC-9/MD-80/MD-90 family: a roughly 100-seat aircraft intended primarily for short-haul routes. The airplane would eventually be designated the MD-95.

In a move that reflected either shrewd marketing or growing desperation, McDonnell Douglas also circulated plans for an entirely different kind of aircraft that it would refer to as the MD-12. The proposed MD-12 would be a four-engine, double-deck airliner capable of carrying 500 or more passengers on intercontinental routes.

The company released artists' renderings of the enormous aircraft and approached several airlines to gauge their interest. Those efforts suggested that McDonnell Douglas was at least seriously studying the concept. Yet the proposal left many industry analysts perplexed. Given the company's financial condition, it was difficult to understand how McDonnell Douglas could afford to develop any clean-sheet commercial aircraft, much less one of unprecedented size, cost, and complexity. It was equally difficult to understand why the company would attempt it.

***

By the early 1990s, Airbus had developed a broad family of aircraft serving nearly every major commercial market segment. The original A300 had evolved into the more capable A300-600, offering greater payload, range, and operating efficiency. The smaller A310 served medium and long-range routes. The A320 family covered the narrow-body market through the A318, A319, A320, and A321. The A330 and A340 addressed medium, long, and ultra-long range widebody missions. Each family occupied a distinct position within Airbus' product portfolio.

One significant gap nevertheless remained at the upper end of the market — the very-large, high-capacity, long-range segment. Only one aircraft had ever successfully occupied that space — Boeing's legendary 747.

In June 1994, Airbus announced that it was beginning serious work on a new aircraft intended to challenge the 747. Initially referred to as the A3XX, the proposed airplane would eventually become the A380. Airbus envisioned a full-length double-deck aircraft capable of transporting approximately 550 to 600 passengers between major international hub airports.

With Boeing and Airbus already dominating the conventional medium and long-range widebody market, there was little strategic logic in McDonnell Douglas attempting to reenter that segment after the disappointing performance of the MD-11. But the 747's underlying design was approaching 30 years of age, while advances in engine technology, avionics, materials, and aerodynamics continued at a rapid pace.

McDonnell Douglas appears to have concluded that the market for very-large aircraft might represent one of the few remaining openings in which it could compete without confronting Boeing and Airbus across their entire product ranges. Most analysts believed that market would remain a relatively narrow niche. Even so, it offered the possibility of entering a segment in which only one established aircraft existed. The problem was financing. How could McDonnell Douglas possibly design, certify, and manufacture an aircraft of such enormous proportions when it lacked sufficient capital to fund far smaller programs?

The company's proposed answer was straightforward. McDonnell Douglas would not develop and manufacture the MD-12 alone. It would recruit international partners to share the design work, production responsibilities, development costs, and financial risk. Major structures and systems would be designed and manufactured by supplier-partners, while McDonnell Douglas would serve primarily as the aircraft's overall architect and integrator. The concept was entirely consistent with Stonecipher's history and management philosophy.

At General Electric, he had been exposed to Jack Welch's emphasis on concentrating corporate resources on those activities that produced the greatest competitive and financial returns. At Sundstrand, Stonecipher had taken the philosophy further by outsourcing the manufacture of numerous subcomponents and repositioning the company around system design, integration, assembly, sales, and customer support.

After arriving at McDonnell Douglas in 1994, he began applying the same model to its commercial aircraft programs. The company shifted additional production work associated with the MD-80, the newly introduced MD-90, and the still-producing MD-11 to outside suppliers.

Stonecipher pursued outsourcing with extraordinary intensity. He viewed it as one of the most effective ways to lower production costs, reduce capital requirements, transfer risk to suppliers, and improve profit margins. If another company could manufacture a component more cheaply or efficiently, Stonecipher saw little reason for McDonnell Douglas to continue producing it internally.

His results-oriented approach extended well beyond the commercial aircraft division. He imposed cost-reduction and process-improvement initiatives across the company's military programs, including the troubled C-17 Globemaster III transport.

The C-17 program had suffered from development delays, production inefficiencies, and significant cost overruns. Because of the aircraft's military mission and the security requirements surrounding many of its systems, McDonnell Douglas could not outsource its work as freely as it could on commercial programs. Stonecipher instead attacked the program's manufacturing processes.

Production workflows were reorganized. Assembly procedures were simplified. Layers of management and review were reduced. Factory operations were scrutinized for waste, duplication, and delay. The objective was to reduce the cost of each aircraft and restore the program to financial health. Those measures produced measurable financial improvements, but they also intensified the company's already difficult relationship with its workforce.

McDonnell Douglas was heavily unionized. The International Association of Machinists and Aerospace Workers, commonly known as the IAM, and several other labor organizations represented much of the company's production workforce. As outsourcing and cost-cutting initiatives eliminated jobs and transferred work to outside suppliers, relations between management and labor became increasingly bitter. Billboards appeared around St. Louis, where McDonnell Douglas maintained its headquarters, carrying messages such as: "Harry Stonecipher, Stop Outsourcing Our Jobs!"

Stonecipher's reputation as a forceful and effective corporate leader remained intact and, in some respects, was enhanced. Wall Street rewarded his cost reductions and improving margins with a rising stock price. Investors increasingly viewed him as the executive who had restored financial discipline to a company long regarded as inefficient and directionless. Within the unions, however, he developed a far darker reputation. Workers saw him as an executive who regarded experienced employees, established production capabilities, and institutional knowledge primarily as expenses to be reduced or eliminated. That reputation would follow him and deepen over the years ahead.

Boeing and Airbus were also increasing their reliance on outside suppliers, but their approach remained more measured. Aircraft manufacturers had always purchased major systems from specialized companies. Engines, avionics, landing gear, doors, hydraulic systems, and numerous secondary structures were routinely supplied by outside firms. Historically, however, manufacturers had generally retained responsibility for the design and production of an aircraft's principal structures, particularly its fuselage and wings. Those were not merely interchangeable commodities. They represented core engineering and manufacturing competencies that helped define the aircraft and the company that produced it.

Outsourcing made sense when a component could be designed and manufactured by several qualified suppliers more efficiently and at lower cost. The danger arose when a company transferred so much design, engineering, and production responsibility to outside firms that it began losing the internal knowledge required to understand, integrate, and control the complete aircraft.

To many inside McDonnell Douglas, Stonecipher was no longer merely removing inefficient or nonessential work. He appeared to be taking a wrecking ball to the company's foundational engineering, design, and manufacturing capabilities.

The MD-95 illustrated both the attraction and the danger of his strategy. McDonnell Douglas was pitching the aircraft as a modern 100-seat short-haul jet, but the company lacked the capital necessary to fund the estimated $1 billion required to design, test, certify, and launch it. Although the MD-95 remained an evolution of the DC-9 platform, developing it still required substantial engineering work and financial investment.

Stonecipher reduced McDonnell Douglas' exposure by assigning major portions of the program to outside partners. Alenia Aeronautica of Italy, for example, would assume responsibility for the wings. By requiring suppliers to invest in the development and production of major components, McDonnell Douglas could transfer a substantial portion of the program's cost and financial risk away from itself. When ValuJet placed an order for 50 MD-95s in 1995, the McDonnell Douglas board quickly approved the program. Such a sizable order appeared to validate Stonecipher's strategy.

Outsourcing became the business model on which Stonecipher publicly pinned the survival of McDonnell Douglas. It allowed the company to introduce new products while conserving cash, reducing capital investment, and avoiding the enormous financial exposure associated with traditional aircraft development. At least, that was the theory presented publicly.

Stonecipher, however, was an exceptionally perceptive and hard-minded executive. He could recognize strategic realities that others either failed to see or preferred not to acknowledge. What he may have seen at McDonnell Douglas was that no amount of outsourcing, marketing, or incremental product development could provide the company with a viable path toward long-term independence.

Despite the supposedly new aircraft McDonnell Douglas was offering to airlines, its narrow-body product line remained rooted in the DC-9, an airplane conceived during the early 1960s. The DC-9 had evolved into the MD-80, then the MD-90, and finally the MD-95, but each remained a derivative of the same underlying design.

The company was developing a reputation within the airline industry as a one-trick pony. It continued to offer variations of a single aging platform while Boeing and Airbus fielded broader and more modern families of aircraft.

McDonnell Douglas' military order book remained substantial, but those programs did not generate the profits investors expected. Stonecipher's cost-cutting initiatives were improving their financial performance, particularly on the C-17, but efficiency alone could not solve the company's larger strategic problem. McDonnell Douglas possessed no credible successor to the MD-11. It lacked the capital to develop a clean-sheet commercial airliner. Its narrow-body products were steadily losing ground. Its military programs were vulnerable to post-Cold War spending reductions and industry consolidation. There was no convincing product strategy that could preserve the company indefinitely as an independent aerospace manufacturer.

Stonecipher nevertheless succeeded in dramatically improving its financial appearance. Through cost reductions, outsourcing, factory restructuring, and tighter operational controls, he helped drive McDonnell Douglas' stock price from just under $19 per share to nearly $70. On the surface, he appeared to be restoring the company to health.

But he was not rebuilding the engineering capabilities or developing the aircraft that would secure its future. He was improving its margins, strengthening its balance sheet, increasing its stock price, and making it more attractive to another corporation. Stonecipher may have understood from the beginning that McDonnell Douglas could not be saved as an independent company. What he could do was maximize its value for shareholders by preparing it for acquisition.

And he knew precisely which company would be interested in acquiring McDonnell Douglas. .

The New and Improved Boeing

The year 1996 would prove to be the zenith of Boeing's commercial dominance. The company commanded the airliner market with its highly successful 737, 747, 757, and 767 families, while its newly introduced 777 was living up to nearly every promise Boeing had made about it. By the end of the year, Boeing had accumulated a staggering 719 orders across its commercial product line — nearly twice the combined total won by Airbus and McDonnell Douglas.

Boeing had not experienced an order year of that magnitude since 1989, when it received 819 commercial aircraft orders. Yet its position in 1996 was arguably even stronger. In 1989, McDonnell Douglas remained a more substantial commercial competitor, and Airbus and McDonnell Douglas together had recorded 510 orders, leaving Boeing with approximately 64 percent of the market. In 1996, Boeing's 719 orders represented roughly 66 percent of all commercial jet orders for aircraft seating more than 100 passengers.

The value of those orders made Boeing's dominance even more pronounced. Its large widebody aircraft — the 747, 767, and new 777 — were substantially outselling their competitors, giving Boeing an even greater share of the market when measured by revenue rather than by the number of aircraft ordered. Its order books were overflowing, its production lines were operating at full pace, and its influence extended across nearly every segment of the aerospace industry. Investors recognized the company's strength. Boeing's stock, which had traded at $24.69 per share in January 1991, had nearly doubled to almost $50 by the end of November 1996.

When Philip Condit assumed the chief executive's office and joined Boeing's board of directors in April 1996, he inherited a company that, by almost every measure, was firing on all cylinders. Its future could scarcely have appeared brighter. It was as though he had been handed the fabled goose that laid the golden eggs. His principal responsibility was to keep it healthy and to act as a faithful steward of a global institution whose reputation for engineering excellence and marketing prowess was unmatched.

That marketing strength was soon reflected in a series of extraordinary exclusivity agreements. Beginning with American Airlines in 1996, followed by similar arrangements with Delta Air Lines and Continental Airlines, Boeing secured long-term contracts under which those carriers agreed to purchase their commercial aircraft exclusively from Boeing for as long as 20 years. (Those exclusivity contracts would eventually have to be torn up to assuage government regulator demands.)

One of Condit's first major actions as chief executive was Boeing's acquisition of Rockwell International's North American aerospace and defense businesses. Boeing had already played a dominant role in the space race of the 1960s and early 1970s. The Rockwell acquisition reinforced its position in space and defense as the company prepared to enter a new century less than four years away.

Across the Atlantic, executives at Airbus headquarters in Toulouse debated how they could counter Boeing's consistent — and increasingly alarming — dominance. As events would soon demonstrate, they need not have worried quite as much as they did.

Sometime in late 1996, Harry Stonecipher met with the McDonnell Douglas board of directors to pitch a plan he had hatched. Although the company's financial condition had improved and its major programs had stabilized, McDonnell Douglas still lacked the resources to finance an entirely new commercial aircraft such as the proposed MD-12.

The company could not fund such a program alone. Its standing within the commercial aviation industry had also deteriorated so badly that it would have been difficult to persuade suppliers or international partners to share the enormous financial risks. Attracting launch customers willing to commit to a new

McDonnell Douglas aircraft — and to place their confidence in it against competing Boeing and Airbus products — would have been equally challenging. Even maintaining the company's small share of the commercial market with its existing aircraft would be difficult, regardless of how many incremental improvements were made to them.

Stonecipher had built his career as a numbers man. No matter how he sliced the data or analyzed the commercial aircraft division's prospects, the conclusion remained the same: its future was bleak. Unless something drastic changed, McDonnell Douglas would eventually be forced to abandon the commercial airliner business altogether.

He also questioned whether the company could remain a leading aerospace corporation into the 21st century. McDonnell Douglas possessed a strong balance sheet, valuable military programs, and a stock price near historic highs. Yet one major product-development failure or other major misstep could still threaten the company's survival. Its elevated share price would also be difficult to sustain if investors began to recognize how limited its long-term growth prospects had become.

Stonecipher nevertheless had a plan that could preserve the company's strongest businesses while delivering a substantial return to its shareholders. After all, in his view, the central obligation of a publicly traded corporation was to reward the investors who had placed their capital and confidence in it.

His plan was to approach Philip Condit and explore whether Boeing might acquire McDonnell Douglas.

Stonecipher knew that Condit wanted to diversify Boeing by expanding its defense and space operations. Although acquiring McDonnell Douglas would be an enormous undertaking, the strategic logic was compelling. The transaction could strengthen Boeing's position in defense almost overnight while giving McDonnell Douglas shareholders a profitable exit from a company whose independent future was becoming increasingly uncertain. The McDonnell Douglas board authorized Stonecipher to explore the possibility.

He also understood that a combination of the two companies would receive intense regulatory scrutiny on both sides of the Atlantic. Government authorities might demand substantial concessions or even attempt to block the transaction. Stonecipher controlled neither those regulators nor Boeing's board. But he recognized one advantage he possessed — the contrast between his personality and Philip Condit's.

Stonecipher was forceful, decisive, and highly persuasive. He was disciplined, analytical, and driven by numbers and facts. Once he had studied the available evidence and reached a conclusion, contrary opinions rarely dissuaded him. Yet despite his ability to immerse himself in detail, he also possessed a talent for seeing the broader strategic picture.

He was shrewd and unusually conscious of appearances — how others perceived him, his decisions, and his words. He projected strength, confidence, and cunning. His demeanor communicated an unmistakable warning: Do not ever cross me.

Philip Condit came from an entirely different mold. He was an engineer and a consensus builder who was often willing to reconsider his position. Where Stonecipher might bulldoze through an obstacle, Condit was more inclined to go around it — or to revise the objective because the obstacle existed. If someone crossed him, he was more likely to retreat, move on, or return later in an effort to repair the relationship.

Stonecipher may have seen opportunity in those qualities. Condit appeared more malleable, more receptive to persuasion, and less confrontational than Stonecipher himself. Perhaps Stonecipher believed that he might eventually succeed Condit. Even if he did not, he may have concluded that he could exercise enormous influence over both Condit and the direction of the combined company.

Stonecipher and Condit already knew each other well — not merely because they led two of the world's largest aerospace manufacturers, but because years of industry interaction had brought them into repeated contact. Stonecipher could easily approach Condit directly and present the idea in pragmatic, strategic terms.

Boeing dominated commercial aviation. McDonnell Douglas remained a formidable military contractor whose F-15, F/A-18, and C-17 programs were among the most important aircraft programs in the United States defense establishment. The defense industry had been contracting and consolidating since the end of the Cold War in 1991. For Boeing, acquiring a company responsible for producing several of the military's most important frontline aircraft and which were expected to remain in production despite defense cutbacks made considerable strategic sense.

Condit had publicly expressed a desire for defense-related operations to account for as much as half of Boeing's revenue. Reaching that target organically — or by acquiring numerous smaller companies — would require years of effort and a series of difficult integrations. Each acquired business would bring its own leadership, culture, operating practices, and internal systems. But by acquiring one enormous defense contractor such as McDonnell Douglas, Boeing could approach Condit's 50 percent objective in a single move.

When they finally spoke, Condit was receptive to Stonecipher's proposal and indicated that he wanted a more formal and detailed discussion after consulting Boeing's board. Stonecipher returned to the McDonnell Douglas board and reported the results of his conversation. He then met with John McDonnell, chairman of the company's board, and together they outlined the principal terms McDonnell Douglas would seek in negotiations.

Among them were several seats on Boeing's board of directors; a senior leadership position for Stonecipher as president and chief operating officer of the combined company; executive positions for other senior McDonnell Douglas officials; an acceptable valuation for McDonnell Douglas shares; and a new corporate name: Boeing-McDonnell.

In December 1996, Stonecipher and Condit met at the Four Seasons Hotel in Seattle and negotiated the details of the proposed combination. Incredibly, the two men reached agreement on most of the major terms within only a few hours. The principal exception was the name. Rather than becoming Boeing-McDonnell, the combined enterprise would retain the Boeing name.

When the agreement was announced publicly on December 16, 1996, it represented a watershed moment for the global aerospace industry. Two industrial giants that had shaped military and commercial aviation throughout the postwar era were preparing to become a single corporation. The end of the Cold War had prompted widespread predictions of massive consolidation throughout the defense and aerospace industries. The Boeing–McDonnell Douglas combination appeared to validate those forecasts but on a scale few had imagined. At the time, it was expected to become the tenth-largest merger in American corporate history and the largest ever undertaken in the aerospace industry.

Condit declared that the combined organization would become the world's largest, strongest, broadest, and most admired aerospace corporation — as well as, by a substantial margin, the largest exporter in the United States.

In a telephone interview shortly after the announcement, Stonecipher said that informal conversations about a possible merger had taken place as far back as 1994, when he first became chief executive of McDonnell Douglas. He acknowledged that, upon taking the position, he had concluded that if McDonnell Douglas were ever to merge with another company, that company should be Boeing.

Although neither man stated it publicly, the sequence of events invited speculation that John McDonnell may have selected

Stonecipher in part because he believed Stonecipher could eventually engineer such a transaction. Stonecipher's actions also suggested that he may have viewed the preparation of McDonnell Douglas for a merger as one of his objectives from the beginning.

If Stonecipher had previously approached Boeing — or Condit personally — the timing may not yet have been right. When Stonecipher joined McDonnell Douglas in 1994, Condit was Boeing's president, but Frank Shrontz remained chairman and chief executive. Shrontz's management style and personality were closer to Stonecipher's than to Condit's, and he would likely have been a more difficult negotiating counterpart.

Everyone within the industry understood that Condit was the presumed successor when Shrontz retired. It therefore appeared possible that Stonecipher and Condit had been waiting for the proper moment — Condit's elevation to the chief executive's office and his eventual appointment as chairman of Boeing's board.

In the meantime, Stonecipher used the intervening years to put McDonnell Douglas' financial house in order. By the time Condit controlled Boeing, McDonnell Douglas had become a more attractive acquisition target, its shareholders were positioned to receive a substantial return, and Stonecipher was ready to negotiate the transaction that would carry him into the leadership ranks of the world's most powerful aerospace corporation.

***

On its face, Philip Condit and Boeing appeared to be the shrewd ones — the predator taking advantage of a weakened prey, McDonnell Douglas. But a closer examination suggested something very different. It was almost as though the wounded prey had lured the predator into a trap and then become the predator itself. Although the merger made sense from a broad strategic perspective, many industry observers and analysts were left wondering why Boeing had agreed to the terms it did.

When the details were scrutinized, it appeared that McDonnell Douglas had dictated much of the agreement, including the price Boeing would pay. Several key leadership positions within the combined company would be occupied by McDonnell Douglas executives. John McDonnell received a seat on Boeing's board of directors, as did Harry Stonecipher. Stonecipher was also named president and chief operating officer of Boeing, placing him directly in the line of succession to the chief executive's office.

Stonecipher also ensured that several of his handpicked lieutenants received highly influential positions within the company. One of the most important — the office of chief financial officer — would, within two years, be occupied by the man who had served as McDonnell Douglas' CFO at the time of the merger.

Michael M. Sears had joined McDonnell in the late 1960s as an engineer after earning an engineering degree from Purdue University. He rose steadily through the company and eventually oversaw several major programs, including development of the F/A-18E/F Super Hornet, the larger and more advanced successor to the original F/A-18. The earlier F/A-18 program had badly strained McDonnell Douglas' profit margins during development and production. Sears, however, managed the Super Hornet program with such discipline that he received industry recognition for its execution. He developed a reputation as a no-nonsense, methodical executive with a keen ability to identify cost-saving opportunities.

Having spent most of his career under John McDonnell and, later, Harry Stonecipher, Sears understood the importance they placed on controlling costs and maximizing profitability. He viewed the aerospace business through much the same lens as Stonecipher. Efficiency and cost containment were not merely financial considerations; they were the foundation upon which higher margins and greater shareholder returns were built.

Both men believed that a public company's ultimate obligation was to increase shareholder value. A higher stock price was the pot of gold at the end of the rainbow, and nearly every management decision was to be judged by whether it moved the company closer to it.

When Stonecipher became chief executive of McDonnell Douglas in 1994, he already knew Sears from earlier encounters within the industry. The two men developed a mutual admiration, perhaps because each recognized similar traits in the other. Both were highly confident, strongly opinionated, and convinced of the soundness of their own judgment. There was little rivalry between them. Stonecipher clearly held the senior position, and Sears was eager to execute his directives. More importantly, Sears shared Stonecipher's interpretation of McDonnell Douglas' circumstances and of the aerospace industry itself. Both regarded efficiency, cost reduction, and improved margins as paramount to all other considerations.

Sears became instrumental in carrying out Stonecipher's strategy for McDonnell Douglas. He helped stabilize the company, improve its financial performance, and prepare it for the merger that Stonecipher increasingly appeared to regard as its ultimate destination.

When the Boeing merger was announced, Sears understood that he would occupy an important position within the combined enterprise. Stonecipher made little secret within his inner circle that Sears was expected eventually to become Boeing's chief financial officer after the departure of its existing CFO, Debra Hopkins.

In the meantime, Sears was assigned an equally consequential role. He became president of Boeing's newly created Military Aircraft and Missile Systems organization. Coming from one of the nation's leading defense contractors, the position placed him on familiar ground. He would be responsible for integrating McDonnell Douglas' most prominent military programs —

including the F-15, F/A-18, and C-17 — into Boeing. The defense business would account for a much larger share of the new company's revenue than it had before the merger. That gave Sears considerable influence over how the combined enterprise would be organized and managed.

In the years leading up to the merger, McDonnell Douglas' defense programs had become increasingly profitable. Stonecipher and Sears believed that this improvement was due in no small part to their emphasis on financial discipline, efficiency, and cost control. They had developed what they considered a proven operating model, and they intended to extend it throughout the merged company. As McDonnell Douglas' military programs were absorbed into Boeing, Sears effectively carried forward many of the management practices that had governed them before the merger. From the perspective of Stonecipher and Sears, there was little reason to abandon a formula they regarded as successful.

Their objective went beyond preserving McDonnell Douglas practices within the inherited military programs. Stonecipher, Sears, and many of the other executives who crossed over in the merger believed that Boeing devoted too much attention to engineering and too little to cost containment. They increasingly viewed the products built by Boeing and McDonnell Douglas as commodities. Under that interpretation, airplanes and weapons systems were not fundamentally different from other manufactured products. The winners would be the companies that could bring them to market faster, produce them more cheaply, and extract the highest possible margins from their sale.

That outlook represented a profound departure from Boeing's traditional culture. For more than eight decades, Boeing had regarded engineering as the central organizing principle of the company. Its aircraft were not merely products to be manufactured and sold. They were complex technological achievements whose commercial success depended upon technical superiority, durability, safety, and performance.

The incoming McDonnell Douglas leadership did not necessarily regard engineering excellence as unimportant. Rather, they believed Boeing had elevated it beyond its proper place. In their view, the company's engineering-centered culture had prevented management from extracting the full financial value of its products.

Beginning with John McDonnell, Harry Stonecipher, Michael Sears, and the executives who reported to them, the former McDonnell Douglas leadership began a systematic effort to persuade Boeing's managers and employees to see their industry differently.

They saw opportunities everywhere to reduce expenses, accelerate production, reorganize the supply chain, and improve margins. Boeing's engineering focus had built one of the most successful aerospace companies in history, but Stonecipher and his allies believed that it had also created inefficiency, excess capacity, and insufficient attention to shareholder returns.

If Boeing's aircraft and defense systems were ultimately commodities, they reasoned, then the company should concentrate less on the internal mastery of every technical detail and more on managing the end-to-end production system. Greater value might be extracted by transferring work to suppliers, forcing competition throughout the supply chain, shortening development schedules, reducing fixed costs, and measuring success increasingly through the lens of financial performance.

That philosophy would gradually reshape Boeing. The company McDonnell Douglas had been unable to defeat in the marketplace would now be transformed from within by the executives of the company it had supposedly conquered.

***

Even before the merger was completed in August 1997, the profound differences between the two corporate cultures had become unmistakable. Worse for Boeing, McDonnell Douglas' cost-cutting, profit-driven philosophy was already beginning to displace Boeing's engineering-first mindset. Harry Stonecipher's vision for the combined company was taking hold through his shrewdness and his ability to read people — Philip Condit in particular. Whatever concerns Condit may have developed as the two companies began integrating their operations, Stonecipher persuaded him to continue moving forward and to embrace increasingly more of the McDonnell Douglas approach. Stonecipher had assessed Condit accurately from the beginning. He had concluded that Condit could be persuaded, and the early months of the merger appeared to confirm it.

The cultural divide was immediately apparent to Boeing's rank-and-file employees, and many of them did not like what they saw. A perception spread that Boeing's hallways were increasingly being patrolled by bean counters — accountants and financially oriented managers, many of them from McDonnell Douglas. Boeing employees were also familiar with the friction Stonecipher and his leadership team had created with the McDonnell Douglas workforce only a few years earlier. They began bracing themselves for the trouble they feared was coming. It did not take long for that trouble to begin.

When the merger was announced in late 1996, the United States and much of the world were experiencing an extraordinary period of economic expansion. Only a few years had passed since the collapse of the Eastern Bloc and the end of the Cold War. Trade barriers were falling, global markets were opening, and a new economic order centered on the expansion of capitalism was driving dramatic increases in international commerce.

That expansion created an insatiable demand for air travel, both for business and leisure. Disposable income was rising across many societies, while countries that had long contained predominantly poor populations — particularly in Asia and the Far East — were beginning to witness rapid growth in their middle and upper classes. The implications for commercial aviation were enormous.

Airbus and Boeing, now the two principal manufacturers of large commercial aircraft, began experiencing a sharp rise in demand. Boeing, still the dominant player, moved quickly to use its size and scale to cement that position. The directive from the executive offices in Seattle was clear: Boeing would rapidly increase production across nearly its entire commercial product line, with particular emphasis on the newly developed 737 Next Generation family.

Short and medium-haul aircraft were expected to experience the greatest demand. Boeing wanted its airline customers to believe that it could deliver new aircraft faster — and at a more competitive price — than Airbus could deliver its rival product. Stonecipher reached into the same playbook he had used at McDonnell Douglas. The plan called for a broad examination of Boeing's production system to identify inefficiencies and eliminate costs.

It began with the supply chain. Suppliers were told to reduce the prices of their components or risk losing their contracts. Many of them were small manufacturers that had worked with Boeing for decades. They felt the pressure immediately and had little leverage with which to resist.

Stonecipher and his team did not stop with outside suppliers. They also examined the components Boeing continued to manufacture internally and began pushing to outsource many of them. These included stamped parts, brackets, fittings, panels, sheet-metal components, tubing, and other items that had traditionally been produced within Boeing facilities. The effort soon expanded beyond individual parts to include complete subassemblies.

Stonecipher and his allies believed Boeing would be more profitable if it evolved from a manufacturer into an integrator. Outside suppliers would produce the components, assemble many of the structures, and deliver increasingly complete sections of the aircraft. Boeing's role would be to bring those pieces together and deliver the finished airplane. That, at least, was the ultimate objective.

Transforming the way an aircraft is manufactured is neither simple nor immediate, particularly when production systems have been operating successfully for years or even decades. Therefore, Outsourcing began incrementally. Stonecipher presented each step to Boeing's legacy leadership as part of a modern and more efficient future. His reasoning remained consistent. If airplanes were commodities, Boeing did not need to manufacture every part itself. Suppliers could build the components as long as they met Boeing's specifications and produced them at a lower cost.

Boeing's value would increasingly come from controlling the supply chain and orchestrating the complex movement of parts and assemblies into a finished aircraft. The company would distinguish itself by delivering that commodity faster and more efficiently than its competitor. Where it could not offer a lower price, it could still protect shareholder value through higher margins.

The shift in thinking had begun. A new philosophy was descending across the company.

Cost reduction, however, was not the only tactic Stonecipher intended to use to preserve Boeing's lead in commercial aviation. Another directive emerged from the executive suite — production would be increased dramatically so that Boeing could deliver 550 commercial aircraft by the end of 1997. That goal represented an extraordinary leap — more than double the company's output from the previous year.

Stonecipher persuaded Condit and Boeing's board that the company should move aggressively to exploit the rapid increase in demand for commercial airliners. The production surge, he argued, would generate enormous profits. There was no immediate need to develop entirely new products or make major changes to existing aircraft. Demand was rising so rapidly that Boeing's principal competitive advantage could simply be its ability to deliver airplanes sooner than Airbus.

Airlines opening new routes or adding capacity to existing ones often could not afford to wait years for new aircraft. If Boeing could offer delivery slots a year earlier — or perhaps even sooner — than Airbus, that advantage might outweigh differences in technology, operating economics, or passenger appeal. Availability itself would become a powerful sales weapon.

The proposed production surge also helped Boeing pressure suppliers into lowering their prices. In exchange, Boeing promised dramatically larger orders. Suppliers were told they could accept lower margins on each component because the increased volume would generate greater total profits. The argument was presented as an opportunity to benefit from economies of scale.

On paper, the strategy appeared compelling. Boeing would increase deliveries, suppliers would gain volume, airlines would receive aircraft sooner, and shareholders would benefit from rapidly rising revenue and profit. Wall Street responded enthusiastically. Boeing's stock price rose approximately 25 percent since the merger.

For a brief time, the strategy appeared to be working exactly as promised. But the euphoria would soon come crashing down.

***

A perfect storm had formed, and it was beginning to wreak havoc across Boeing. Unlike the destructive forces of nature, however, this storm was arguably man-made and largely avoidable.

It had been created by overconfidence, unrealistic expectations, and the outsized egos of the executives now directing the company.

Boeing was simultaneously attempting to absorb and integrate the vast and complex McDonnell Douglas organization, reorganize its own operations to accommodate the merger, force hundreds of suppliers to increase production, and accelerate its assembly lines to support an extraordinary increase in aircraft deliveries. It did not take long for the entire system to begin unraveling.

For all the intellectual and engineering talent within Boeing, its leadership appeared to have overlooked an obvious problem — the hundreds of suppliers expected to increase production could not all expand at the same rate. Some possessed the equipment, employees, capital, and capacity to respond quickly. Others did not. As a result, the thousands of parts required to build Boeing's airplanes began arriving at different times and in insufficient quantities.

Bins positioned at workstations throughout the factories were supposed to contain the parts needed to complete the tasks assigned to each station. Increasingly, those bins were empty or held only a portion of what was required. Aircraft had to move forward without work that should have been completed at earlier stations. Each unfinished task followed the airplane down the line, creating still more disruption at subsequent workstations. Eventually, aircraft emerged from the factory in varying stages of completion and were parked outside.

Some lacked only a few relatively minor items. Others were missing more substantial parts or contained numerous incomplete installations. Aircraft requiring only minor work might proceed through flight testing and then be parked until the remaining items arrived. Those with more serious deficiencies sat on the tarmac like abandoned houses — outwardly recognizable as airplanes but nowhere near ready for delivery.

The missing work then had to be completed through what Boeing called traveled work. Mechanics, engineers, tools, instructions, and parts had to follow the aircraft to wherever it had been moved and complete tasks that should have been performed earlier on the assembly line. Traveled work was not new to aircraft manufacturing. It had traditionally occurred in limited circumstances, usually because of an isolated parts shortage or an unexpected production problem. What was different in 1997 was its scale. It was becoming an accepted stopgap for coping with systemic failures across Boeing's production system.

The practice was inefficient and expensive. Employees performed tasks out of sequence, often in locations poorly suited to the work. Access to the aircraft became more difficult after surrounding components had already been installed. Tools and parts had to be transported repeatedly, and work already completed sometimes had to be removed so that an earlier operation could be performed. Rather than flowing smoothly through the factory, airplanes were becoming collections of deferred jobs.

The disruption was magnified by another problem: much of Boeing's commercial production infrastructure had become antiquated. The company had incorporated advanced digital technologies into the design and production of the 777, but large portions of its broader production and information systems remained rooted in methods developed decades earlier. Paper documents still moved through layers of approval. Production information was fragmented. Administrative processes were cumbersome, and small disruptions could take far too long to identify and resolve. The company that had used computers to design one of the most technologically advanced airliners in history was still relying upon an aging manufacturing bureaucracy to build much of the rest of its product line. Why had Boeing not modernized its factories more aggressively?

Part of the answer lay in complacency and in a narrative Boeing's leadership had repeated for years about its competitors. During much of the preceding two decades, McDonnell Douglas had been Boeing's largest American competitor in commercial aviation. Yet Boeing could see, as could nearly everyone else in the industry, that McDonnell Douglas was gradually losing its ability to compete. Its commercial product line was aging, its market share was shrinking, and its capacity to finance new aircraft was disappearing. That left Airbus.

As Airbus steadily increased its market share throughout the 1980s and early 1990s, Boeing executives often attributed its success to European government support. Boeing had initially regarded Airbus as little more than a government-sponsored job-creation project — an industrial experiment that would never become a genuine threat to American aerospace leadership.

Even as Airbus accumulated customers and introduced increasingly competitive aircraft, Boeing continued to explain much of its success through subsidies. After McDonnell Douglas ceased to be an independent competitor, that interpretation became even more deeply embedded within Boeing.

Airbus undeniably benefited from government-backed financing that reduced the amount of private capital it needed to raise when developing new aircraft. But concentrating almost exclusively on the source of Airbus' financing allowed Boeing to discount what Airbus was doing with the money.

Airbus was producing aircraft that were increasingly comparable to Boeing's and, in many respects, more technologically advanced. Its extensive use of cockpit automation, digital flight controls, and standardized systems across aircraft families gave airlines operational advantages. Airbus could offer its airplanes at prices competitive with Boeing's and, in many campaigns, could undercut it.

It was easier for Boeing to attribute those advantages entirely to subsidies than to examine whether Airbus had created a more modern and efficient industrial system. Airbus had directed substantial investment toward contemporary production methods, tooling, automation, and manufacturing infrastructure. Most of its aircraft families were comparatively new. With the exception of the original A300, Airbus' commercial product line in 1997 had been developed during the preceding 15 years. Its factories and production systems had therefore been created around newer technologies and manufacturing concepts. Even the A300 had benefited from continuing modernization.

Boeing, by contrast, had become caught in a cycle of producing new variants of aircraft originally designed decades earlier. Those derivatives remained commercially successful, but many continued to be assembled in factories and on production systems developed for their predecessors.

Before the merger, McDonnell Douglas had been building the MD-80 family on a production system whose origins reached back to the DC-9 of the 1960s. The MD-11 was an extensively modified derivative of the DC-10, and much of its tooling and manufacturing infrastructure traced its lineage to the earlier aircraft.

Boeing faced a similar situation with several of its most important products. The 747-400 was assembled in the same enormous Everett facility built for the original 747, using production concepts, jigs, and tooling whose origins extended back to the late 1960s.

The 737 Next Generation family — the -600, -700, -800, and forthcoming -900 — was being built in Renton, where the original 737 had entered production three decades earlier. Although the aircraft itself had been substantially modernized, much of the surrounding industrial system remained tied to the legacy production line.

The 757, 767, and especially the 777 benefited from more recent tooling, automation, and manufacturing processes because they had been introduced later. But across Boeing's broader production network, aging infrastructure and outdated information systems remained widespread.

The emergence of Airbus should have created a sense of urgency to modernize those capabilities. Instead, Stonecipher and the financially oriented leadership surrounding him believed Boeing could achieve greater efficiency by extracting more productivity from what it already possessed. Completely renovating the factories would have required enormous capital investment. New tooling, computerized production systems, automated machinery, redesigned factory layouts, and upgraded supplier networks would have consumed billions of dollars and reduced near-term earnings. It was easier — and appeared cheaper — to retain the existing system and force it to operate faster. That decision proved disastrous.

Boeing was attempting to more than double production in approximately a year while relying upon factories, information systems, suppliers, and manufacturing practices that were not equipped to absorb such rapid growth. By the late summer of 1997, bottlenecks had appeared throughout its commercial aircraft operations.

The 737 was experiencing the greatest demand. It was also being produced on a system with roots extending back to the original aircraft's introduction in the 1960s. As Boeing forced the Renton operation toward unprecedented output, the line began to stumble.

The problems were particularly severe on the 737 Next Generation and 747 programs, although parts shortages, jobs behind schedule, and out-of-sequence work affected nearly every Boeing commercial aircraft assembled in the Puget Sound region.

Tasks initially expected to fall behind by only a few days slipped by weeks and then months. Airline customers became increasingly angry. Many contracts contained provisions requiring

Boeing to compensate carriers for late deliveries, and customers began threatening to invoke them.

Those penalties added to the financial pressure already confronting management. But perhaps the greater concern was that missed delivery targets would cause Boeing to miss its earnings projections. That, in turn, threatened the company's stock price and the compensation packages of executives whose bonuses depended upon financial performance.

Management continued hoping that the disruption could be contained. But by early October, the unfinished aircraft parked outside the factories and the chaos inside them could no longer be dismissed as ordinary growing pains. On October 3, 1997, Boeing announced an aggressive production recovery plan.

The company stopped portions of the 747 production line for 20 manufacturing days so employees could complete jobs that had fallen behind. On the 737 Next Generation line, aircraft already in production continued moving through the factory, with missing parts installed whenever they became available. However, no additional Next Generation 737s entered final assembly for 25 days.

At the time of the announcement, Boeing acknowledged that parts shortages were the company's most pressing production problem. The shortages affected all of its commercial programs but were especially severe on the 747 and 737 Next Generation. Boeing had expanded output from approximately 18 commercial aircraft per month only 18 months earlier to roughly 40 per month, and the production system had failed to keep pace. The pause allowed employees to reduce the backlog of unfinished work while giving both internal operations and outside suppliers an opportunity to catch up.

The 747 recovery effort reduced recorded parts shortages from more than 500 to fewer than 100 and cut jobs behind schedule by approximately 80 percent. The 737 rebalancing effort reduced both parts shortages and delayed jobs from their peaks, although

Boeing acknowledged that its factories might not return to proper operating balance until well into 1998.

The production crisis coincided with the Asian financial crisis, which began in 1997 and weakened several of the fastest-growing aviation markets in the world. Airlines throughout Asia began reassessing expansion plans, delaying deliveries, and reducing expected capacity growth. The downturn eventually eased some of the pressure on Boeing's production system by reducing near-term demand.

When production finally resumed, it did so under intense scrutiny from managers, airline customers, analysts, and investors. The damage, however, had already been done.

Boeing reported a net loss of $178 million for 1997, compared with earnings of approximately $1.8 billion the previous year. It was the company's first annual in nearly 40 years. The result reflected both the enormous cost of the commercial production breakdown and charges associated with writing down assets and inventory acquired in the McDonnell Douglas merger. The company took billions of dollars in pretax charges related to parts shortages, delayed production, excess labor, traveled work, late deliveries, and the restructuring of inherited McDonnell Douglas programs.

This disaster was not the result of a minor manufacturing interruption. It was a major corporate failure. In the language of sports, it was an unforced error — one produced almost entirely by Boeing's own decisions. The failure was especially striking because it occurred during one of the largest commercial aircraft-ordering booms the industry had ever experienced. Boeing possessed a record backlog. Airline demand was surging, its factories were full, and its principal American competitor had just been absorbed. The company should have been earning record profits. Instead, it stumbled over its own success.

A palpable sense of unease settled across Boeing and the aerospace industry. This was the world's largest aerospace company, an organization renowned for its engineering expertise, manufacturing skill, and ability to manage projects of extraordinary complexity. Yet in less than a year, it had turned unprecedented demand into a production catastrophe and its strongest commercial position in decades into its first annual loss in decades as well. How could so much have gone wrong so quickly — particularly after the extraordinary success of 1996?

Let the Games Begin

By 1997, only two companies remained capable of competing globally in the market for commercial airplanes seating 100 passengers or more, and the term "duopoly" was beginning to enter the commercial aviation lexicon. Russia's Tupolev had developed the Tu-204, and later the closely related Tu-214, as a rough counterpart to Boeing's 757. But neither aircraft achieved certification from U.S. or European aviation authorities, severely limiting its ability to compete for airline customers outside Russia and other traditional Soviet-aligned markets. For all practical purposes, commercial aviation had become a two-horse race between Boeing and Airbus.

At the beginning of 1997, with Boeing in the process of absorbing McDonnell Douglas, many wondered how Airbus could possibly fare against the enormous aerospace company Boeing was becoming. By the end of the year, after Boeing's production-line debacle and Airbus' continued advances in the marketplace, fewer people were wondering whether Airbus could compete.

A useful starting point for measuring Airbus' ascent is 1984. Only seven years had passed since Frank Borman's Eastern Airlines placed its landmark order for the A300, helping save both the aircraft and Airbus from possible extinction. But 1984 was also the year Airbus demonstrated that its foundation had become stable enough to support an expansion of its product line by formally launching development of the A320. In that sense, 1984 marked Airbus' transition from a multinational experiment intended to preserve Europe's commercial aircraft industry into a serious contender for aircraft orders throughout the world.

That year, Airbus captured 48 orders, compared with 135 for McDonnell Douglas and 182 for Boeing. Airbus accounted for only 13 percent of the combined orders placed with the three manufacturers, while McDonnell Douglas captured 37 percent and Boeing claimed the remaining 50 percent.

However, thirteen years later, the balance had changed dramatically. In 1997, the year Boeing completed its merger with McDonnell Douglas, Airbus recorded 460 firm orders while Boeing reported 568, including orders for 17 legacy Douglas aircraft. Airbus had captured nearly 45 percent of the combined market, compared with approximately 55 percent for Boeing. Boeing remained in front, but the margin separating the two companies had narrowed to a degree that would have seemed almost inconceivable in 1984. Airbus would never again be regarded as merely a distant challenger.

By the end of 1999, only three years after Boeing had decisively defeated both Airbus and McDonnell Douglas in the annual orders race — and had absorbed one of those competitors — Airbus surpassed Boeing. It captured approximately 54 percent of the combined orders placed with the two manufacturers.

The year 1999 was a watershed. Airbus did not merely approach parity with Boeing; it won the annual orders race decisively. Over the following quarter-century, it would finish ahead of Boeing in most years and gradually reverse their respective positions in the marketplace. The company that had once struggled for survival would eventually command roughly 60 percent of the market in many measurements, leaving Boeing with approximately 40 percent. It was nothing less than a stunning reversal in the established order of the world's commercial aircraft industry.

The biennial Paris and Farnborough airshows had long served as the aerospace industry's most important international showcases. The Paris Air Show originated with an aviation exhibition held in 1908 and became a separate event the following year. Britain's principal industry airshow moved to Farnborough in 1948. Paris eventually settled into the odd-numbered years and Farnborough into the even-numbered years, ensuring that one of the two major European shows would take place every year.

By the 1960s, the rapid advancement of commercial aviation, military aircraft, missiles, and space technology had transformed the two events into global gatherings for the aerospace industry. Manufacturers used them to display new aircraft, unveil proposed programs, demonstrate prototypes, and meet with airlines, governments, suppliers, and prospective customers from around the world. By the 1990s, the shows had also become theaters in the increasingly public competition between Boeing and Airbus.

Airbus recognized that an aircraft order represented more than a sale. Announced at the right moment and before the right audience, it could also become a powerful demonstration of market momentum. As Airbus gained ground against Boeing, it increasingly used Paris and Farnborough to showcase that progress.

Rather than announce every agreement immediately after it was signed, Airbus sometimes accumulated orders and commitments in the months leading up to an airshow. It could then unveil several deals in rapid succession before the thousands of journalists, airline executives, government officials, and industry analysts gathered at the event. The effect could be dramatic.

Boeing might announce orders steadily from January through June and appear to be building a commanding lead for the year. Airbus would then arrive at Paris or Farnborough and announce commitments covering scores — or even hundreds — of airplanes within a matter of days. The orders had not necessarily been negotiated at the airshow, but announcing them there created the impression that Airbus was sweeping the market before the assembled aerospace world.

Airbus was playing the competitive game on its own terms, and in doing so it was capturing the spotlight. Its publicity operation reinforced the message. Each announcement emphasized that another airline had selected an Airbus aircraft because of its efficiency, technology, passenger appeal, operating economics, or suitability for the carrier's network. Taken together, the announcements portrayed Airbus as energetic, flexible, and increasingly irresistible to the world's airlines.

Boeing generally followed a more traditional approach. It usually announced orders after agreements were completed and the customer authorized their disclosure. Boeing might hold an announcement briefly when a transaction was finalized immediately before an airshow, but it was less inclined to accumulate months of business merely to produce a dramatic week of publicity.

That approach may have been more straightforward, but it surrendered control of the airshow narrative. Journalists arriving at Paris or Farnborough were looking for news, and Airbus repeatedly supplied it. An airline order announced months earlier by Boeing was old news by the time the show opened. An Airbus agreement announced from a crowded press conference at the show became an immediate headline.

As a result, Airbus could dominate coverage even during periods when Boeing's total orders for the year remained competitive. The impression conveyed by the daily headlines was that Airbus was surging forward while Boeing was standing still.

Boeing insisted that it would not manipulate the timing of its announcements merely to manufacture publicity at the European airshows. But the company's refusal to engage Airbus on the same terms did little to improve its image. Instead, it reinforced the growing perception that Boeing was an old and increasingly cumbersome giant — still formidable, but slower to react and less adept at presenting itself to a changing market.

Airbus understood that the contest was being fought not only in airline boardrooms but also before the press, investors, governments, and the broader aviation industry. By concentrating its announcements at Paris and Farnborough, it transformed routine sales disclosures into major international events. Airbus was not simply winning orders. It was making certain the world watched it win them.

Airbus' come-from-behind victory during the late 1990s was not entirely unexpected. The company had methodically and consistently increased its share of the market, and it was not as though executives in Seattle had failed to notice the threat emerging from across the Atlantic.

Whenever Boeing lost an important sales campaign to Airbus — such as its highly publicized loss of the United Airlines order in 1992 — the explanation offered in Seattle frequently centered on the government subsidies Airbus received from its European partners. Those subsidies, Boeing argued, allowed Airbus to develop aircraft with less financial risk and to sell them at prices Boeing could not economically match.

Price was indeed an important factor in United's decision. Boeing acknowledged that it had been unable to match Airbus' offer for the 50 A320s United agreed to lease, with options for 50 more. But after the initial protests surrounding the loss subsided, Boeing and Airbus moved on to the next airline competition, and the issue once again receded into the background.

By the late 1990s, however, Airbus was no longer merely winning occasional sales campaigns. It was threatening to gain the upper hand in the worldwide competition. As that threat became unmistakable — and particularly after Boeing's disastrous performance in the 1999 orders race — the refrain of "government subsidies" grew steadily louder in Seattle.

Around this same time, Boeing also began explaining some prominent sales losses by asserting that it was placing profitability ahead of market share. The company maintained that it would not discount its airplanes merely to win orders and that protecting profit margins would ultimately produce greater value for its shareholders.

That argument may have sounded reassuring to some analysts on Wall Street, but it did little to satisfy aviation-industry observers who watched Boeing lose important customers and strategically significant sales campaigns. In several cases, experts were left wondering whether Boeing was demonstrating financial discipline or simply failing to understand how aggressively Airbus was reshaping the marketplace.

One of the clearest examples came in 2002, when Boeing went to sell its planes to an existing customer in the United Kingdom.

***

Although airline deregulation originated in the United States, the movement spread across the Atlantic during the 1980s and accelerated after the collapse of the Soviet bloc. Europe began liberalizing its airline industry in stages, culminating in a largely deregulated market by the late 1990s.

One of deregulation's inherent advantages was that it gave airlines greater freedom to experiment with their business models and organize their operations according to market forces rather than government mandates. One such experiment, pioneered in the United States by Southwest Airlines, proved enormously successful and eventually transformed commercial aviation throughout the world.

At the heart of Southwest's model was simplicity. The airline flew only one type of airplane: the Boeing 737. Operating a single aircraft type substantially reduced the complexity and cost of its operations. Pilots, mechanics, flight attendants, ground personnel, and cabin-service crews needed to be trained on only one type of aircraft. Maintenance procedures could be standardized, spare-parts inventories simplified, and aircraft and flight crews moved throughout the network with far greater flexibility.

Southwest also operated primarily through a point-to-point network. Rather than funneling passengers through major connecting hubs, it emphasized direct service between cities. Whenever practical, it served secondary airports near major metropolitan areas rather than the congested airports used by the larger carriers. These airports generally charged lower fees and allowed Southwest to move its airplanes through the gate more quickly, reducing both operating costs and turnaround times.

The resulting savings allowed Southwest to offer fares substantially lower than those charged by the established airlines. Its prices were often so low that when Southwest entered a new market, competing airlines reduced their fares to avoid losing passengers. The phenomenon became known as the "Southwest Effect."

Southwest had begun flying in 1971, seven years before Congress deregulated the U.S. airline industry. Because it initially operated solely within the state of Texas, it was not subject to the Civil Aeronautics Board's regulation of interstate airline routes and fares. Once deregulation took effect nationally, Southwest was free to expand its model beyond Texas and across the continental United States.

It was a scrappy carrier whose success was built on simplicity — one aircraft family, open seating, limited cabin service, rapid aircraft turnarounds, and a point-to-point network. Flight attendants often assisted with cleaning the cabin so the aircraft could be boarded again as quickly as possible.

Southwest's business model was designed primarily to attract travelers who might otherwise drive rather than fly. Herb Kelleher, one of the airline's founders, frequently explained that Southwest's true competition was not necessarily another airline but the automobile. His objective was to persuade a family considering a five or six-hour drive to visit relatives that flying could be almost as affordable and considerably faster.

It was a brilliant business model that produced decades of growth and an extraordinary record of profitability. By the end of

2019, Southwest had reported 47 consecutive profitable years, a streak that continued through recessions, fuel-price shocks, and the disruption that followed the terrorist attacks of September 11, 2001. It did not report an annual loss until 2020, when the COVID-19 pandemic brought much of the world's airline industry to a standstill.

Southwest's success did not go unnoticed. Its model was soon emulated in other parts of the world, helping place air travel within reach of millions of people who previously could not afford it. Many of the airlines that copied Southwest attempted to reproduce its operating model in as much detail as possible. That frequently included choosing the same aircraft.

When Southwest prepared to begin operations in 1971, the Boeing 737 and the McDonnell Douglas DC-9 were the principal aircraft it considered for the short-haul markets it intended to serve. The MD-80, the enlarged and improved derivative of the DC-9, would not become available until the end of the decade. Southwest ultimately acquired three Boeing 737-200s that had originally been intended for other airlines, receiving favorable pricing and financing terms from Boeing.

With Southwest's selection of the 737, the aircraft became inseparable from the operating model. And that operating model would eventually come to be known as that of the low-cost carrier, or LCC. As other low-cost airlines began operations around the world, many selected the 737 because Southwest had already demonstrated how the aircraft could be used to support an efficient, high-utilization operation.

Throughout much of the 1990s, Boeing remained the preferred supplier to the expanding low-cost-airline sector. The 737 Classic family — the 737-300, -400, and -500 — was still rolling off the assembly line in Renton in substantial numbers. As a result, many emerging low-cost carriers began flying the 737 even though

Airbus' newer and more technologically advanced A320 had already entered service.

Most of these airlines started with relatively small fleets, sometimes numbering fewer than a dozen aircraft, but their business plans anticipated rapid expansion. Once successful, they could add dozens of airplanes as they opened new bases and expanded their route networks.

The 737 Next Generation family — the 737-600, -700, -800, and -900 — began entering airline service in late 1997. Boeing planned to end production of the Classic series by 2000 as customers transitioned to the newer models. One of Boeing's most important promises to its existing 737 customers was that the Next Generation aircraft would share a common pilot type rating with the earlier 737 models. Although the new airplanes had more efficient engines, redesigned wings, modernized cockpits, and performance comparable to that of the Airbus A320 family, pilots already qualified on the 737 could transition to the newer models with relative ease and minimal familiarization training.

A pilot could, therefore, fly a 737 Classic on one trip and a 737 Next Generation aircraft on another. The two generations also shared substantial systems and parts commonality, mitigating the disruption to maintenance operations, technical training, and spare-parts inventories when an airline upgraded its fleet.

Boeing was confident that low-cost carriers operating the 737 Classic would naturally transition to the 737 Next Generation as their fleets expanded. If an airline were to consider switching to the Airbus A320 family, it would require the airline to introduce a second aircraft type, at least during the years-long period in which its older Boeing aircraft were being phased out. That would mean separate training programs, pilot pools, maintenance procedures, spare-parts inventories, and technical-support systems — the very kind of complexity the low-cost model was designed to avoid.

Boeing had good reason to believe that its existing customers were unlikely to switch manufacturers. But Boeing may have become too confident.

One of the airlines that adopted Southwest's model was easyJet, which began flying from London's Luton Airport in 1995. Like Southwest, easyJet built its early operations around the Boeing 737. By 2002, particularly after its acquisition of Go Fly, the airline controlled a fleet of approximately 50 aircraft that included both 737 Classics and newer 737 Next Generation models.

EasyJet was preparing for a massive expansion throughout Europe and announced its intention to acquire more than 100 additional airplanes. Almost everyone expected the order to go to Boeing. EasyJet's chief executive, Ray Webster, later acknowledged that Boeing was the natural choice.

Nevertheless, Webster was determined to negotiate the best possible terms. So, he invited Airbus to submit a proposal for the A320 family, both to evaluate the aircraft but primarily to place additional pressure on Boeing. Boeing understood why Airbus had been invited, but its representatives apparently believed that the European manufacturer had little chance of winning. EasyJet already operated dozens of Boeing aircraft. The enormous cost of introducing an entirely different airplane would outweigh any discount Airbus could possibly offer on the purchase price.

And Webster clearly understood the economics. Even if Airbus offered the A319 at a substantially lower price than Boeing's 737-700, easyJet would still incur the expense of operating two aircraft families during the years-long transition until the Boeings were phased out. It would need additional pilot and mechanic training, new maintenance programs, different spare-parts inventories, and a separate technical-support structure.

But easyJet was a low-cost airline, and Webster had an obligation to ensure that the largest aircraft purchase in the company's history up until then was negotiated as aggressively as every other part of the airline's operation. Boeing would not receive the order merely because it was the incumbent supplier.

Boeing had additional reasons to feel secure. Southwest Airlines and Ireland's Ryanair — two of the world's most prominent low-cost carriers — had placed enormous orders for the 737 Next Generation. EasyJet was already operating approximately 50 Boeing aircraft. From Boeing's perspective, abandoning the 737 would be commercially irrational.

What followed sent a shock through the commercial aviation industry.

Airbus had been watching the rapid expansion of low-cost airlines throughout the world. In Asia and other developing regions, growing middle classes were creating vast new markets for inexpensive air travel. Low-cost carriers were likely to account for a large share of future narrow-body demand, yet Boeing maintained a commanding position in that market.

Airbus was determined to break Boeing's hold. So, it offered easyJet an extraordinarily aggressive package. The precise price paid for the aircraft was never publicly disclosed, but contemporary estimates suggested that easyJet obtained the A319s for approximately half their published list price. Deep discounts alone, however, would not eliminate the cost of introducing a second aircraft family.

Airbus went further. It agreed to support easyJet in meeting some of the costs associated with introducing the A319, reducing the financial burden of pilot training, technical preparation, and operating a dual fleet during the transition. EasyJet concluded that the value of Airbus' overall proposal more than compensated for the complexity of operating two aircraft types.

On October 14, 2002, easyJet announced that it had selected Airbus for 120 A319s, with price-protected options for another 120 aircraft. The A319 was the smaller member of the A320 family and competed most directly with Boeing's 737-700.

The announcement stunned Boeing as well as most of the aviation community. Airbus had persuaded one of Europe's most prominent Boeing operators to abandon the assumption that a low-cost carrier needed to follow the Southwest model by flying the 737.

Airbus may have accepted little profit on the easyJet order, and it may even have been willing to sustain losses on parts of the package. But the immediate profit on those 120 airplanes was not the only consideration. Airbus had broken the Boeing-only pattern that had shaped the low-cost-carrier market throughout the world, and it had secured a strategically important customer with enormous growth potential.

Boeing responded by arguing that Airbus' government support allowed it to offer prices that a privately financed American manufacturer could not economically match. Boeing also maintained that it had submitted a highly competitive proposal but would not reduce its price to a level that diminished the long-term value of the 737 or sacrificed an acceptable return. That explanation reinforced the impression that Boeing had placed margin preservation ahead of defending a strategically important market position.

Airbus had an obvious reason to price its aircraft aggressively — it needed to establish itself in the low-cost-carrier market. But Boeing had an equally compelling reason to compete aggressively. It needed to protect the position it already held.

Had Boeing retained easyJet, it might have supplied the airline's aircraft needs for decades. Instead, the 2002 agreement began a relationship under which Airbus became easyJet's principal — and exclusive — supplier of large commercial aircraft. The original order was followed by repeated purchases as easyJet expanded, replaced its early A319s, and introduced larger A320 and A321 variants.

Whatever profit Airbus may have surrendered on the original transaction was dwarfed by the strategic value of the relationship it established, not to mention the profit it made on subsequent orders from the airline.

The easyJet agreement proved to be a watershed for Airbus and helped establish the methods it would use to penetrate other strategically important fleets. The company's approach was straightforward: Price the initial aircraft aggressively, help the airline absorb the costs of introducing them, and establish an operational foothold. Once the aircraft had been integrated into the fleet, the same economics of commonality that had once protected Boeing would begin protecting Airbus.

An airline operating A320-family aircraft would need more airplanes as it grew. It would eventually need replacements for its earliest aircraft. It might acquire another airline or be acquired itself, creating further opportunities to expand or standardize the fleet. The initial sale was therefore not an isolated transaction. It was the beginning of a relationship that could produce orders for decades.

Airbus also worked closely with easyJet to demonstrate that a large and efficient low-cost carrier could be built around the A320 family. It would then use easyJet as proof that the Southwest model did not require the Boeing 737. Airbus was developing its own playbook for the low-cost-airline market.

One might have expected Boeing to reassess its approach after losing easyJet. Instead, the company appeared to reinforce its emphasis on maintaining pricing discipline and protecting shareholder returns. That philosophy had become increasingly visible since the merger with McDonnell Douglas and reflected the financial priorities associated with Harry Stonecipher and the new Boeing leadership team.

The strategy was simple: Sell airplanes at acceptable margins rather than pursue market share at any cost. If Boeing lost an occasional sales campaign, there would be other opportunities in a rapidly expanding commercial-aircraft market.

But the loss of easyJet demonstrated the weakness in that reasoning. An aircraft sale was not merely a one-time transaction. Winning the initial order could determine which manufacturer supplied an airline for the next several decades. Airbus was willing to accept a smaller immediate return to capture that continuing stream of business. Boeing appeared more concerned with short-term profitability.

The easyJet campaign was not the first time Airbus had demonstrated the effectiveness of that approach. A few years earlier, Airbus had broken Boeing's grip on British Airways. The airline had become such a prominent Boeing customer that some people jokingly referred to it as "Boeing Airways." British Airways did operate a small number of A320-family aircraft inherited through its acquisition of British Caledonian, but Boeing remained its dominant aircraft supplier.

In 1998, British Airways placed an order for 59 A320-family aircraft, with options for as many as 129 more. The decision gave Airbus a major victory at the flag carrier of the country that had once been one of the founding partners in the Airbus consortium. Boeing again pointed to European government support, arguing that subsidies enabled Airbus to offer prices it could not otherwise sustain.

Other losses followed. Frontier Airlines, based in Denver, was another emerging low-cost carrier operating an all-Boeing fleet. As part of its expansion plans, it requested proposals from both manufacturers. Airbus ultimately won the competition, and Frontier began taking delivery of A319s in 2001. It later became the launch customer for the smaller A318 and eventually retired its remaining 737s, becoming an all-Airbus operator.

Once again, Boeing attributed the loss in part to Airbus' ability to offer unusually aggressive financial terms. It also emphasized that Boeing intended to preserve the value of its products and generate acceptable returns for shareholders.

To the airline industry, however, a pattern was emerging. With every prominent sales campaign Airbus won during the late 1990s and early 2000s, its market share increased and another customer was introduced to its product line. Every successful introduction strengthened its global training, maintenance, financing, leasing, and technical-support network. The argument that airlines would hesitate to operate Airbus aircraft became progressively less persuasive because more airlines were already operating them successfully.

Boeing countered that Airbus might be winning more annual orders, but Boeing continued to manufacture and deliver more airplanes. The distinction mattered. Airlines paid most of an aircraft's purchase price as it approached delivery, not when the order was first announced. Boeing also entered this period with a much larger backlog accumulated during earlier years, allowing it to maintain higher production and delivery totals even as Airbus began winning more new business. Boeing could therefore continue calling itself the world's largest commercial-aircraft manufacturer based on the number of airplanes delivered.

But the company's leaders surely understood that this advantage could not continue indefinitely. If Airbus consistently booked more orders than Boeing, those orders would eventually move through its production system and become deliveries. At some point, Airbus would catch Boeing in the measurement Boeing preferred as well.

Perhaps Boeing expected Airbus to stumble under the strain of its expansion. Perhaps it believed that its own next major aircraft program would restore the competitive advantage it had enjoyed for decades. But while Boeing waited, Airbus continued selling airplanes.

***

Did government subsidies actually allow Airbus to sell its airplanes at lower prices?

The answer was more complicated than Boeing's public statements sometimes suggested. Government support unquestionably played an important role in allowing Airbus to launch new aircraft. The principal mechanism was known as launch aid, or later as repayable launch investment. Under the terms established by the European governments backing Airbus — principally France, Germany, the United Kingdom, and Spain — government financing could cover as much as one-third of the development cost of a new aircraft program.

If the development of a new airplane cost $3 billion, for example, Airbus might receive as much as $1 billion from its sponsoring governments. Instead of repaying the financing through fixed principal-and-interest installments, Airbus generally made royalty payments based on the number of airplanes it sold.

The precise repayment terms varied from program to program. But payments would continue after the original amount advanced by the governments had been recovered with each sale of an aircraft. This meant that a successful program could produce a positive return for the governments involved. Airbus could therefore argue that launch aid was not necessarily a taxpayer giveaway and could actually prove beneficial to the taxpayer. When an aircraft sold well, the government could recover its original investment and earn a return long into the future. That argument was particularly persuasive when applied to a successful program such as the A320 family which, 40-plus years on since its launch with government launch aid, it is still selling incredibly well and continuing to provide a healthy return to its sponsor governments.

But launch aid was still considerably more advantageous to Airbus than an ordinary commercial loan would have been. The financing was generally provided on terms that Airbus could not have obtained from private lenders. Repayment depended heavily on the commercial success of the aircraft. If sales fell below expectations, the amount repaid could be reduced, delayed, or left unpaid. Airbus was not necessarily required to make up the shortfall from the profits of its other aircraft programs.

Nor did the financing carry the same rigid repayment schedule as conventional commercial debt. If development problems delayed an aircraft's entry into service, Airbus did not face the same immediate principal-and-interest payments it might have faced under a conventional bank loan or bond offering. The practical effect was that European governments assumed a meaningful share of the financial risk associated with developing a new airplane.

From Boeing's perspective, this distorted the market. Launching an entirely new commercial aircraft was one of the largest and riskiest investments any industrial corporation could undertake. Boeing had to commit billions of dollars years before the first airplane would be delivered, with no guarantee that airlines would order enough aircraft to recover the development cost.

Airbus faced those same engineering and commercial risks, but launch aid transferred part of the financial exposure from Airbus to its sponsoring governments. If an aircraft succeeded, Airbus repaid the financing through royalties. If the program failed to generate the expected sales, the governments bore part of the loss.

Boeing had a legitimate argument that launch aid enabled Airbus to undertake aircraft programs on terms that were not commercially available to Boeing. It reduced Airbus' cost of capital, limited its downside exposure, and allowed it to develop aircraft sooner — or with less financial risk — than it otherwise might have been able to do.

The World Trade Organization would later validate much of that argument. In the case brought by the United States against the European Union, the WTO found that Airbus had received launch aid and other subsidies on noncommercial terms. It concluded that the financing shifted a significant portion of the risk of launching new aircraft from Airbus to the governments providing the money.

The WTO went further. It found that Airbus would not have been able to launch and bring to market the same aircraft when it did without the government assistance it received. The subsidies therefore contributed materially to Airbus' emergence as a full-scale competitor to Boeing and to sales Airbus won at Boeing's expense.

But the WTO did not validate every part of Boeing's argument. Most importantly, it rejected the claim that the United States had demonstrated that launch aid allowed Airbus to reduce the prices it charged airlines during the period examined. The subsidies strengthened Airbus' ability to develop aircraft and compete in the market, but the WTO did not find sufficient evidence that they directly caused Airbus to offer lower prices. That distinction was critical.

Government financing helped place Airbus in the competition. It reduced the risk and cost associated with creating the aircraft Airbus was offering. But that did not automatically prove that Airbus' price in a particular sales campaign — easyJet, British Airways, Frontier Airlines, or any other customer — was lower because of launch aid.

By the early 2000s, Boeing also argued that the original industrial-policy justification for supporting Airbus had expired. Airbus was no longer a struggling European consortium attempting to establish itself against dominant American manufacturers. It had reached approximate parity with Boeing in annual orders and had access to international capital markets. Boeing maintained that Airbus should finance future projects on ordinary commercial terms rather than continue relying on government-backed launch investment.

Airbus countered that Boeing had little standing to complain about government assistance. For decades, Boeing had received billions of dollars through U.S. military, space, and research contracts. These contracts often included substantial research-and-development work involving advanced aerodynamics, materials, manufacturing methods, electronics, propulsion, and systems integration. Some of the knowledge and technical capabilities developed through government-funded programs had also benefited Boeing's commercial aircraft business.

Airbus argued that the American government had, in effect, paid for research and technological development from which Boeing later profited in the commercial market. Because many defense and space contracts were restricted to American companies, Airbus and its European predecessors had no opportunity to compete for them.

Boeing also received federal tax benefits and state and local incentives. Washington State and municipalities near Boeing's manufacturing facilities provided tax reductions and other incentives connected with aircraft production. Other states offered their own packages to persuade Boeing to locate assembly plants and related operations within their jurisdictions.

In Airbus' view, those benefits reduced Boeing's costs just as European government financing reduced the financial burden on Airbus.

The dispute eventually produced two enormous and complicated cases before the World Trade Organization: one brought by the United States against European support for Airbus and another brought by the European Union against American support for Boeing.

Between 2011 and 2012, in separate rulings the WTO found merit in both sides' complaints. It ruled that Airbus had received subsidies through launch aid, government equity contributions, infrastructure support, and certain research programs. Some of those subsidies caused adverse effects to Boeing by contributing to lost sales and the displacement of Boeing aircraft from certain markets.

In the parallel case, the WTO found that Boeing had also received prohibited or otherwise actionable subsidies. These included federal tax treatment, research-and-development assistance associated with NASA and the Department of Defense, and state and local tax incentives — most notably those offered by Washington State.

The WTO's ruling effectively found both manufacturers had benefited from forms of government support that violated international trade rules or caused adverse effects to the other.

But it would be inaccurate to say that the WTO determined that the two companies had benefited to precisely — or even approximately — the same extent. The two disputes involved different forms of assistance, different periods, different aircraft programs, and different market consequences. The WTO assessed whether particular measures violated trade rules and caused adverse effects; it did not produce a single balance sheet declaring the subsidies to Airbus and Boeing economically equivalent.

The scale of the authorized retaliation also differed. In 2019, the WTO authorized the United States to impose countermeasures of approximately $7.5 billion annually because of the effects of European subsidies to Airbus. In 2020, it authorized the European Union to impose countermeasures of approximately $4 billion annually because of American subsidies to Boeing. Those figures were measures of permitted retaliation under the circumstances examined, not a complete accounting of every benefit either company had received throughout its history.

The broader conclusion was nevertheless unmistakable: Neither side entered the competition with entirely clean hands. Airbus benefited from government financing that reduced the risk of developing new airplanes and enabled it to bring aircraft to market when it otherwise could not have done so. Boeing benefited from tax incentives and government-funded research that strengthened its technical capabilities and reduced some of its costs.

Boeing's complaints about launch aid had a legitimate foundation. Airbus had received valuable financial assistance on terms unavailable in the private market. But Boeing's repeated suggestion that every lost sales campaign resulted from Airbus using subsidies to underprice it went beyond what the evidence — and eventually the WTO — established.

Indeed, the WTO specifically rejected the claim that the United States had demonstrated that launch aid allowed Airbus to suppress aircraft prices during the period it examined. Airbus' government support helped make the company and its airplanes formidable competitors. It did not excuse Boeing's failure to recognize the strategic importance of customers, respond creatively to Airbus' sales packages, or address its own production costs and organizational weaknesses.

Subsidies helped Airbus reach the negotiating table. They did not necessarily explain why Boeing walked away from it empty-handed.

***

As the world approached the dawn of a new century, Boeing and Airbus — now increasingly regarded as equals — turned their attention toward what commercial flight might look like in the decades ahead.

During the first half of the 1990s, Airbus had filled out its product portfolio with the twin-engine A330 and four-engine A340. Together with the A320 family and the original A300 and A310, those aircraft allowed Airbus to compete with Boeing across nearly every major segment of the commercial-aircraft market, from short-haul narrow-body routes to long-haul international service. Only one important segment remained beyond Airbus' reach: the very-large, high-capacity, long-range market dominated by Boeing's 747.

The 747 had become, in many respects, a cash cow for Boeing. If an airline needed an aircraft capable of carrying approximately 400 passengers over long distances, there was effectively only one airplane available. Boeing could charge a premium because it faced no direct competitor.

Aircraft designed for that part of the market came to be described as very large aircraft, or VLAs. They were intended to carry unusually large numbers of passengers over high-density, long-haul routes, particularly between major international hubs. Their economic appeal rested on the ability to spread the cost of a flight among hundreds of passengers.

The 747 was the only established VLA in commercial service. Moreover, most of the original aircraft's development cost had been recovered decades earlier. Although Boeing had continued investing in improved versions, the basic program had long since crossed the financial threshold that had made the original 747 such an enormous corporate gamble. Each additional sale had the potential to generate an attractive return.

For all the progress Airbus had made against Boeing, its leaders believed that the company would not possess a genuinely comprehensive product line until it could compete in the market occupied by the 747. As early as 1990, Airbus was publicly discussing the possibility of developing an aircraft larger than anything in its existing portfolio. The idea evolved into what Airbus initially called the A3XX.

By the middle and late 1990s, global air travel was expanding rapidly. Airbus forecast that many of the world's largest hub airports would eventually reach the limits of their runway and gate capacity. Few entirely new airports were being constructed near the cities experiencing the greatest increases in traffic, while adding runways to existing airports was often politically difficult or physically impossible because urban development had surrounded them.

Many of the most important airports would therefore become increasingly constrained. At slot-constrained airports, an airline could not simply add another flight whenever passenger demand increased. Takeoff and landing opportunities were finite, particularly during the most desirable hours of the day. Airbus concluded that larger airplanes would provide the most practical means of increasing passenger capacity at those airports.

The principle was straightforward. If an airport could accommodate 20 aircraft movements during a given period and each airplane carried 250 passengers, those flights could move a combined total of 5,000 people. If the same number of aircraft carried 500 passengers each, the airport could move 10,000 people without adding another takeoff or landing slot.

Airbus envisioned a future in which international traffic would continue to flow through a relatively small number of enormous hubs. Airlines would use very large airplanes to move passengers between those hubs and then distribute them to their final destinations aboard smaller aircraft.

Boeing examined the same market and reached a fundamentally different conclusion. Its analysts believed that air travel was becoming increasingly fragmented. Passengers preferred flying directly to their destinations rather than traveling through a major hub and changing airplanes. Advances in aircraft range and efficiency were allowing airlines to open direct routes between cities that previously could not generate enough traffic to support a large four-engine airplane.

Boeing's 767 had played an important role in that transformation. As regulators approved extended-range operations by twin-engine aircraft, airlines were able to use the 767 across the Atlantic and on other long overwater routes that had previously required airplanes with three or four engines. The 777 carried the trend further by combining long range, substantial capacity, and twin-engine efficiency. These aircraft allowed airlines to bypass traditional hubs and connect a growing number of cities directly.

Boeing believed this fragmentation of the market would reduce the need for extremely large airplanes. Instead of concentrating ever-greater numbers of passengers aboard a few enormous aircraft flying between major hubs, airlines would divide those passengers among a larger number of direct routes served by smaller long-range twin-engine airplanes. Fragmentation would not eliminate congestion at the largest airports, but it could redirect some traffic away from them. A passenger traveling from a secondary American city to a secondary European or Asian city might no longer need to connect through New York, London, Frankfurt, Tokyo, or another major international gateway.

The two manufacturers had arrived at sharply different visions of the future. Airbus believed continued growth would reinforce the hub-and-spoke system and create demand for airplanes substantially larger than the 747. Boeing believed passengers and airlines would increasingly favor direct service aboard smaller, efficient twin-engine aircraft.

Which company had interpreted the market correctly? In truth, neither forecast was absolute. Airbus was not claiming that every international passenger would travel through a hub, nor was Boeing arguing that the market for very large aircraft would disappear entirely. Both companies recognized that the future would contain elements of each model. The disagreement concerned which trend would dominate and how large the market for a new VLA would ultimately become.

Airbus also had a strategic reason to favor the development of a very large airplane. It wanted to complete its product portfolio and challenge Boeing in the last segment where the American company retained an uncontested monopoly.

Boeing, for its part, was not entirely dismissive of the VLA market. It could see that certain densely traveled routes would continue to require aircraft capable of carrying 400 or more passengers. It also knew Airbus was moving steadily toward launching an airplane that would enter the 747's territory. Boeing's concern was not that the market for very large airplanes would vanish. It was that the market would remain too small to justify the enormous cost of developing an entirely new aircraft.

A clean-sheet VLA could require an investment of between $8 billion and $10 billion in mid-1990s dollars. Depending on the price and profitability of each aircraft, several hundred airplanes might have to be sold before the manufacturer recovered its development cost. Boeing's forecasts suggested that demand might not be sufficient to support two competing aircraft programs of that size, let alone one.

The recent performance of the 747 appeared to reinforce Boeing's skepticism. Boeing had formally launched the 747-400 in October 1985, and the first aircraft entered airline service with Northwest Airlines in February 1989. It became the most commercially successful version of the 747 and benefited from strong replacement demand among airlines operating earlier variants.

Many customers purchased the 747-400 to replace aging 747-100s, -200s, and -300s rather than to open entirely new markets. The aircraft also benefited from the rapid growth of Asian airlines and international traffic during the late 1980s and early 1990s. Orders for the 747 reached their historic peak in 1990, when Boeing recorded 122 net orders for the family. But that surge proved temporary. Demand declined sharply during the recession and airline downturn that followed the Gulf War, and it never returned to its 1990 level.

The booming airline market of the middle 1990s produced a temporary recovery. In 1996, Boeing recorded 75 gross orders for the 747, although cancellations reduced the net total to 53. Even that improvement remained far below the extraordinary demand recorded in 1990, and orders fell dramatically during the following years.

The slowdown suggested that the market for very large passenger aircraft was limited, especially since Boeing believed fragmentation was replacing much of the need for VLAs. It also suggested that many airlines had already replaced the older 747s they intended to replace.

At the same time, the 747's underlying design was growing old. Although the 747-400 itself had been in service for less than a decade, it remained an evolution of an airframe conceived during the 1960s. By the middle of the 1990s, the basic architecture of the airplane was approximately 30 years old. Its systems, engines, cockpit, and aerodynamics had been substantially improved, but the fundamental configuration remained unchanged.

But as the old business adage warned, it was better for a company to render its own product obsolete than to allow a competitor to do it. Therefore, Boeing examined ways to preserve its position at the high end of the market without assuming the expense and risk of developing an entirely new aircraft.

In 1996, the company unveiled studies for two much larger 747 derivatives, designated the 747-500X and 747-600X. Boeing presented the proposals publicly at the Farnborough International Airshow that September.

The 747-500X was intended to combine substantially greater range with a modest increase in passenger capacity. Depending on the cabin configuration under discussion, it would have carried approximately 460 passengers and flown as far as 8,500 nautical miles. The longer 747-600X would have carried approximately 540 to 550 passengers and offered a range of roughly 7,700 nautical miles. It would have approached the capacity Airbus was proposing for the A3XX while retaining the familiar 747 configuration.

Both aircraft would have incorporated a new wing using technology derived from the 777, along with updated engines, flight-deck systems, and passenger accommodations. The proposed changes were so extensive that the program would have been considerably more than a routine stretch of the 747-400. Boeing conducted aerodynamic testing of the designs during 1996 as it attempted to determine whether sufficient airline interest existed.

Boeing estimated that developing the two derivatives would cost more than $5 billion — far less than the probable cost of an entirely new Airbus VLA, but still a substantial investment. Because Boeing would be building upon an existing aircraft, it believed it could bring the enlarged 747s to market sooner and at lower cost than Airbus could develop the A3XX. The derivatives also gave Boeing a potential defensive weapon. If airlines preferred an enlarged and modernized version of the proven 747, Airbus might conclude that there was insufficient room in the market for its proposed aircraft. Keeping Airbus out of the VLA market would preserve Boeing's monopoly without requiring it to undertake a completely clean-sheet design.

But airline reaction to the 747-500X and 747-600X was lukewarm. Some potential customers questioned whether the derivatives offered enough improvement over the 747-400 to justify their cost. Others were dissatisfied as Boeing repeatedly modified the proposals. Lufthansa, one of the airlines most closely involved in evaluating the aircraft, publicly criticized changes that moved the design away from what it wanted.

Boeing failed to secure sufficient commitments and suspended the 747-500X and 747-600X studies in early 1997. From Boeing's perspective, the lack of airline enthusiasm confirmed its assessment that the market for very large aircraft was too limited to support the investment. The airlines had been offered a larger, longer-range version of an airplane they already knew, yet none had been willing to make the commitments necessary to launch it.

Airbus interpreted the same evidence differently. Its executives believed the decline in 747 orders did not necessarily mean airlines had lost interest in very large aircraft. Instead, they concluded that the market had not yet matured and that airlines were waiting for a substantially better airplane.

In Airbus' view, the greatest capacity constraints at major airports were still years away. Airlines did not yet need hundreds of

VLAs, but continuing growth would eventually create that need. When it did, the industry would require an aircraft designed specifically for the demands of the twenty-first century rather than another derivative of an airplane conceived during the 1960s.

Airbus believed that airport congestion could best be addressed by moving more passengers on each flight, while Boeing increasingly believed traffic growth would be distributed among a larger number of direct routes. Those competing forecasts became the foundation of the manufacturers' next major product decisions. Airbus itself later described the A380 as its answer to overcrowded hubs and the need to transport more passengers with fewer aircraft movements.

Time would determine which company had interpreted the future more accurately. In the meantime, Airbus and Boeing set off along divergent paths in search of commercial aviation's next great airplane. The contest would involve more than engineering. Each company was going to be wagering billions of dollars on its prediction of how the world would choose to fly. But, in the meantime, both companies used every opportunity to tout their vision of the future.

Building a Better Mousetrap

As the page turned on the old century and a new one began, Boeing appeared to have finally digested the enormous pill that was McDonnell Douglas. The acquisition had given Boeing additional products, facilities, defense programs, and technical capabilities, helping make an already dominant aerospace company even larger. But in commercial aviation, Airbus had emerged as a direct and increasingly dangerous threat. By the end of the 1990s, it had surpassed Boeing in annual orders, capturing the symbolic crown in the yearly sales race even though Boeing continued to deliver more airplanes.

Airbus' product lineup — and, in particular, the technological architecture around which its aircraft were built — was resonating with airlines around the world. Its highly automated, fly-by-wire cockpits offered substantial commonality among different models, allowing pilots trained on one Airbus aircraft to transition more easily to another. Airlines operating several Airbus types could reduce training expenses, simplify crew scheduling, and standardize many of their cockpit procedures.

By the early 2000s, Airbus was preparing to phase out the A300, its original widebody aircraft dating back to the 1970s. Its principal product families were now the A320, A330, and A340, all of which shared similar design philosophies, cockpit layouts, flight-control systems, and operating characteristics. Yet each served a different segment of the market.

The single-aisle A320 family included the smaller A318 and A319, the standard A320, and the stretched A321. The twin-engine, widebody A330 was offered as the smaller A330-200 and the longer A330-300. The four-engine A340 family included the A340-200, A340-300, ultra-long-range A340-500, and stretched A340-600. Through these derivatives, Airbus could offer airlines a broad range of passenger capacities, operating ranges, and performance capabilities without forcing them to abandon the common technological platform that tied the aircraft together.

Boeing's commercial lineup consisted of the 737, 747, 757, 767, and 777. Boeing offered more distinct models than Airbus, along with numerous shortened, stretched, and extended-range derivatives. But each Boeing family had been developed largely independently, during a different technological era. With the exception of the 777, Boeing's principal aircraft had been conceived at least two decades earlier. The 737 dated to the 1960s, while the 747, 757, and 767 traced their origins to the 1960s and 1970s. Although all had been substantially updated over the years, their underlying designs still reflected the periods in which they had originated.

The leadership in Seattle, now dominated by Philip Condit and Harry Stonecipher, had a significant challenge on its hands — and a less-than-stellar image to overcome. Airline executives were publicly complaining that Boeing appeared to be slowing down and was no longer the innovative powerhouse it had once been. The company that had repeatedly transformed commercial aviation now seemed to be relying on derivatives of existing products while Airbus captured an increasing share of the market with a newer and more integrated family of airplanes.

Clearly, Boeing needed to do something to counter both Airbus and the growing perception that it had lost its technological edge. It needed a new product that could retake the initiative and demonstrate that Boeing remained the great innovator of commercial aviation — that it still possessed the ingenuity and daring that had made it a legend.

Boeing had anchored the future of its commercial-airplane business around the 777. From the beginning, the aircraft had been designed with enough structural capacity, aerodynamic efficiency, and technological flexibility to remain adaptable for decades. Its design allowed Boeing to introduce progressively larger, heavier, and longer-range versions as engines and other technologies improved. The company had wagered that the future of long-distance air travel would revolve around large twin-engine airplanes, and it had designed the 777 to remain relevant well into the twenty-first century.

But Airbus also offered a modern twin-engine widebody in the form of the A330. Moreover, because the competing manufacturers were forced to satisfy many of the same operating requirements and within the same aerodynamic constraints, their airplanes inevitably performed many of the same missions. And to the untrained or unfamiliar eye, there was often little obvious difference between competing Boeing and Airbus models parked alongside one another on an airport ramp.

The differences that mattered most to airlines were generally found beneath the surface — acquisition price, fuel consumption, maintenance costs, crew-training requirements, fleet commonality, reliability, financing arrangements, delivery availability, and the long-term support offered by the manufacturer. Boeing and Airbus could each emphasize particular advantages, but when two competing aircraft offered similar capabilities, aggressive pricing and financing could become decisive.

The year 2000 would become a watershed for both manufacturers. Airbus was betting that the next major phase of international air travel would favor very large aircraft operating between the world's largest hub airports. Since the mid-1990s, it had been accelerating work on its very large aircraft program that it had tentatively designated the A3XX, with the "XX" serving as a placeholder for the numerical designation that would eventually be assigned.

Airbus' market studies had predicted demand for approximately 1,200 very large aircraft over the following twenty years. As the program advanced, its forecasts would grow even more optimistic, eventually approaching 1,500 to 1,700 aircraft. According to Airbus, increasing congestion at major international airports would compel airlines to carry more passengers on each available takeoff and landing slot. The solution was not simply to operate more flights, because many of the world's largest airports would not have the capacity to accommodate them. The solution, Airbus believed, was to build a substantially larger airplane.

On December 19, 2000, Airbus formally launched the A3XX program. Several customers had already agreed to support the new aircraft, giving Airbus sufficient confidence to proceed with what would become the most ambitious commercial airplane development program in European history.

In Seattle, Boeing's leadership was left scratching its collective head. Orders for the existing 747-400 had slowed considerably, and Boeing's proposals for larger derivatives — collectively referred to as the 747X — had generated little enthusiasm among airlines. Based on that experience, Boeing concluded that Airbus was wildly exaggerating the size of the market for very large aircraft, perhaps in part to justify the enormous development expenditure and government-backed financing that the program would require.

Boeing's own analysis projected demand for only about 300 very large aircraft over the next twenty years. The difference between the two forecasts was enormous. Airbus saw a market large enough to justify an entirely new aircraft program. Boeing saw a limited and declining niche that could be served by improved versions of the 747. Both visions could not be correct. One of the two manufacturers was making a monumental strategic mistake.

Airbus expected much of the demand to come from Asia. Air travel throughout the region was expanding rapidly as economic growth created a larger middle and upper class with the means to travel internationally. China and India, with their enormous populations and rapidly developing economies, appeared likely to generate vast numbers of passengers traveling to destinations throughout Asia, Europe, and the Americas. Congested airports in cities such as Tokyo, Hong Kong, Singapore, and eventually Beijing and Shanghai appeared to provide precisely the operating environment for which the new aircraft was intended.

Airbus would officially designate the airplane the A380, leapfrogging the unused A350, A360, and A370 designations. The number eight was considered auspicious in China and elsewhere in Asia, the region Airbus believed would become one of the aircraft's most important markets. The designation also visually suggested the aircraft's distinctive double-deck cross-section.

Yet the customers that initially committed to the A380 were not Chinese carriers. They included Emirates, then an ambitious and rapidly expanding Middle Eastern airline; Air France; Qantas; Singapore Airlines; Virgin Atlantic; and International Lease Finance Corporation, one of the world's largest aircraft-leasing companies. Their commitments gave Airbus a diverse launch group representing Europe, Asia, Australia, the Middle East, and the global aircraft-leasing industry.

With the launch of the A380, Airbus dominated the world's aviation headlines. The company portrayed the new aircraft as the twenty-first-century equivalent of the 747. Just as the 747 had become the defining commercial aircraft of the jet age, Airbus intended the A380 to become the defining aircraft of the new century.

Airbus' marketing ensured that even people with little interest in aviation knew that an enormous new airplane would soon be arriving at the world's major airports. With two passenger decks extending nearly the entire length of the aircraft, the A380 was designed to carry approximately 550 passengers in the representative configuration Airbus initially promoted, although airlines could configure it for considerably fewer passengers in premium-heavy layouts or for more than 800 passengers in an extremely dense, all-economy arrangement.

Its immense interior volume also allowed Airbus and prospective operators to imagine amenities that had rarely, if ever, been practical aboard a commercial airliner. Promotional concepts included private suites, lounges, bars, sleeping areas, shops, showers, and even casinos or gaming rooms. Airlines would ultimately adopt only some of these ideas, but the speculation helped reinforce the image of the A380 as something more than another passenger airplane. It was being presented as a flying destination in its own right.

Boeing, meanwhile, found itself increasingly overshadowed. Its market share was shrinking, its recent strategic decisions had produced disappointing results, and Airbus was commanding the world's attention with the launch of the largest commercial aircraft ever attempted. Boeing was not merely losing individual orders. It was losing the initiative — and, with it, the reputation for developing bold, leading-edge aircraft that airlines believed would shape the future of commercial flight.

Something had to be done. Boeing needed more than another derivative of an existing airplane. It needed a dramatic new idea that could seize the headlines from Airbus, restore confidence among its airline customers, and prove that the company that had created the modern jet age had not lost its ability to envision the future.

***

Looking back over the first century of powered flight, commercial airplanes had generally evolved along two basic measurements — size and speed. Often, the two had advanced together, as increasingly powerful engines allowed aircraft not only to fly faster but also to carry more passengers over greater distances. Boeing believed that the pursuit of ever-greater capacity had reached the point of diminishing returns. The company saw limited demand for airplanes substantially larger than the 747. Speed, however, was another matter.

As the aviation industry had discovered during the 1960s, supersonic passenger flight held enormous promise. But environmental concerns, sonic-boom restrictions, and the tremendous amount of fuel required to fly faster than the speed of sound made supersonic service commercially impractical for most routes. Concorde had proved that supersonic passenger travel was technologically possible, but it had also demonstrated how difficult it was to make such an airplane economically viable.

Modern commercial jetliners, regardless of manufacturer or design, typically cruised between approximately Mach 0.80 and Mach 0.85, or 80 to 85 percent of the speed of sound. Most could fly a little bit faster if necessary, but that range provided an efficient balance among speed, fuel consumption, engine performance, and aerodynamic drag.

As an airplane approached the speed of sound, or Mach 1.0, however, shock waves began forming around portions of its wings and airframe. Those shock waves produced a sharp increase in aerodynamic drag. The closer the airplane came to the speed of sound, the more power its engines had to produce — and the more fuel they had to consume — to overcome that drag.

But what if continuing advances in aerodynamics, propulsion, computer-aided design, and lightweight materials made it possible to build an airplane that could fly approximately 15 percent faster than conventional jetliners without breaking the sound barrier and without incurring a corresponding increase in fuel consumption?

On short flights, a 15 percent increase in speed would produce relatively modest time savings. On long intercontinental routes, however, it could shorten a flight by two or even three hours, depending on the distance traveled, prevailing winds, and routing. A long transpacific flight that required approximately twelve hours aboard a conventional airliner might be completed in closer to ten.

Passengers would undoubtedly appreciate reaching their destinations sooner, but Boeing believed airlines would benefit as well. Faster flights could allow an aircraft to complete more trips over a given period, increasing its utilization and revenue-producing potential. Shorter block times might also reduce some crew and operating expenses. Boeing's argument was that even if flying near the sound barrier consumed somewhat more fuel, the added expense could be offset by higher aircraft utilization and the premium passengers might be willing to pay for significantly shorter journeys. From that reasoning emerged Boeing's hoped-for next great advancement in commercial aviation — and the airplane that might restore the company's reputation as the industry's technological leader: the Sonic Cruiser.

Boeing publicly revealed the Sonic Cruiser concept on March 29, 2001. If nothing else, it was a stunning visual declaration that Boeing had not lost its ability to think boldly. The airplane's design looked nothing like the conventional cylindrical fuselage, swept wings, tail assembly, and underwing engines that had become familiar at airports throughout the world.

The Sonic Cruiser was expected to carry approximately 200 to 250 passengers at speeds between Mach 0.95 and Mach 0.98 — roughly 15 to 20 percent faster than conventional jetliners. Boeing also contemplated a family of different-sized versions, potentially carrying anywhere from 100 to 300 passengers.

Its configuration was unlike anything then in commercial service. The airplane featured a long, tapered fuselage; a distinctive cranked-delta wing; small forward-mounted canards; rear-mounted engines blended into the wing structure; and twin vertical tail surfaces. Its unconventional shape made the Sonic Cruiser appear less like an evolutionary successor to the 767 than an airplane from an entirely different era.

The Sonic Cruiser had not been conjured up overnight. Its origins could be traced to an internal Boeing airplane-development study established in 1995 to explore how advances in propulsion, aerodynamics, materials, and manufacturing might shape the next generation of commercial aircraft. One of the concepts emerging from that work was known internally as Project Glacier, a design that by late 2000 had begun to resemble what would soon be presented publicly as the Sonic Cruiser.

The program fell under the leadership of Alan Mulally, president and chief executive officer of Boeing Commercial Airplanes. Mulally had previously played a central role in the development of the 777, ultimately leading the program that had become Boeing's greatest recent commercial success. He now had the opportunity to oversee an airplane that promised to be even more revolutionary.

One of the Sonic Cruiser's essential enabling technologies was its extensive use of advanced composite materials. Conventional commercial aircraft were constructed primarily from aluminum, titanium, steel, and other metallic alloys. Carbon-fiber-reinforced polymer, commonly abbreviated CFRP, offered a much higher strength-to-weight ratio than many traditional materials. Properly engineered, composite structures could be made lighter, more resistant to corrosion, and capable of being formed into complex aerodynamic shapes that would be difficult to manufacture efficiently from aluminum.

The aerospace industry had been expanding the use of carbon-fiber composites since the 1960s, when they began appearing in limited space and military applications. Their use grew substantially during the following decades as NASA, the Department of Defense, aircraft manufacturers, and materials suppliers funded research into composite structures and production methods.

Boeing benefited directly from those government-supported programs. Under NASA contracts during the 1970s, the company evaluated composite aircraft components as a means of reducing structural weight and improving performance. During the 1980s and 1990s, Boeing participated in military programs as a subcontractor involving extensive composite applications, including the B-2 Spirit stealth bomber, V-22 Osprey, and F-22 Raptor. Those programs expanded the company's knowledge of how advanced composites could be designed, manufactured, inspected, repaired, and incorporated into primary aircraft structures.

The Sonic Cruiser offered Boeing an opportunity to transfer that accumulated knowledge into commercial aviation on an unprecedented scale. By 2002, the company envisioned a design composed of approximately 60 percent composite materials. Boeing estimated that the resulting weight savings, combined with advanced aerodynamics and more efficient engines, would allow the Sonic Cruiser to fly 15 percent faster while consuming approximately the same amount of fuel per passenger as the best contemporary twin-engine widebodies.

Boeing's senior leadership gave Mulally authorization to present the concept to airlines and gauge their interest. On June 19, 2001, at the Paris Air Show, Mulally unveiled a detailed three-dimensional model of the Sonic Cruiser inside Boeing's pavilion. It gave the aviation world its first full view of the radically unconventional airplane Boeing was proposing.

The concept generated considerable attention. Several major airlines expressed interest, including American Airlines and Virgin Atlantic Airways. Virgin Atlantic chairman Richard Branson was particularly enthusiastic about the possibility of an airplane that could significantly reduce long-distance flight times. Yet expressions of interest were not firm orders, and Boeing had not formally launched the program. The company still had to determine whether enough airlines would commit to the airplane — and whether they would pay enough for its speed — to justify the immense development cost.

By the summer of 2001, the battle lines appeared to have been drawn for commercial aviation in the twenty-first century. Boeing and Airbus, now roughly matched in the annual competition for orders, were advancing fundamentally different visions of the industry's future.

Airbus was proceeding with the A380 because it believed international aviation would increasingly revolve around large hub airports. Enormous airplanes would carry hundreds of passengers between major global hubs, after which smaller aircraft would distribute them to their final destinations. Although Airbus acknowledged that some market fragmentation would occur as airlines introduced more nonstop routes, it believed the greatest growth would remain concentrated among the world's largest cities and airports.

With the A380 added to its product lineup, Airbus could offer aircraft for virtually every capacity and range requirement, from the single-aisle A320 family to the mid-sized A330 and A340 widebodies and, finally, the immense A380. In Airbus' vision, airport congestion and limited takeoff and landing slots would force airlines to carry more passengers aboard each flight.

Boeing saw a different future. It believed long-range, twin-engine aircraft would allow airlines to bypass traditional hubs and connect an increasing number of cities directly. Major hubs would not disappear, but congestion, limited capacity, and the traveling public's preference for nonstop service would push airlines toward more point-to-point routes linking secondary cities.

Instead of carrying 500 or more passengers through a massive hub, an airline might carry 200 to 300 passengers directly between cities that previously could not support nonstop international service. For Boeing, the future did not belong to ever-larger four-engine airplanes. It belonged to smaller, highly efficient twin-engine aircraft capable of flying farther, faster, and more frequently between a much larger number of city pairs.

Boeing needed to regain the initiative and reaffirm its leadership in commercial aviation. The Sonic Cruiser appeared perfectly suited to that objective. It was a futuristic-looking twin-engine airplane that promised faster travel, longer range, and more direct service. More important, it presented Boeing as the company defining the next stage of commercial flight rather than reacting to the vision being promoted by Airbus.

Just as Boeing regarded the A380 as a strategic folly, Airbus viewed the Sonic Cruiser with similar skepticism. Airbus executives argued that relatively few routes were long enough for the Sonic Cruiser's speed to produce time savings substantial enough to justify its likely acquisition and operating costs. They questioned whether passengers would consistently pay higher fares merely to arrive one or two hours earlier. There were even suggestions within Airbus that the Sonic Cruiser was little more than an elaborate diversion — a dramatic concept Boeing had introduced to draw attention away from the A380 while it determined what kind of airplane it actually wanted to build.

Boeing, however, was hedging its bet. While Mulally's organization continued refining the Sonic Cruiser, another group was studying how the same emerging technologies could be incorporated into a more conventional commercial airplane. Instead of using lightweight composites, improved aerodynamics, and advanced engines primarily to increase speed, this alternative design would use them to achieve dramatically lower operating costs.

The tradeoff was straightforward. The Sonic Cruiser would carry roughly the same number of passengers as a 767 while flying approximately 15 percent faster and consuming about the same amount of fuel per passenger. The more conventional alternative would fly at approximately the same speed as existing jetliners but consume substantially less fuel, and cost less to maintain. Boeing believed the new technologies might eventually reduce operating costs by as much as 20 percent. The broader technology effort became known as Yellowstone. It encompassed three general aircraft categories intended to provide the technological foundation for the eventual replacement of Boeing's entire commercial product line.

Yellowstone 1, or Y1, focused on the single-aisle market then served primarily by the 737 and Airbus A320 families. Y2 addressed the middle of the market occupied by the 757 and 767. Y3 contemplated a future large-aircraft family that would eventually succeed the 777 and 747. The precise boundaries among the projects would evolve, but together they represented Boeing's long-term plan for transferring advanced composites, advanced aircraft systems, improved aerodynamics, and a new generation of engines throughout its commercial lineup.

Y1 presented perhaps the most difficult economic challenge. Single-aisle aircraft carried fewer passengers over shorter distances. Because their flights consumed less fuel and incurred lower total operating costs than long-haul widebody services, the absolute savings available from new technologies were correspondingly smaller. Boeing's early studies suggested that the available technologies might produce an operating-cost improvement of less than 10 percent in a new single-aisle airplane. Company leadership doubted whether that would be sufficient to persuade airlines to abandon relatively young and highly efficient fleets of 737s and A320s — or to justify the billions of dollars Boeing would have to spend developing an entirely new replacement.

Y3 posed a different problem. The 777 had been in service for only a few years, and its sales were gaining momentum. Boeing expected the airplane to remain the foundation of its long-range product line for decades. Launching a new large-aircraft family too soon could cannibalize 777 sales before Boeing had recovered the full value of its investment.

Nevertheless, the Y3 studies helped validate the technologies that could someday be incorporated into a high-capacity, long-range twin-engine airplane. Boeing believed such an aircraft would eventually provide much of the capacity and range traditionally associated with the 747 — but with two engines rather than four.

The success of the 777 was already weakening demand for the 747. Airlines had discovered that large twin-engine aircraft could perform many long-range missions more economically while offering greater scheduling flexibility and presenting less risk than filling a very large four-engine airplane. From Boeing's perspective, that trend demonstrated why the market for an aircraft as large as the A380 would remain confined to the extreme upper end of commercial aviation.

Airbus intended to serve the market for aircraft carrying 100 or more passengers through three principal categories: the A320 family, the A330 and A340 families, and the A380. Boeing was preparing to organize its future product line along a comparable three-tier structure built around Y1, Y2, and Y3.

There was, however, one fundamental difference between the two strategies: Boeing saw no need for an airplane approaching the size of the A380.

This Ain't Your Father's Boeing

By the time Harry Stonecipher took the helm at McDonnell Douglas in 1994, he had learned to loathe the enormous cost of developing new aircraft. He was not alone in his aversion. John McDonnell shared it. The cost of launching a clean-sheet commercial airplane had climbed dramatically over the decades, even after accounting for inflation. During the late 1960s and 1970s, McDonnell Douglas invested approximately $1 billion in developing, producing, and bringing the DC-10 to market — roughly equivalent to $4.4 billion in 2001 inflation adjusted dollars. However, by 2001, the estimated cost of developing an entirely new commercial airplane had climbed substantially higher, between $10 to $12 billion.

Rapid technological advancement had made aircraft more capable and safer, but it had also made them more complex and expensive to develop. The greater the investment, the longer it took for a manufacturer to recover its development costs. Capital remained tied up for years, sometimes decades, while the risks multiplied — technical problems, production delays, economic downturns, changing airline requirements, and the possibility that a competitor might introduce a better airplane before the original investment had been recovered.

This was one of the principal reasons McDonnell Douglas' commercial airplane business had withered. Its reluctance to invest in new programs — particularly when capital was scarce — had led the company to repeatedly modify existing aircraft instead. Each derivative required less money and carried less immediate financial risk, but the strategy gradually left McDonnell Douglas attempting to sell increasingly antiquated products against newer Boeing and Airbus designs.

Philip Condit and Harry Stonecipher recognized that Boeing would eventually have to develop a new airplane if it hoped to blunt Airbus' advances. True to form, however, Stonecipher was appalled by the projected cost of launching another clean-sheet program. He was also a committed proponent of outsourcing. He had seen the savings that outsourcing could produce at General Electric, Sundstrand, and McDonnell Douglas, although those companies had generally used it to reduce the cost of existing production activities.

But what if outsourcing were incorporated into an airplane program from its inception? What if Boeing did not merely purchase components from suppliers but transferred to them much of the responsibility for designing, financing, manufacturing, and assembling entire sections of the airplane?

As chief operating officer and senior board member, Stonecipher wanted to take the concept further by borrowing elements of the industrial model Airbus had used from its beginning. Airbus had originally been organized as a consortium in which manufacturers from participating European nations assumed responsibility for major portions of each aircraft. On the A320, facilities in France produced such components as the forward fuselage, center wing box, engine pylons, and landing-gear components, while Toulouse performed final assembly on many aircraft. German facilities manufactured other fuselage sections, the vertical tail, flaps, and portions of the empennage, and Hamburg later became another major A320-family final-assembly center. Spain produced much of the horizontal stabilizer, while Airbus' British operations designed and manufactured the wings.

The resulting components were transported from manufacturing facilities across Europe and brought together on final-assembly lines to produce completed airplanes. Airbus had demonstrated that a commercial aircraft did not have to be designed and manufactured almost entirely within one vertically integrated company.

Stonecipher believed Boeing could adapt that model and go considerably further. Responsibility for major portions of the new airplane — including its wings, fuselage sections, empennage, systems, and major subassemblies — would be assigned to strategic partners. Those companies would not merely manufacture parts according to Boeing drawings. They would help design the components, invest their own capital in their development, manage networks of lower-tier suppliers, and deliver completed sections ready to be joined together.

The theory appeared compelling. Because each strategic partner would concentrate on a particular section or system, it could develop specialized expertise and continually improve the efficiency with which that component was produced. Boeing would retain responsibility for the airplane's overall configuration, systems architecture, certification, marketing, and final assembly — or, as it increasingly came to be called, final integration.

As Stonecipher saw it, such an arrangement offered the only financially prudent way to launch another clean-sheet commercial airplane. The investment and risk could be distributed among Boeing and its strategic partners rather than borne almost entirely by Boeing. The company expected to contribute only a portion of the total development cost, with suppliers financing much of the remainder.

Stonecipher appears to have convinced Condit of the viability of this model, and so they presented it to the board as more than a way to reduce Boeing's financial exposure. They argued that it would also shorten development time, reduce the amount of capital tied up in factories and tooling, and allow Boeing to bring new airplanes to market faster.

For the moment, however, Boeing was still promoting the Sonic Cruiser. Throughout 2001, the company presented the unconventional aircraft as the future of long-range travel. Designed to fly at approximately Mach 0.95 to Mach 0.98, it promised to reduce travel times on long routes substantially while offering fuel efficiency per passenger comparable to that of the most efficient widebody airplanes then in service. American Airlines, Virgin Atlantic Airways, and several other carriers expressed interest, but none was prepared to make a firm commitment until Boeing provided more detailed performance guarantees, pricing, and development schedules.

Then, on the morning of September 11, 2001, the course of history changed.

Whatever transformation Boeing had undergone during the four years following its merger with McDonnell Douglas, the events of that morning accelerated it. The fact that all four airplanes hijacked by terrorists were Boeing aircraft reflected Boeing's historic dominance of the world's commercial-airliner fleet. For nearly half a century, Boeing had produced more of the world's jetliners than any other manufacturer, and its airplanes still dominated airport ramps around the world.

The terrorist attacks brought commercial aviation to an immediate standstill. Demand for new aircraft had already begun weakening during 2001 as the global economy slowed, but after September 11, airline traffic collapsed and orders for new airplanes nearly disappeared. When commercial flights resumed, passenger volumes remained far below their previous levels. Airlines grounded aircraft, reduced schedules, laid off employees, deferred deliveries, and canceled or renegotiated orders with both Boeing and Airbus. With more airplanes than they needed and no certainty about when passengers would return, carriers had little appetite for purchasing an unconventional new aircraft built primarily around the promise of greater speed.

The airline industry's difficulties did not end with the decline in passenger traffic. Oil prices initially fell as the attacks and the weakening global economy reduced demand, but the decline proved temporary. Over the following years, geopolitical instability, disruptions to Iraqi oil production, constrained refining capacity, and rapidly increasing worldwide demand from emerging economies helped drive oil prices steadily upward.

For airlines struggling to recover from the collapse in passenger traffic, fuel became an increasingly serious burden. Speed was no longer their principal concern. They wanted airplanes that consumed less fuel, required less maintenance, and could operate profitably even when passenger demand and ticket prices were uncertain.

Whatever enthusiasm airlines had shown for the Sonic Cruiser quickly evaporated. Boeing continued studying and promoting the concept through much of 2002, but airline executives increasingly told the company that they would prefer an airplane offering dramatically better operating economics rather than one offering dramatically greater speed.

By December 2002, Boeing had placed the Sonic Cruiser on hold. The company redirected the technology developed for it — including advanced composite structures, advanced systems, improved aerodynamics, and new manufacturing techniques — toward a more conventional aircraft. Boeing brought forward the Yellowstone Y2 concept, a proposed replacement for the 767 that would carry approximately 200 to 300 passengers over long distances while consuming roughly 20 percent less fuel than comparable existing aircraft.

The mandate from Boeing's leadership was clear: determine whether the world's airlines wanted such an airplane, and whether Boeing's new global network of risk-sharing partners could make it possible.

***

For a number of reasons, 2001 was a watershed year for Boeing — and not just because of the events of September 11. On March 21, 2001, Philip Condit announced that the company was moving its headquarters out of Seattle. The announcement rocked the Seattle area as though a bomb had been detonated over the city. Boeing was synonymous with Seattle, and Seattle was synonymous with Boeing. After 85 years, why the sudden desire to move the company's headquarters?

When the announcement was made, Philip Condit said, "We are a global enterprise. Our headquarters team needs to be located where it can best reach out to our customers, partners, and operations around the world — not necessarily where our largest business unit happens to be located." Beyond his public statement, there were three primary reasons Condit wanted to move Boeing's headquarters out of Seattle.

The first was that he wanted the company's corporate leadership to operate independently of any one of its manufacturing divisions. As he later explained, headquarters needed to focus on strategy, global growth, and customer relations rather than the day-to-day management of factories. By Condit's reasoning, remaining in Seattle meant Boeing's leadership would remain physically and culturally tied to the commercial-airplane division, which had always been — and still remained — the heart of the company's identity.

Condit wanted the headquarters to reflect the broader Boeing that he was trying to create, one that included space, defense, communications, financing, and related services. Moving headquarters away from Seattle, where the commercial airplane division dominated, would ostensibly place those other businesses on more equal footing. Their respective headquarters were scattered across other regions of the country, and Condit believed that Boeing's corporate center should be separate from all of them.

The second reason involved Boeing's global image and accessibility. Condit wanted Boeing to be seen as a global corporation rather than merely a Seattle-based airplane manufacturer. He believed the headquarters should be located in a centrally situated American city with convenient access to Washington, D.C., the financial centers on the East Coast, Boeing's dispersed business units, and international markets.

The third reason was cultural transformation. Although Boeing's acquisition of McDonnell Douglas had been completed in 1997, the two corporate cultures had never been fully integrated. Moving headquarters out of Seattle would symbolically break the combined company away from the traditional Boeing culture and help establish a new identity that would reflect the company's broader scope and focus.

The move away from Seattle was a logical extension of Condit's dream of diversifying Boeing and making it less dependent on commercial aviation and more heavily oriented toward defense, space, and services. But it was impossible not to see the influence wielded by former McDonnell Douglas executives — Harry Stonecipher in particular — in the decision to break with Seattle and 85 years of the company's storied history.

Many questioned the wisdom of moving the headquarters out of Seattle. Everyone understood Condit's desire to diversify into space and defense, but the fact remained that the largest share of Boeing's revenue still came from commercial aviation, and that business had always been the driving force behind the company's success. Moving the headquarters away from Seattle meant that Boeing's most senior executives would be physically detached from the rank and file and from whatever was happening on the production floor. The common fear was that the company's leadership would lose its ability to keep a finger on Boeing's pulse.

But there was something else about the decision that reflected the mindset taking hold within the company: Philip Condit's approach to leadership. When Condit became CEO in 1996, he could have doubled down on the formula that had made Boeing a resounding success, a commercial-aviation behemoth, and the undisputed king of the aerospace industry. Instead, he took the company in a substantially different direction — one that some believed created an enormous and unnecessary distraction.

The decision to move Boeing's headquarters away from Seattle only accelerated Condit's lurch — and the company's — into the realm of "What were they thinking?" It was the kind of decision that would become legendary, but for all the wrong reasons.

When Condit announced the plan to leave Seattle, he identified three cities that Boeing's board was considering for its new headquarters: Chicago, the Dallas–Fort Worth metropolitan area, and Denver. All three were centrally located within the United States, at least relative to Seattle, and all three were considered comparatively neutral because none of Boeing's three largest operating divisions was headquartered in them. Each also had a major airport offering convenient access to domestic and international destinations.

Chicago's strength as a candidate was its powerful corporate environment. It was home to numerous Fortune 500 companies, had long been recognized as an international financial and business center, and offered extensive air service through both O'Hare and Midway airports. It also possessed the cultural and urban appeal considered important when entertaining visiting dignitaries, customers, investors, and corporate executives.

Dallas–Fort Worth was attractive because of its business-friendly environment, relatively low taxes, lower cost of living, and rapidly growing metropolitan economy. Its location also offered convenient access to the southern and eastern United States and to many of Boeing's military customers. The region's enormous airport provided extensive domestic and international connections, and Texas officials were eager to add the headquarters of the world's largest aerospace company to the state's expanding corporate roster.

Denver was considered because of its quality of life, outdoor amenities, central location, and established aerospace presence. Numerous aviation, defense, satellite, and space companies already maintained operations in Colorado. Denver also projected an image of technology and innovation — attributes that resonated with Boeing's desire to present itself not simply as an airplane manufacturer, but as a forward-looking global aerospace pioneer.

If Condit had previously come across as a mild-mannered, reserved consensus builder, he became a seasoned showman after announcing the planned relocation. He remained tight-lipped about which city might be leading, but periodically made off-the-cuff remarks that kept public interest piqued. The drama intensified when he made what were supposedly low-key and unannounced visits to each of the three cities, only for word of his arrival to leak almost immediately. Some observers joked that Condit's visits were the worst-kept secrets in corporate America.

The three cities eagerly competed for Boeing's favor. Political leaders, business executives, civic organizations, and cultural institutions courted Condit and the company's selection team. They offered financial incentives, prospective office locations, and elaborate demonstrations of what their cities could provide. The process began to resemble less a corporate site selection and more a competition to host the Olympic Games.

The final act came with the announcement of the winning city. Boeing revealed in advance that its selection would be made public on May 10, 2001. But rather than simply issue the customary press release, Condit wanted the announcement to have the drama of an opera's finale.

On the morning of May 10, Condit and other Boeing executives boarded one of the company's corporate 737s in Seattle. The airplane would take off and land in the city Boeing had selected as its new home. Reports circulated that the flight crew had prepared three possible flight plans — one for Chicago, one for Dallas–Fort Worth, and one for Denver — and that the destination would not be revealed until the airplane was in the air.

Whether three flight plans were actually filed became a matter of dispute. But Boeing did nothing to discourage the public impression that the destination remained a closely guarded secret, and the story of the three possible flight plans added considerably to the spectacle.

Once the airplane was airborne, Boeing announced that it was headed for Chicago. Condit's corporate 737 landed at Midway International Airport, where Chicago and Illinois officials had literally rolled out a red carpet. Condit emerged from the airplane and announced that Chicago would become Boeing's new world headquarters. He and other Boeing executives then boarded a helicopter for an aerial tour of the city and of the building along the Chicago River that would become their new corporate home.

It was obvious to even the most casual observer that much of it had been staged for show. Boeing was not simply relocating a corporate office; Condit was presenting the move as a grand public demonstration that the old Boeing was being transformed into something new.

The analogy to an opera's finale was particularly fitting. Condit was known for his love of opera and symphonic music, as was Harry Stonecipher, and Chicago was internationally renowned for both. The city was home to the Chicago Symphony Orchestra and the Lyric Opera of Chicago, whose opera house would be within walking distance of Boeing's new headquarters.

Some insiders suggested that Chicago's cultural stature had appealed personally to Condit and helped distinguish it from Dallas and Denver. Although it would be difficult to establish that opera and symphonic music were decisive factors in a multibillion-dollar corporation's headquarters decision, they certainly fit Condit's tastes and the image he wanted Boeing's new home to project. Chicago possessed the international prestige, corporate sophistication, cultural institutions, and urban grandeur that Condit believed should surround the headquarters of the world's largest aerospace company.

This was the same Philip Condit whose personality and personal tastes had increasingly become intertwined with the privileges of his position. Coupled with the previous reports that alleged he had the executive suite at the Four Seasons Hotel altered at company expense only reinforced the impression that Condit had become increasingly comfortable using Boeing's resources to shape his surroundings according to his own tastes.

Fewer than 500 Boeing employees would initially move into the new headquarters, but they included the company's most senior executives and the corporate staff responsible for setting Boeing's overall direction. The number of jobs transferred was small compared with the tens of thousands who remained in the Puget Sound region. However, the significance of the move was not how many people left Seattle; it was who left.

Boeing's leaders had intentionally placed themselves in a city where the company designed, manufactured, and assembled virtually nothing. The nearest commercial-airplane assembly plant would be approximately 1,700 miles away. Senior executives could visit Boeing's factories whenever they wished, but visiting a production line was not the same as living near it, interacting routinely with its engineers and managers, and absorbing the everyday realities of building airplanes.

Physical separation would inevitably make it easier for troubling information to be filtered as it traveled upward through layers of management. Factory problems could be reduced to charts, statistics, and carefully worded presentations. Engineering concerns could be translated into financial risks. Manufacturing delays could become schedule variances. Quality problems could be discussed as metrics rather than as physical defects affecting actual airplanes.

The move was meant to free Boeing's corporate leadership to think strategically. It also freed that leadership from having to confront the daily consequences of its strategies.

However, if Condit's effort to move Boeing's headquarters out of Seattle was intended to improve the company's diversification or strengthen its ability to win enormous defense contracts, the results would prove bitterly disappointing. Boeing's first opportunity to demonstrate the wisdom of its defense-oriented transformation would come almost immediately — and it would end in a misstep of enormous proportions.

***

Before its 1996 merger agreement with Boeing, McDonnell Douglas had produced some of America's most successful frontline military aircraft, including the A-4 Skyhawk, F-4 Phantom II, F-15 Eagle, and F/A-18 Hornet. The F-15 and later variants of the F/A-18 would remain central to the fighter forces of the U.S. Air Force and Navy for decades, while also serving with air forces around the world. Meanwhile, the company's production lines were building the Air Force's newest strategic airlifter, the C-17 Globemaster III. McDonnell Douglas' ability to design innovative military aircraft was nearly unmatched within the defense industry — as was the influence of the executives and lobbyists who accompanied it into Boeing.

In late 1993, the Department of Defense began the effort that would eventually become the Joint Strike Fighter program. Its goal was to develop a new generation of combat aircraft for the Air Force, Navy, and Marine Corps. Rather than procure three entirely separate designs, the services would operate variants of a common aircraft built around shared systems and components. The new fighter would incorporate advanced sensors, an adaptable systems architecture, and the latest stealth technology.

The concept was driven as much by cost as by capability. As the expense of developing and acquiring advanced weapons systems continued to rise, the Pentagon believed that a common platform produced in large numbers could generate substantial economies of scale. Shared development, production, parts inventories, maintenance procedures, and training programs were expected to reduce costs across all three services. In theory, standardization would make the next generation of fighter aircraft both more capable and more affordable.

On paper, the Joint Strike Fighter concept made sense. From a technological standpoint, however, the obstacles were formidable. The aircraft had to satisfy the Air Force's tactical and operational requirements, withstand the unforgiving demands of Navy carrier operations, and provide the Marine Corps with a short-takeoff-and-vertical-landing capability.

Designing one basic aircraft to satisfy all three missions required compromises that would test the limits of engineering. The Air Force wanted a relatively lightweight conventional fighter. The Navy required a stronger airframe, larger wings, folding wingtips, reinforced landing gear, and an arresting hook for carrier operations. The Marine Corps needed an aircraft capable of taking off from short runways and landing vertically aboard amphibious assault ships and at austere forward bases.

In late 1996, the Department of Defense awarded demonstration contracts to Boeing and Lockheed Martin. Each company would develop aircraft representing a conventional-takeoff-and-landing version and a short-takeoff-and-vertical-landing version, commonly abbreviated STOVL. Each contractor received approximately $750 million to build and demonstrate its concept. The Pentagon would then evaluate the competing aircraft before awarding the development and production contract in what was effectively a winner-take-all competition.

This would be no ordinary defense contract. At the time the program was announced, the United States was expected to purchase approximately 3,000 Joint Strike Fighters, with the possibility of thousands of additional orders from allied nations. By

2001, estimates of the total cost of acquiring, operating, and maintaining the aircraft over its anticipated service life approached or exceeded $1 trillion. It promised to become the largest and most expensive military aircraft program in history.

This was one reason Boeing's acquisition of McDonnell Douglas had electrified the aerospace industry. Boeing had, in effect, purchased additional defense credibility. McDonnell Douglas brought with it generations of experience designing and producing fighters, attack aircraft, transports, and missiles. The acquisition dramatically expanded Boeing's military-aircraft capabilities. Winning the Joint Strike Fighter contract would have further transformed the company and could have made Boeing the world's dominant defense contractor.

Boeing and Lockheed Martin designed, built, and flew their demonstrator aircraft by 2000. Boeing's competing aircraft was designated the X-32, while Lockheed Martin's was designated the X-35. Both companies then subjected their aircraft to a demanding series of flight demonstrations and evaluations conducted by the Department of Defense and the participating military services.

In October 2001, the Pentagon announced that Lockheed Martin had won the competition. The decision did not come as a surprise to much of the aerospace industry.

Losing the contract was a severe blow to Boeing, but company officials had seen the writing on the wall during the evaluation period, as had many outside observers. Although the Department of Defense publicly maintained that both aircraft had met or exceeded the program's stated requirements, privately there was a widespread perception that the competition had not been particularly close.

The unfortunate truth was that Boeing's X-32 looked like an abomination assembled from mismatched portions of several different aircraft. It appeared to be an obese, out-of-shape fighter whose best days had long since passed. Lockheed Martin's X-35, by comparison, looked like a sinister and svelte championship fighter ready to take on any opponent in the ring.

The X-32's most conspicuous feature was its enormous, distorted-looking engine intake beneath the forward fuselage. The gaping opening quickly acquired the nickname "The Monica" — a crude reference to Monica Lewinsky and former President Bill Clinton's White House sex scandal of the late 1990s.

Appearance alone did not decide the competition, but the X-32's awkward shape reflected the compromises Boeing had made while attempting to satisfy the program's competing requirements. If Boeing had hoped to demonstrate its newly acquired prowess as one of the world's largest defense contractors, the X-32 instead blew up in the company's face.

Philip Condit's much-vaunted plan to diversify Boeing, along with his decision to move the company's headquarters away from Seattle, had delivered no discernible advantage in the competition. Losing the Joint Strike Fighter contract rattled the company to its core, but it also exposed, in a particularly glaring way, the inner workings of Boeing and what had happened to the company since its acquisition of McDonnell Douglas.

The X-32 became a classic example of how a company could lose an enormous must-win contract because of competing internal priorities, excessive compromise, and decision-making by committee. It was as though each major portion of the aircraft had been designed in isolation and optimized for a particular requirement without sufficient consideration for how it would affect the airplane as a whole.

The result was an aircraft that was not only gut-wrenchingly strange in appearance but aerodynamically and operationally compromised. Its thick, broad wing and deep fuselage were intended to satisfy requirements for fuel capacity, weapons carriage, structural strength, and vertical lift. But those features produced an ungainly shape and made it more difficult for Boeing to achieve the degree of stealth and performance offered by Lockheed Martin's more conventional design.

Lockheed Martin's X-35 outperformed Boeing's X-32 most decisively during the STOVL demonstrations. Lockheed Martin developed a novel shaft-driven lift-fan system that redirected engine power to a vertically mounted fan behind the cockpit. Combined with a swiveling rear exhaust nozzle and smaller roll-control ducts in the wings, the system provided balanced vertical lift while allowing the aircraft's main engine to remain optimized for conventional flight.

Boeing instead chose a direct-lift system derived conceptually from the arrangement used by the AV-8B Harrier II, an aircraft in which McDonnell Douglas had played a central role. The Harrier directed engine thrust downward through rotating exhaust nozzles. Boeing's X-32 similarly relied on redirected exhaust from its main engine to produce vertical lift.

The approach appeared less mechanically complex than Lockheed Martin's lift fan, but it created significant design problems. The X-32 had to be built around the demands of the direct-lift system, particularly the enormous volume of air required by the engine while hovering. This contributed to the aircraft's oversized forward intake, deep fuselage, and unconventional overall shape.

The design also encountered difficulties during testing. Boeing's engineers had to make late changes to the lift system, adding weight to an aircraft that was already struggling to meet its performance targets. During vertical-flight demonstrations, a section of the front intake structure had to be removed to improve airflow and reduce the risk of the engine ingesting hot exhaust gases reflected from the ground.

The X-32 did demonstrate the ability to take off over a short distance, accelerate through supersonic flight, and land vertically.

But it did not complete those tasks in a single uninterrupted mission. Lockheed Martin's X-35B, by contrast, performed what became known as Mission X — a short takeoff, supersonic acceleration, and vertical landing during one continuous flight sequence. The demonstration provided the Pentagon with a vivid illustration of how much more mature Lockheed Martin's concept was than Boeing's.

What Defense Department officials saw in the X-35 was a relatively nimble aircraft that had convincingly demonstrated the principal capabilities demanded by the contract. What they saw in the X-32 was an aircraft that had struggled through testing and whose configuration was already destined for substantial redesign. During the competition, Boeing acknowledged that the production aircraft it intended to build would look markedly different from the demonstrator. The X-32's large delta wing would be replaced by a more conventional swept wing and separate horizontal tail surfaces. The inlet, fuselage, and lift system would also be modified.

Prototype demonstrators were not expected to be identical to eventual production aircraft, and Lockheed Martin's design would also undergo extensive changes. But Boeing was asking the Pentagon to select an airplane whose final configuration had not been fully demonstrated. The very characteristics that would distinguish the production version from the X-32 were therefore characteristics that had not yet been flown and tested.

The Department of Defense evaluators were not convinced. In the X-35, they saw an aircraft whose fundamental configuration already worked and had performed to their expectations. In the X-32, they saw a concept that would require extensive redesign, introducing additional technical, financial, and scheduling risks. Awarding the contract to Lockheed Martin became the obvious choice.

This was certainly not the Boeing of the Bill Allen years — a company that had taken calculated risks and repeatedly won enormous military and commercial opportunities. Many industry analysts were left wondering how the second-largest defense contractor in the world, and the largest aerospace company in the world, could have come up so short in such an important and lucrative competition.

The doubts had begun when the X-32 was first unveiled. Industry observers were aghast at its appearance, and many questioned whether such an awkward design could credibly evolve into the next generation of American fighter aircraft. Combat effectiveness could not be judged solely by appearance, of course, but fighters had traditionally embodied speed, agility, menace, and technological sophistication. The X-32 projected none of those qualities.

It was evident that Boeing had designed the aircraft around the demands of the STOVL system. Because the company had committed itself to direct lift, the airframe had to accommodate the enormous airflow requirements and exhaust arrangement needed for vertical flight. The rest of the airplane was then shaped around that solution.

Lockheed Martin approached the problem from the opposite direction. It first designed the X-35 as a more conventional stealth fighter and then integrated the lift-fan system into that basic configuration. The result was an aircraft whose STOVL equipment appeared to have been designed around the fighter rather than a fighter designed around its vertical-lift machinery.

For all the public explanations Boeing offered for losing the Joint Strike Fighter program, the most obvious one went largely unmentioned — something was wrong with the leadership structure that had allowed such a compromised prototype to become the company's entry in the most important military-aircraft competition in decades.

If Boeing had become infiltrated by bean counters following the acquisition of McDonnell Douglas, it had also become burdened by decision-making that was reviewed and reconsidered by committees ad nauseam. Technical compromises accumulated, but no one at the top appeared willing or able to step back, examine the aircraft as a whole, and ask whether Boeing was truly presenting its best possible work.

The optics were equally damaging. Condit had pushed Boeing toward defense and space, acquired McDonnell Douglas to strengthen the company's position in those markets, and moved the corporate headquarters out of Seattle as part of his effort to transform Boeing into a broader global enterprise. Yet when the most lucrative defense opportunity of the era came before the company, Boeing's entry was comprehensively outclassed.

In 2002, Harry Stonecipher stepped down as Boeing's president and chief operating officer. At 67, he was two years beyond the company's customary retirement age for senior executives, making it unlikely that he would succeed Condit as chief executive. He nevertheless retained his seat on Boeing's board of directors — and his influence over the company was far from over.

***

Fresh from losing the Joint Strike Fighter contract and still trying to assess the fallout from the September 11 attacks, Boeing's comedy of errors — many of them unforced — continued.

In 2002, Michael Sears, the former chief financial officer of McDonnell Douglas who had joined Boeing through the acquisition, became Boeing's chief financial officer. Like Harry Stonecipher, Sears was a no-nonsense, take-no-prisoners executive who possessed great confidence in his own judgment and believed he knew better than most. Unfortunately, his leadership would soon cast another pall over the company. This time, there would be no outside event to blame. The scandal that followed would be the product of the toxic culture that had taken hold within Boeing.

When Boeing acquired McDonnell Douglas in 1997, Sears was appointed president of Boeing's Military Aircraft and Missile

Systems division. The position brought him into direct contact with major U.S. Air Force procurement programs.

Darleen Druyun was the Principal Deputy Assistant Secretary of the Air Force for Acquisition and Management, making her one of the most powerful civilian procurement officials in the service. She exercised enormous influence over many of the Air Force's largest acquisition programs, including the C-17 and F-15 — both inherited by Boeing through its acquisition of McDonnell Douglas.

The precise date of Sears and Druyun's first meeting is unclear, but they almost certainly encountered one another in 1997 or 1998 during discussions concerning Air Force acquisitions and contract management. Their professional interactions would have involved major Boeing programs under Druyun's authority. At some point, however, those conversations crossed an ethical and legal boundary and began to include discussions about Druyun's future employment.

Druyun expected to retire from government service within a few years and understood the influence she could wield over a prospective employer while she still controlled billions of dollars in Air Force procurement decisions. Sears understood that influence as well.

In October 2002, while Druyun was still serving as a senior Air Force acquisition official, Sears met with her privately at a hotel near Washington, D.C., to discuss the possibility of her joining Boeing after she retired from government service. The conversation was not disclosed to the Air Force, nor was Druyun properly removed from matters affecting Boeing.

The discussions were not merely unethical. They violated federal conflict-of-interest laws. Druyun was negotiating future employment with a company whose contracts she was still in a position to influence. The promise of a lucrative position at Boeing came with an obvious, if initially unspoken, expectation — she would use her authority to favor the company in major procurement decisions. And she did so in spectacular fashion.

After the September 11 attacks, as the United States prepared for sustained military operations in Afghanistan and elsewhere, the Air Force turned its attention to its aging fleet of KC-135 aerial-refueling tankers. Many of the aircraft dated to the Eisenhower administration and were among the oldest operational assets in the American arsenal.

The Air Force had considered replacement options for years, but other procurement priorities repeatedly took precedence. Now, with the United States anticipating prolonged military operations overseas while still maintaining global readiness, pressure mounted to augment and eventually replace the KC-135 fleet.

The timing appeared ideal for Boeing. Orders for commercial aircraft had collapsed after September 11, leaving the company with excess production capacity and a dwindling backlog for the 767. What better way to support the domestic aerospace industry than to order new tankers from the only American manufacturer capable of producing them in the quantities required?

The 767 was the logical aircraft to convert. It offered sufficient capacity and range, was based on a mature design, and remained in production. Its assembly line could be sustained and potentially accelerated without requiring Boeing to develop an entirely new airplane. The aircraft was also not that much larger in size to the KC-135 that it would require the Air Force to radically alter many of its existing hangars, ramps, maintenance facilities, and ground-handling procedures. The Air Force estimated that 100 new tankers would be needed to maintain or even broaden its ability to sustain military operations. Purchasing 100 tankers outright was initially estimated to cost approximately $15 billion — about $27.2 billion in 2026 dollars.

In late 2001, still licking its wounds from the loss of the Joint Strike Fighter program, Boeing proposed a more novel arrangement: the Air Force could lease the aircraft for six years and then exercise an option to purchase them at the end of the lease. The lease proposal was presented as a way to deliver new tankers quickly while spreading the cost across multiple fiscal years. It also allowed the Air Force to avoid some of the immediate budgetary constraints associated with a traditional procurement program.

But the lease arrangement was extraordinarily expensive. The General Accounting Office — later renamed the Government Accountability Office — estimated that if the Air Force leased the tankers and then purchased them at the end of the term, the total cost could exceed $26 billion, or approximately $47.2 billion in 2026 dollars. During the term of the lease for each tanker, the Air Force would pay Boeing almost what the outright purchase cost would be for each plane, yet when the lease for each plane would expire the Air Force wouldn't own them outright and would still have to buy them.

Boeing undoubtedly understood that the Air Force would be under enormous pressure to exercise the purchase option at the end of the lease. After operating the aircraft for six years, training crews and maintenance personnel, constructing support infrastructure, and incorporating the tankers into worldwide operations, returning them would have significantly reduced the Air Force's refueling capability, not to mention the service would have nothing to show for the billions it would have spent on the planes in the first place. The lease appeared to offer Boeing the long-term economics of a sale combined with the additional revenue generated by years of lease payments.

In January 2002, Congress formally authorized the Air Force to pursue the lease and formal negotiations between the service and Boeing commenced shortly afterwards. However, the proposal provoked an uproar. The loudest and most persistent criticism came from Senator John McCain of Arizona, who questioned both the lack of competition and the extraordinary cost of the arrangement. The Pentagon's inspector general, government auditors, and other members of Congress also challenged the proposal's transparency and financial logic.

Critics objected not only to the cost but also to the idea that the Air Force would not own one of the most important assets in its inventory after paying billions of dollars for it. Aerial-refueling tankers were essential to nearly every major American military operation around the world. Under the proposed arrangement, however, the Air Force would lease these indispensable aircraft from a private corporation.

Most troubling of all was the apparent disparity between the cost of an outright purchase and the total cost of the lease-and-purchase arrangement. But Boeing maintained that the lease would allow the service to obtain aircraft more rapidly and distribute the expense across several budget cycles. Critics saw something else: a financially advantageous arrangement for Boeing that had been structured with little competition and insufficient public scrutiny. Why had such an offer received serious consideration in the first place?

It did not take long for those questions to produce investigations. What they uncovered became the largest scandal in Boeing's history up to that time.

Darleen Druyun retired from the Air Force in November 2002 and began working for Boeing in January 2003. She accepted a highly compensated position overseeing the company's missile-defense operations — the same company whose contracts she had influenced while serving as one of the Air Force's most powerful procurement officials.

Investigators soon discovered that Druyun had negotiated her Boeing employment while she was still overseeing matters directly affecting the company. Under questioning, she eventually admitted that she had favored Boeing in several procurement decisions because she wanted to secure employment for herself and members of her family.

Among the matters she acknowledged improperly influencing was the proposed 767 tanker lease. She admitted that she had agreed to a higher price for the aircraft than she believed was justified, describing it as a parting gift to Boeing before joining the company. She also acknowledged favoring Boeing in connection with other programs while seeking employment for herself, her daughter, and her future son-in-law.

Michael Sears, by then Boeing's chief financial officer and one of the company's highest-ranking executives, was found to have actively participated in the illicit employment discussions. Rather than reporting Druyun's approach and ensuring that she recused herself from Boeing-related decisions, Sears secretly negotiated the terms of her future employment.

The scandal began to emerge during the summer of 2003, but it broke publicly on November 24, 2003, when Boeing fired Darleen Druyun and chief financial officer Michael Sears for concealing employment negotiations that had taken place while Druyun was still overseeing Air Force contracts with Boeing. Both were later convicted of federal crimes and sentenced to prison. The tanker lease was abandoned, destroying what Boeing had hoped would become a highly lucrative program at a time when the company desperately needed new work for the 767 production line and a new lucrative defense program contract to confirm its decision to merge with McDonnell Douglas.

The consequences reached far beyond the loss of a single contract. Boeing's name was now associated with corruption at the highest levels of military procurement. A company once admired as one of America's greatest engineering institutions renowned for winning contracts because of its excellence in engineering and innovation had been caught negotiating employment with a government official while that official controlled billions of dollars in contracts affecting the company. The scandal left an indelible stain on Boeing's reputation and storied history.

But the tanker affair was not the only scandal carrying Boeing's name in 2003. Boeing and Lockheed Martin were competing for highly lucrative Air Force launch contracts under the Evolved Expendable Launch Vehicle program. Boeing offered its new Delta IV rocket, while Lockheed Martin competed with the Atlas V.

In 1996, a Boeing engineer and EELV manager recruited a Lockheed Martin engineer working on the EELV program. In return for offering up Lockheed Martin EELV proposals and related proprietary material, the Lockheed Martin engineer was offered a job at Boeing and at a higher salary.

The engineer left Lockheed in January 1997 and joined Boeing's EELV organization, bringing a trove of proprietary Lockheed Martin EELV related documents with him. Those documents included sensitive pricing, technical, and strategic information that could provide a significant advantage during the contract-bidding process. Boeing employees were in possession of approximately 22,000 pages of proprietary Lockheed Martin documents, and many of them were used in preparing Boeing's competitive strategy as well as to prepare and price its competing Delta IV bids. The Air Force ultimately awarded Boeing 19 of the first 28 anticipated launches, giving the company's Delta IV program an enormous advantage over Lockheed Martin's Atlas V.

In 2002, after learning that Boeing had acquired its proprietary material, Lockheed Martin notified the government. The Air Force and the Department of Justice launched an investigation. In July 2003, while the tanker controversy was already attracting intense scrutiny, the Air Force announced severe sanctions against Boeing. It stripped the company of seven launch contracts worth approximately $1 billion, transferred several of them to Lockheed Martin, and temporarily barred Boeing from competing for additional launch business.

The two scandals reinforced one another. In the tanker case, Boeing had corrupted the government procurement process by negotiating employment with a senior acquisition official. In the launch-contract case, Boeing employees had used a competitor's proprietary information to gain an advantage in government bidding.

These were not isolated production failures or unfortunate business judgments. They represented serious breakdowns in ethics, oversight, and corporate governance.

What had been promoted in 1996 as an acquisition that would cement Boeing's position as the world's greatest aerospace company now appeared, in hindsight, to have been a poison pill. Boeing had lost the Joint Strike Fighter competition, forfeited launch contracts because of industrial espionage, and watched the tanker proposal collapse amid criminal prosecutions. The company had also struggled to keep its commercial-airplane production system functioning during the first years after the McDonnell Douglas acquisition.

The accumulating failures gave new ammunition to those who argued that Boeing had lost its way — and that McDonnell Douglas had, in effect, acquired Boeing with Boeing's own money.

The fallout culminated on December 1, 2003, when Philip Condit tendered his resignation as Boeing's chairman and chief executive officer. Condit publicly maintained that he was stepping down voluntarily for the good of the company and that he had not personally participated in the misconduct. Few observers, however, believed he had any meaningful choice. He was under intense pressure from the board to step down because just six years after taking the helm of the world's preeminent aerospace company, that company was now in disarray and stained with scandal.

Since Condit's ascent to the chief executive's office, the once-invincible Boeing had been toppled from its commanding position in commercial aviation. Airbus had emerged as an equal — and in some respects superior — competitor. Boeing's much-ballyhooed acquisition of McDonnell Douglas had failed to produce the anticipated dominance in defense and space. The Joint Strike Fighter contract had gone to Lockheed Martin, the tanker proposal had collapsed into scandal, and Boeing had been punished for possessing a competitor's proprietary launch data. The two scandals would cost Boeing over $615 million in penalties and damages that it had to pay the Justice Department.

The relocation of the company's headquarters from Seattle to Chicago increasingly appeared less like a strategic masterstroke and more like a gimmick — an elaborate corporate gesture that had delivered few tangible benefits while separating Boeing's senior leadership from the company's principal engineering and manufacturing operations.

And now Boeing was facing back-to-back scandals that had generated criminal prosecutions, congressional investigations, canceled contracts, humiliating headlines, and steep fines.

There was one other development in 2003 that would eventually prove to be another poison dart, although it generated few headlines and its effects would not be fully felt for more than a decade. In October, shortly before the scandals reached a fever pitch, Boeing announced that it would end production of the 757.

The decision surprised many industry observers. Boeing had only a small number of unfilled 757 orders remaining, but many expected demand to recover once global air travel returned to pre-September 11 levels. The 737 Next Generation could perform many missions once assigned to the 757, but the larger aircraft still occupied a unique position in the market that no other airplane could completely fill.

The 757 combined the economics and operational flexibility of a single-aisle aircraft with exceptional runway performance, substantial payload capacity, and the range to fly transcontinental and, in some configurations, long transatlantic routes. It could carry more than 200 passengers, operate from relatively short or high-altitude runways, and fly routes that were beyond the practical capabilities of most other narrow-body aircraft of its era.

The market for such an airplane was not enormous, but the 757 faced virtually no direct competition. Once passenger volumes recovered, any airline requiring that particular combination of capacity, range, and performance would have had few alternatives.

Three months before the formal cancellation, Condit told shareholders that Boeing was evaluating the program's future because of what he called changing "market dynamics." To critics, the possibility that Boeing would voluntarily abandon an uncontested market segment offered still more evidence that the McDonnell Douglas philosophy had taken control of the company.

The decision reflected the same reluctance to invest that had caused McDonnell Douglas' commercial-aircraft business to wither. Rather than improve the 757, develop a successor, or at least sustain the line until demand recovered, Boeing elected to end production and rely on smaller 737 derivatives to cover portions of the market.

At the time, the decision could be defended. The production backlog had nearly disappeared, airlines were financially weakened, and keeping an assembly line open without sufficient orders was expensive. Yet the long-term consequences would prove far more significant than Boeing's leadership appeared to understand.

The company was surrendering a strategically valuable segment for which it had no direct replacement. In the years ahead, airlines would discover new uses for long-range narrow-body aircraft, particularly on thinner and longer transatlantic routes. Boeing would eventually attempt to stretch the 737 into portions of the market once occupied by the 757, but the smaller aircraft could never fully duplicate the 757's combination of capacity, range, payload, and runway performance. The decision to end the 757 program would echo through Boeing's halls for years because it would give Airbus another opening to advance on a segment of the market that Boeing once dominated.

Unfortunately, the turbulence that began with Philip Condit's ascent to the chief executive's office was only the beginning. Far greater turbulence awaited Boeing even after he stepped down.

Attempting to Write the Future

With Condit stepping down, Boeing's board needed to move quickly to restore order and credibility. The company needed strong, decisive leadership from someone with deep aerospace experience and an intimate understanding of its sprawling operations. The board considered three internal candidates: Alan Mulally, president and chief executive officer of Boeing Commercial Airplanes, who had played a pivotal role in developing the enormously successful 777; James Albaugh, president of Boeing's Space and Communications division; and Gerald Daniels, president of Boeing's Military Aircraft and Missile Systems division.

Of the three, Mulally appeared to be the strongest candidate. He had risen through Boeing's ranks after joining the company directly out of graduate school with a master's degree in aeronautical and astronautical engineering. He represented the "old" Boeing — the engineering-centered company that had existed before the McDonnell Douglas acquisition — and understood both its culture and the formula that had produced its greatest successes.

Mulally also had a clear grasp of where commercial aviation was headed. Having led the Sonic Cruiser studies and after becoming deeply involved in the Yellowstone family of next-generation aircraft concepts, he understood the technological and commercial opportunities awaiting Boeing. More than either of the other candidates, Mulally appeared positioned to repair the cultural and organizational damage caused by the merger and return Boeing to the engineering discipline and product leadership that had once defined the company.

So the board did what boards sometimes do: it doubled down on one of the principal causes of the company's recent, headline-grabbing failures. It waived Boeing's mandatory retirement age and brought Harry Stonecipher back to lead the company as its new chief executive officer.

Contemporaneous accounts indicated that the directors unanimously agreed that Boeing's next CEO had to be capable of stabilizing the company and restoring its credibility in the marketplace. They also wanted a leader who could rebuild Boeing's relationship with the Pentagon after the loss of the Joint Strike Fighter competition and the scandals involving the tanker lease and EELV programs. The board plainly believed that repairing the defense side of the business had to take precedence over nearly every other consideration.

That judgment was not difficult to understand. Six years after the McDonnell Douglas acquisition, Boeing was still struggling with the organizational and cultural upheaval the merger had unleashed, while its defense business faced a damaged relationship with one of the company's most important customers. Yet many observers would have argued that Boeing's more fundamental problem was not simply its standing with the Pentagon. It was the extent to which the company had abandoned the engineering-centered culture and product discipline that had once defined it. From that perspective, the board's overriding priority should not have been to attempt to restore confidence in the existing leadership model, but to return Boeing to its roots.

If there was ever an example of a board that could not see the forest for the trees, Boeing's board of directors in 2003 was it. The directors unquestionably had their hands full. Their immediate priority was to contain the scandals engulfing the company, restore confidence among customers and government officials, and move the crises into the rearview mirror as quickly as possible. In concentrating on those urgent problems, however, they appear to have lost sight of the larger question: How had Boeing, the world's largest and most successful aerospace company in 1996, become a symbol of corporate scandal, managerial dysfunction, and strategic ineptitude in only a few years?

Two prominent members of the board likely helped shape the decision to return Harry Stonecipher to the chief executive's office. The first was John F. McDonnell, whose family had founded McDonnell Douglas and who, for obvious reasons, would have been favorably disposed toward the executive who had led the company before its acquisition by Boeing. The second was James McNerney.

McNerney, then chairman and chief executive officer of 3M, had joined Boeing's board in 2001. He had spent nearly two decades at General Electric during the years when Jack Welch was transforming the company and had emerged as one of Welch's most prominent protégés. McNerney therefore viewed corporate management through much the same lens — rigorous financial discipline, numerical performance targets, shareholder returns, and an intense focus on the bottom line.

McNerney was not an engineer. He held a bachelor's degree from Yale University and an MBA from Harvard Business School. His expertise was in management, finance, and organizational performance rather than aircraft design or aerospace engineering. He was, above all, a numbers-driven executive — precisely the type of leader inclined to judge a company by its financial metrics rather than by the less easily quantified engineering capabilities, institutional knowledge, and technical culture upon which Boeing's long-term success depended.

As the board weighed the candidates who might lead Boeing out of the morass in which it had become mired, McNerney and John McDonnell likely made a persuasive case for the executive they believed could stabilize the company most quickly. Their priorities were readily apparent — restore Boeing's battered credibility, revive a share price that had fallen sharply since the McDonnell Douglas acquisition, and improve the company's profit margins.

Unfortunately, those priorities favored a leader capable of addressing Boeing's immediate financial and reputational problems rather than one prepared to confront the deeper causes of its decline. The board did not select the candidate best equipped to repair the cultural, organizational, and engineering fractures that had weakened the company's foundation. Instead, it chose the man whose management philosophy had helped create many of them.

When Harry Stonecipher returned to Boeing, his mandate was clear: stabilize the company quickly and restore its battered reputation. The irony was difficult to miss. Many inside and outside Boeing believed that Stonecipher — and the cadre of McDonnell Douglas executives who had followed him into the company after the acquisition — had helped implement the very strategies that had driven Boeing off course. Their objective had been to remake Boeing in McDonnell Douglas' image, emphasizing financial performance and managerial control at the expense of the engineering-centered culture that had long distinguished the company.

Stonecipher had also supported bringing Michael Sears to Boeing from McDonnell Douglas and had recommended his elevation to chief financial officer. Now, with the full backing of the board and relatively few members of Boeing's old guard still occupying positions of influence, Stonecipher was in an even stronger position to impose his vision on the company. He likely felt empowered to take whatever actions he considered necessary, with little internal resistance and no meaningful obstacle standing in his way.

He wasted little time. His first priority — widely regarded by insiders and outside observers alike as the most urgent — was repairing Boeing's battered ethical culture. Several members of the board were assigned responsibility for overseeing the creation of a new company-wide code of conduct. Management repeatedly informed employees that ethical violations, regardless of their severity or the seniority of the person involved, would not be tolerated and would carry serious consequences.

Formal ethics training was instituted throughout the company and made mandatory for continued employment and advancement. The new standards were intended to apply uniformly, from the most junior production worker to the highest-ranking executive in the C-suite. Boeing's leadership wanted to leave no doubt that the conduct exposed by the recent scandals would no longer be accepted as part of doing business.

Stonecipher's second major task was determining how Boeing should position its businesses for the future. On the government side, the McDonnell Douglas acquisition had already provided Boeing with a substantial and relatively dependable stream of revenue from the Department of Defense through the military programs and contracts it had inherited. After the terrorist attacks of September 11, 2001, placed the United States on a war footing, that revenue stream expanded as military spending surged to support the wars in Afghanistan and Iraq.

Boeing benefited from the continued development and production of the F/A-18E/F Super Hornet for the U.S. Navy, production of the C-17 Globemaster III strategic airlifter for the U.S. Air Force, and continued sales of the F-15 Strike Eagle family to foreign air forces. Its defense portfolio also included development and production of the AH-64D Apache Longbow attack helicopter for the U.S. Army and Joint Direct Attack Munition kits, which converted tens of thousands of conventional free-fall bombs into precision-guided weapons.

Beyond the established programs Boeing had inherited through the merger, the company began developing a 737-based airborne early-warning and control aircraft for international customers and remained deeply involved in the construction and integration of major components for the International Space Station. Taken together, these programs gave Boeing a broad and increasingly lucrative government portfolio at precisely the moment when its commercial-aircraft business was struggling.

There was also potential good news on the tanker front because the Air Force still needed a new aerial-refueling aircraft. In the aftermath of the tanker-lease scandal, Congress and the Pentagon insisted that the next procurement be conducted through a fully open and competitive bidding process. Airbus was the only manufacturer from an allied nation with an aircraft large and capable enough to fill the role, and it had already developed and begun marketing a tanker version of its popular A330 passenger aircraft. Shortly after the tanker-lease program was canceled in 2003, the Air Force announced that it would prepare a competitive procurement in which Boeing and Airbus would pitch their respective proposals.

The commercial side of the company, however, demanded a far more fundamental reckoning. Airbus was winning an increasing number of sales campaigns and, by the beginning of 2004, had overtaken Boeing as the world's leading seller of commercial airliners. It appeared to be only a matter of time before Airbus would also surpass Boeing in annual deliveries, making it the undisputed leader of the global commercial-aircraft industry.

Such a reversal would carry consequences extending well beyond prestige or market share. Producing and delivering more airplanes would allow Airbus to generate greater revenue and profits from its commercial aircraft business, strengthen its relationships with airlines and suppliers, and reinvest more heavily in the development of future aircraft. Boeing therefore faced more than a temporary sales downturn. It confronted the prospect that Airbus was building a durable competitive advantage that could become increasingly difficult to reverse.

With oil prices climbing sharply in the aftermath of the U.S. invasions of Afghanistan and Iraq, the Sonic Cruiser was effectively dead on arrival. Airlines were no longer interested in an aircraft that traded greater speed for higher fuel consumption. What they wanted instead was a more efficient airplane that could reduce operating costs at a time when fuel was becoming one of their greatest financial burdens to their business.

With Boeing's share price near historic lows and Airbus seemingly eating its lunch, there was a growing sense that the company needed another game-changing airplane — something comparable to the breakthroughs Boeing had repeatedly delivered from the 707 through the 777. Throughout the jet age, nearly every new Boeing model had established the standard for the market it was designed to serve. The company had built its reputation not merely by responding to the industry, but by anticipating where commercial aviation was headed and developing aircraft that compelled the rest of the industry to follow.

Yet Boeing now appeared increasingly averse to launching an all-new aircraft, preferring instead to develop updated derivatives of its existing product line. That approach was not inherently misguided. A clean-sheet design required enormous capital, years of development, and the assumption of substantial technical and commercial risk. Derivatives could be brought to market more quickly and at far lower cost.

The problem was that Boeing's caution no longer seemed entirely strategic. Since the McDonnell Douglas acquisition, the company had become distracted and increasingly reluctant to make the long-term investments that had once defined it. While Boeing struggled with internal cultural and organizational upheaval, Airbus appeared more focused, more agile, and more responsive to the changing needs of its airline customers.

The pressure was considerable when Stonecipher came out of retirement to lead Boeing. Unlike McDonnell Douglas, whose commercial-aircraft product line had already become anemic by the time he took control of that company, Boeing remained one of the world's two dominant manufacturers of commercial airliners. Stonecipher now had access to engineering talent, manufacturing resources, customer relationships, and a legacy of product innovation unlike anything he had commanded before. His challenge was to marshal those resources, restore Boeing's historical leadership in commercial aviation, and wrest the initiative back from Airbus.

***

Stonecipher had his own ideas about how that should be accomplished, and those ideas would become increasingly apparent in the months ahead. For the moment, Boeing kept moving forward with maturing and refining the Yellowstone project — specifically Y2, the concept for a new twin-engine, twin-aisle widebody aircraft. The central question was whether airlines would embrace it.

From the outside, the airplane would appear broadly conventional. Beneath the surface, however, nearly every major aspect of its design, construction, systems architecture, and production would depart from convention. Boeing was not merely looking at proposing another new airliner. It was preparing to gamble on a fundamentally different way of designing, building, and operating one.

Boeing's engineers believed that the core technologies developed for the Sonic Cruiser could be incorporated into a conventionally configured airliner. Combined with a new generation of engines then under development, those technologies promised to reduce operating costs by an estimated 20 to 25 percent per seat-mile compared with existing aircraft in the same class.

The new airplane would be constructed primarily from carbon-fiber-reinforced polymer, or CFRP, supplemented by titanium and other lightweight alloys in areas where composites were unsuitable. By weight, approximately 50 percent of the aircraft would consist of composite materials. Because the wings and fuselage represented such a large portion of the airframe's structural mass, replacing traditional aluminum alloys with CFRP offered the potential for substantial weight savings.

Those savings were only one of several advantages Boeing expected to gain from composite construction. CFRP was not only lighter than aluminum but also highly resistant to corrosion and fatigue, potentially reducing maintenance requirements and extending the service life of major structural components. It also allowed engineers greater freedom to shape large, aerodynamically efficient structures without relying on the thousands of metal panels, fasteners, and joints traditionally used to assemble a commercial airliner.

CFRP is a high-performance composite material made from carbon fibers embedded within a polymer matrix, typically an epoxy resin. The carbon fibers provide exceptional strength and stiffness at relatively low weight, while the resin binds the fibers together, protects them from damage, transfers loads between them, and allows the material to be formed into complex structural shapes.

Compared with the aluminum alloys traditionally used in commercial-aircraft construction, CFRP offers a much higher strength-to-weight ratio. Because the material can carry substantial loads with less mass, engineers can reduce not only the weight of the primary structure but also, in some areas, the amount of secondary reinforcement required to support it. Those savings compound throughout the aircraft, contributing to lower fuel consumption and operating costs.

CFRP also resists fatigue and corrosion far better than conventional aluminum structures. That promised lower maintenance requirements, longer inspection intervals, and potentially longer service lives for major structural components. It could also provide engineers with greater design flexibility and allow wider safety margins without imposing the same weight penalties associated with traditional metals.

Corrosion, one of the persistent challenges confronting any metal aircraft repeatedly exposed to moisture, temperature changes, and atmospheric contaminants, would largely cease to be a concern in the composite portions of the airframe. It would not disappear entirely, because metallic components, fasteners, and joints that would still be utilized would still require protection, but the extensive corrosion-control burden associated with a conventional aluminum fuselage could be substantially reduced.

From a manufacturing standpoint, CFRP required an entirely new generation of tools, facilities, and production techniques. However, once those systems were established, composite structures could be manufactured with a high degree of automation and with substantially less material waste than many traditional metal components.

Boeing's engineers came up with a method for producing large fuselage sections as single-piece composite barrels. Narrow ribbons of resin-impregnated carbon fiber would be laid in overlapping, crisscrossing patterns around a massive cylindrical mold, known as a mandrel, shaped to match a section of the aircraft's fuselage. The mandrel rotated slowly while a computer-controlled fiber-placement head moved along its length, precisely applying successive layers of composite material at carefully calculated angles.

Once the required layers had been applied, the entire assembly was cured under heat and pressure, producing a rigid fuselage barrel with far fewer panels, joints, and fasteners than a conventional aluminum structure. In some respects, the process resembled an early form of large-scale additive manufacturing, not unlike 3D printing, where the structure was built gradually, layer by layer, under computer control. Unlike modern 3D printing, however, the placement head deposited continuous carbon-fiber tape rather than extruding a liquid or molten material.

Once the required thickness had been built up around the mandrel, the entire assembly would be placed inside a massive pressurized oven known as an autoclave. There, the composite material would be heated to approximately 350°F, or 177°C, and subjected to pressures of roughly 95 pounds per square inch, or 655 kilopascals. Those conditions would be maintained for several hours before the structure entered a carefully controlled cooling cycle.

The combination of heat, pressure, and vacuum caused the resin to flow and cure while consolidating the individual layers into a single rigid structure. The process also helped remove trapped air and minimize voids between the layers. This was critically important because such voids could weaken the surrounding material, reduce its ability to carry loads, and potentially create sites where cracks or delamination might begin under repeated pressurization and flight stresses.

After curing, the mandrel would be removed and the completed fuselage barrel would undergo inspection, trimming, drilling, and installation of frames, floors, doors, windows, and other structural elements. The separate barrel sections could then be joined together to form the completed fuselage.

The wings and empennage would also make extensive use of composite construction, although they were not manufactured in precisely the same manner as the cylindrical fuselage barrels. Their spars, skins, and other major components were laid up on specially shaped tools and molds, cured, inspected, and then assembled into the finished wing and tail structures.

One of the greatest promises CFRP offered commercial-aircraft manufacturing was a dramatic reduction in the number of individual parts required to build the airframe. Fewer parts meant fewer suppliers, smaller inventories, less handling, fewer assembly operations, and fewer opportunities for manufacturing defects.

A conventionally constructed airliner depended on thousands of separate components to form its fuselage, wings, and tail. Aluminum began as raw material that had to be cast, rolled, forged, or extruded before being cut, machined, chemically milled, bent, and formed into the required shapes. Much of this work was performed by a broad network of suppliers, each producing different skins, frames, stringers, fittings, clips, and reinforcement pieces.

Once those components reached the assembly plant, they still had to be positioned, drilled, aligned, and joined to the airframe with enormous numbers of rivets and bolts. The fuselage of an aircraft such as the 767 contained hundreds of thousands of fasteners, each of which required a hole to be drilled, inspected, sealed, and secured. Every joint added labor, weight, complexity, and another potential location for fatigue, corrosion, or manufacturing error.

CFRP offered the possibility of replacing many of those separate panels and supporting pieces with much larger, integrated structures. Instead of assembling a fuselage from thousands of relatively small metal parts, manufacturers could produce entire barrel sections as single composite pieces. The result promised not only a lighter airframe, but also a far simpler bill of materials and a substantial reduction in the inventory, tooling, labor, and assembly time required to build each airplane.

The use of CFRP in aircraft construction was not new. Military aircraft had incorporated steadily increasing amounts of composite material for decades, and Boeing had accumulated considerable experience with it through programs such as the B-2 stealth bomber, on which the company served as a major subcontractor. Composite structures had also appeared in selected components of commercial aircraft, including tail sections, control surfaces, and other secondary structures.

What was unprecedented was the scale of Boeing's ambition. CFRP had never before been used so extensively in a commercial airliner, nor had it been employed to produce a large passenger fuselage that would be subjected to tens of thousands of pressurization cycles over its service life. If Boeing could make the concept work reliably and economically, it would represent a genuine technological breakthrough.

A composite airliner promised airlines lower fuel consumption, reduced maintenance requirements, and potentially longer structural service lives. It also held out the possibility of lowering manufacturing costs by reducing the number of individual parts, fasteners, assembly operations, and labor hours required to build each aircraft. Whether those savings would actually materialize depended on the cost of the composite materials, specialized tooling, autoclaves, quality-control systems, and the production network required to manufacture the aircraft. But from Boeing's perspective, the commercial attraction was obvious: if it could reduce the cost of building the airplane while selling it at prices comparable to conventionally constructed competitors, the difference would flow directly into higher profit margins.

The manufacturing process represented only one of the many opportunities Boeing's engineers saw to improve the new airplane. Determined to extract as much efficiency from the design as possible, they also examined whether many of the pneumatic systems traditionally used aboard commercial aircraft could be eliminated.

On a conventional jetliner, a portion of the highly compressed air flowing through the engine is diverted before it reaches the combustion chamber. This air, known as bleed air, is extracted through ports in the engine's compressor section and routed through a network of ducts, valves, heat exchangers, and control equipment to power several aircraft systems.

Bleed air was commonly used to pressurize and air-condition the cabin, prevent ice from forming on the wings and engine inlets, pressurize hydraulic reservoirs, and maintain pressure within the potable-water system so that water could be delivered to the galleys and lavatories without relying entirely on electric pumps. It was a proven and dependable arrangement, but it required extensive ductwork and supporting equipment and diverted compressed air that the engines had already expended energy producing.

The bleed-air system had been an ingenious solution when it emerged at the dawn of the jet age, but it imposed an unavoidable efficiency penalty. The engine had to expend energy compressing air that was then extracted before reaching the combustion chamber, reducing the amount available to produce thrust. Depending on the operating conditions and the demands being placed on the system, supplying bleed air could increase fuel consumption by several percent.

Eliminating most of the pneumatic system would allow the engines to devote more of their compressed airflow to producing thrust, contributing to the new airplane's overall fuel-efficiency gains. It would also remove much of the extensive network of high-temperature ducts, valves, pre-coolers, regulators, and related equipment required to distribute and control bleed air throughout a conventional aircraft. Those components added weight, occupied valuable space, and required regular inspection and maintenance because they operated under high temperatures and pressures.

Boeing proposed replacing many pneumatic functions with electrically powered systems. Cabin pressurization and air conditioning would be supplied by electrically driven compressors, while other traditionally pneumatic or hydraulic functions would increasingly rely on electric motors and pumps.

The advantage was not simply that electrical systems were inherently lighter, because the larger generators, wiring, cooling equipment, and power electronics they required also added considerable weight. Rather, electrical power could be generated and distributed more precisely according to demand. A conventional pneumatic system continuously extracted compressed air and then regulated or discarded whatever was not required. An electrical system could deliver power only when and where it was needed, reducing wasted energy and allowing the aircraft's systems to operate more efficiently.

CFRP's strength, fatigue resistance, and low weight also allowed Boeing's engineers to make changes to the cabin environment that promised a more comfortable experience for passengers. One of the most important was the ability to maintain a lower cabin altitude.

Although conventional airliners are pressurized at cruise, their cabins are not maintained at sea-level pressure. Doing so would create a much greater pressure differential between the inside and outside of the aircraft, placing additional stress on the fuselage during every flight. Instead, most conventional jetliners maintain a cabin altitude of approximately 8,000 feet, or 2,400 meters, when operating at their highest cruise altitudes.

The cabin air is also extremely dry. Outside air at cruising altitude is very cold and contains little moisture, and conventional aircraft generally avoid adding significant humidity because condensation could promote corrosion within the aluminum structure and behind interior panels. The resulting low-humidity environment helps protect the airframe, but it can also contribute to dry eyes, irritated nasal passages, dehydration, and a general sense of fatigue during long flights.

The 787's composite fuselage offered greater resistance to both repeated pressurization cycles and corrosion. This would allow Boeing to design the aircraft around a cabin altitude of roughly 6,000 feet and to maintain higher humidity levels (via air humidifiers) than were typical aboard conventional aluminum airliners. Passengers would be able to disembark their flights after a long intercontinental journey feeling far more energetic and refreshed than they would on a comparable flight on a conventional commercial aircraft.

Boeing's engineers pushed the advantages of CFRP even further. Because the composite fuselage could accommodate larger structural openings without incurring the same weight penalties as a conventional aluminum design, they proposed passenger windows substantially larger than those found on comparable airliners — roughly 40 percent larger in overall area.

The larger windows would bring more natural light into the cabin and allow passengers seated farther from the fuselage to see more of the world outside. They would also help create a greater sense of openness, making the cabin feel less confined during long flights.

Larger windows, however, would also require larger shades. Since the new aircraft was already being designed around a more-electric systems architecture, Boeing explored an alternative to conventional pull-down shades. The engineers came up with a design where each window would incorporate an electronically dimmable panel controlled by buttons beneath it.

The technology relied on an electrochromic layer laminated within the window assembly. When a small electrical charge was applied, the layer changed its optical properties and progressively darkened. Passengers could select among several levels of tint, from nearly clear to heavily shaded. Even at its darkest setting, the window would continue to provide some visibility outside while blocking most of the glare and heat from direct sunlight. The system would also allow the cabin crew to control the windows collectively. During certain phases of flight or when creating a darkened cabin for rest, the crew could dim every window simultaneously, while still allowing individual passengers some control within preset limits.

For the first time in decades, Boeing appeared to be designing a commercial airliner whose passenger experience was itself a central selling point. Since deregulation, airlines had increasingly competed on price, schedules, frequent-flier programs, and service levels rather than on the intrinsic comfort of the aircraft they operated. The new Boeing promised something different — an airplane that airlines might promote not merely as transportation, but as a noticeably better way to travel. At least in theory.

***

Throughout 2004, an intense internal debate continued at Boeing over whether to launch Y2. On one point, however, there was broad agreement: if Boeing moved forward with the new airplane, Airbus would find it exceedingly difficult to respond with a competing design.

Airbus had already committed itself to its own vision of the future — the enormous A380. The program was consuming a substantial share of the company's financial resources, engineering talent, and management attention. By 2004, it was also beginning to fall behind schedule and incur significant cost overruns, placing even greater strain on Airbus' ability to undertake another major clean-sheet aircraft program.

Boeing believed Y2 could dominate the broad middle of the long-range market then served by the 767 and Airbus' A330 and A340 families. The 767 was already more than two decades old and badly needed either a major redesign or a replacement, especially since it was rapidly losing ground to Airbus' A330. The A330 and A340, by contrast, were still relatively young programs. Airbus had only recently begun to recover the enormous investment required to develop them, and many of their most profitable years were still expected to lie ahead.

Launching a direct competitor to Y2 would place Airbus in a difficult position. It would require another massive financial commitment immediately after the A380 while simultaneously threatening sales of the A330 and A340 — the very aircraft Airbus still needed to sell in large numbers to earn an adequate return on its investment. In effect, Airbus would be forced to cannibalize its own product line in order to prevent Boeing from sweeping the market.

In Seattle, the prevailing belief was that Airbus would not be able to launch an all-new aircraft to compete directly with Y2 for at least another decade. Boeing's market analysts believed the timing could hardly have been better. By the time the aircraft would enter commercial service, global air travel was expected not only to have recovered from the collapse that followed the September 11 attacks, but to have surpassed its previous record levels. If the Y2, now renamed 7E7, with the 'E' denoting efficiency, could genuinely reduce operating costs by 20 to 25 percent compared with existing aircraft in its class, airlines would have a powerful financial incentive to acquire it.

Persistently high fuel prices would only strengthen that incentive. Fuel had become one of the largest and most volatile expenses confronting airlines, and an aircraft capable of delivering such substantial efficiency gains could transform fleet economics.

Boeing believed carriers might be willing to replace even relatively young widebody aircraft, including the Airbus A330 and A340, if the savings promised by the 7E7 were sufficiently large. Boeing saw the potential to reset the economics of the entire mid-sized widebody market and compel airlines to reconsider aircraft that otherwise might have remained in service for many more years.

With each passing week, Boeing released more details about the 7E7 it was considering launching, and the aviation industry began to take notice. So did Airbus, which now faced the task of dampening the growing enthusiasm surrounding its rival's proposed aircraft. Predictably, Airbus publicly questioned Boeing's claims and emphasized the technical risks involved in constructing a commercial airliner so extensively from CFRP.

No one denied that the engineering challenges were considerable, but Boeing appeared to have an answer for each one. Among the most frequently cited concerns was lightning protection. Unlike aluminum, CFRP does not readily conduct electrical current. Since commercial airliners are struck by lightning on average about once a year, critics warned that a direct strike could damage the composite structure at the point of contact and send electrical surges into the aircraft's sensitive systems.

Boeing's engineers addressed the problem by proposing to incorporate a thin conductive mesh — made primarily of copper — into the outer layers of the composite skin. The mesh would provide a low-resistance path across the surface of the aircraft, allowing the electrical energy from a lightning strike to disperse over the airframe and exit safely, much as it did through the aluminum skin of a conventional airliner. Additional bonding, grounding, shielding, and surge-protection measures would protect the aircraft's electrical and electronic systems.

The solution would not make the composite airframe immune to lightning damage, nor would it eliminate the need for inspection after a significant strike. It did, however, demonstrate that the problem was manageable and that a commercial aircraft made primarily from composites could be designed to meet the same lightning-protection standards as its aluminum predecessors.

The concern over "ramp rash" was more difficult to dismiss. Commercial aircraft routinely suffer minor damage while parked at airport gates, where baggage carts, belt loaders, catering trucks, and other ground-service vehicles occasionally scrape against or strike the fuselage. On a conventional aluminum aircraft, such encounters often leave behind little more than a scratch or shallow dent.

Critics argued that CFRP behaved differently. Aluminum could deform under impact, providing visible evidence of damage that, if severe enough to necessitate a repair, would provide a visual cue that required attention. Composite material, by contrast, might retain its outward shape while sustaining cracking, crushed fibers, or delamination beneath the surface. Because such damage could be difficult to detect visually, there was concern that an apparently minor impact might compromise the strength of the structure without leaving an obvious dent.

Moisture intrusion was another concern. If the outer protective layers were penetrated, water could enter the damaged area and potentially worsen delamination through repeated temperature and pressure cycles. In a pressurized fuselage, any undetected weakening of the laminate demanded careful inspection because the structure had to withstand thousands of cycles over the aircraft's service life.

Boeing's engineers countered that the fuselage skin consisted of many layers of composite material and was designed to tolerate the kinds of minor impacts commonly encountered during ground operations. Superficial scratches would not normally penetrate deeply enough to affect the load-bearing fibers. More significant damage could be identified through specialized inspection techniques and repaired by removing the damaged material and bonding a composite patch into place.

Boeing also argued that many repairs could be completed without replacing large sections of the airframe. Airlines would be provided with approved materials, procedures, and portable repair equipment for use at maintenance stations throughout their networks. Nevertheless, the process would not be quite as simple as filling a scratch with polymer on the ramp. Composite damage first had to be assessed accurately, and repairs affecting the pressurized structure would require controlled preparation, curing, and inspection before the aircraft could safely return to service.

The comparison with aluminum was therefore not entirely one-sided. Metal aircraft could continue flying for some time with certain dents or scratches that fell within published limits, but those imperfections could also become sites for fatigue cracking or corrosion. CFRP eliminated some of those traditional vulnerabilities while introducing a different set of inspection and repair challenges.

But just to prove the durability of their proposed new "plastic plane" — a derogatory term that seems to have emanated at some point in Toulouse — Boeing's sales executives who went pitching this new plane to potential launch customers took along a suitcase-sized piece of CFRP material that had been fabricated during test runs of the manufacturing process, along with a heavy sledgehammer. They challenged any executive they were talking with to take the sledgehammer and slam it down as hard as they could on the CFRP sample. In each instance whenever the challenge was taken up, or even if the Boeing sales executive did it himself, there was no noticeable damage from the blows.

It was a clever sales pitch that had the intended effect because those same executives knew that a sledgehammer blow to a sheet of aluminum the same shape and size as the CFRP sample would cause immense damage.

Back in Seattle, a palpable sense of excitement was beginning to build around the 7E7. In some quarters, there was a growing belief that the aircraft might be precisely what Boeing needed to rekindle the spirit of the company's earlier years and restore its reputation for setting the direction of commercial aviation.

Given the number of leading-edge technologies being considered for the aircraft — and the dramatic gains in efficiency they promised — some even suggested that the 7E7 could become the 21st century's equivalent of the 707 — an airplane that would not merely compete in an existing market, but redefine the standards by which future commercial aircraft would be judged.

Yet, one formidable obstacle still stood between the 7E7 and a formal launch: Harry Stonecipher.

Although preliminary feasibility work on the 7E7 had begun under Condit as early as April 2003, the program already faced resistance from two influential figures on Boeing's board. John McDonnell and Harry Stonecipher, who had remained a director after stepping down as president and chief operating officer, both expressed concern about the cost of launching the aircraft. They warned that the project could be canceled if its development expenses became excessive.

The meaning of "excessive," however, depended largely on what Stonecipher and McDonnell believed Boeing should be willing to invest. By then, virtually everyone in the aerospace industry understood that developing an all-new commercial airliner such as the 7E7 would require an investment of at least $8 to $10 billion, probably more. A clean-sheet aircraft incorporating an unprecedented composite airframe, new engines, a more-electric systems architecture, and an entirely new production process was never going to be inexpensive.

The underlying message appeared unmistakable: Boeing's leadership was no longer prepared to commit the sums traditionally required to develop a genuinely new commercial aircraft. The company wanted the technological and competitive benefits of a revolutionary airplane, but Stonecipher and McDonnell seemed determined to obtain them within financial limits more appropriate to a derivative program. That tension — between Boeing's engineering ambitions and its leadership's reluctance to fund them fully — would shape nearly every major decision that followed.

Seasoned veterans of Boeing and the aerospace industry were aghast. To them, such messages suggested that the executives now pulling the levers of power intended to transform Boeing into another McDonnell Douglas, complete with the same underinvestment, shrinking product ambitions, and eventual decline in commercial aviation. Some went even further, concluding that Stonecipher had little genuine interest in Boeing's commercial-aircraft business and was focused primarily on a single customer: the United States government.

The thinking of Stonecipher and the board, however, was not entirely opaque. It eventually emerged that Boeing had established extraordinarily aggressive financial targets for the new aircraft. The company wanted to develop the 7E7 for approximately 40 percent of what it had cost to develop the 777, while reducing its manufacturing cost to roughly 60 percent of the 777's.

To many experienced observers, those targets bordered on fantasy. The 7E7 was supposed to incorporate an unprecedented composite airframe, new engines, an extensively electrical systems architecture, and revolutionary manufacturing processes, yet somehow cost dramatically less to develop and build than its far more conventional predecessor.

Alarm began to spread among those who believed Boeing's future depended upon launching a bold new commercial airliner. They feared that the company was attempting to obtain the benefits of a technological revolution without accepting the investment, risk, and institutional commitment such a revolution required.

Stonecipher and his supporters appeared to regard commercial aircraft manufacturing as an increasingly commoditized business — capital-intensive, cyclical, highly competitive, and burdened by thin margins. Defense contracting, by contrast, seemed to promise healthier returns, steadier revenues, and greater insulation from economic downturns. Government programs were rarely free of risk, but long-term contracts, political support, and the barriers to entry for competition made them appear far more predictable than gambling billions of dollars on the next commercial jetliner.

Boeing's most influential leaders were not necessarily opposed to developing a new commercial airliner. They simply wanted to ensure that the company did not assume the enormous financial risk traditionally associated with launching one. Outsourcing appeared to offer a way to reconcile those competing objectives. If Boeing moved forward with the 7E7, outside suppliers would play a far greater role than they had on any previous company program.

This would not be a conventional aircraft-development effort in which Boeing designed most of the airplane, manufactured its principal structures, and hired suppliers primarily to produce components to company specifications. Instead, the new program would be designed from the outset around outsourcing as much of the development, engineering, manufacturing, and financial burden as possible. Outsourcing would not be merely one element of the 7E7's production strategy. It would become the organizing principle around which virtually the entire program was constructed.

Manufacturing outsourcing had traditionally occurred only after a company had already established and refined a production process internally. Engineers would design the component, develop the tooling, perfect the manufacturing sequence, and optimize it as far as practical. Only then, usually under pressure to reduce costs and improve profit margins, would the company transfer production to an outside supplier capable of making the part more cheaply.

That work was often sent to regions where labor, facilities, and raw materials cost less. The supplier, however, generally inherited a mature design and a prescribed manufacturing process. Its task was not to reinvent the component, but to reproduce it according to the original company's drawings, specifications, tooling requirements, and quality standards.

Stonecipher and his lieutenants envisioned something far more ambitious. They believed greater savings could be achieved by giving selected suppliers responsibility not merely for manufacturing components, but also for designing them and determining how best to produce them. Freed from Boeing's established methods and internal cost structure, those partners could, at least in theory, devise more efficient processes and drive costs even lower.

That was precisely the strategy Boeing adopted for the new airplane. The company planned to outsource approximately 70 to 75 percent of its design and manufacturing work to a relatively small group of major partners. These companies would not function simply as conventional suppliers. They would become risk-sharing partners, responsible for financing, engineering, producing, and delivering entire sections and systems of the aircraft.

Boeing would increasingly cease to act as the manufacturer of most of the airplane's major components. Instead, it would oversee the architecture of the aircraft, manage the global supplier network, integrate the completed sections, and perform final assembly.

***

Working under a mandate to cap development costs if the new jetliner program was to be launched, Alan Mulally, still president and chief executive officer of Boeing Commercial Airplanes, and Mike Bair, then executive vice president of Boeing's Commercial Aviation Services organization, developed a plan intended to meet the aggressive financial targets established for the 7E7. Their strategy became known as the "risk and revenue-sharing partnership" model.

Under it, entire sections of the new aircraft would be assigned to major supplier-partners, which would assume responsibility not only for manufacturing those sections but, in many cases, for designing, engineering, and financing them as well. In return, the partners would receive a share of the revenues generated by aircraft sales.

This represented a fundamental departure from Boeing's traditional production system. In the past, suppliers had delivered parts and subassemblies to Boeing plants, where company employees combined them into major structures such as fuselage sections, wings, and tail assemblies before bringing those structures together on the final assembly line.

Under the new model, the supplier-partners themselves would produce substantially completed major sections of the aircraft. Those sections would then be transported to Boeing for systems integration, joining, testing, and final assembly. Boeing would therefore shift much of its role from designing and manufacturing the airplane to coordinating a global network of partners and integrating the large structures they delivered.

Boeing's involvement in the design of many major structures would be substantially reduced. The company would establish the aircraft's overall architecture, define performance and safety requirements, and specify the interfaces through which each section would connect with the rest of the airplane. The supplier-partners would then be responsible for much of the detailed engineering, tooling, and manufacturing needed to produce their assigned sections.

Once completed, those structures would be shipped to Boeing for final assembly. Even the meaning of "final assembly," however, was about to change. Under the proposed system, the major sections would arrive in Everett in a finished — or nearly finished — condition. They were expected to be delivered fully "stuffed," with as many systems, fittings, and interior components installed as practical before shipment.

In theory, fuselage sections would eventually arrive with much of their wiring, plumbing, insulation, floor structure, and interior equipment already in place. Some plans even contemplated the installation of items such as overhead bins and passenger interiors before the sections reached Boeing. Boeing workers would then join the major sections, connect the systems crossing the interfaces, install remaining equipment, test the completed aircraft, and prepare it for delivery.

The transformation would be profound. Boeing would no longer manufacture and assemble most of the airplane in the traditional sense. Its principal role would become that of systems architect, supply-chain manager, and final integrator of a commercial aircraft largely designed and built by others.

The concept was not entirely unlike the production system Airbus had used throughout much of its history. Airbus' industrial partners in France, Germany, Britain, and Spain were each responsible for designing and manufacturing major sections of an aircraft, whether fuselage sections, wings, empennage components, or other structural assemblies. Those sections were then transported to a central location for final assembly.

The critical difference, however, was that Airbus' partners were not independent outside suppliers in the conventional sense. They formed the industrial foundation of the Airbus consortium and later became integrated divisions of the Airbus organization. Although work was geographically dispersed, the participating companies remained part of the same broader enterprise and operated within a common corporate and engineering structure.

Airbus therefore retained direct institutional control over the design, manufacture, and integration of its aircraft. Whatever role government financing and subsidies played in reducing its exposure, Airbus and its shareholders ultimately bore the financial and commercial risk of launching a new airplane.

Boeing's proposed model for the 7E7 was fundamentally different. Major sections of the aircraft would be designed and manufactured by independent companies that Boeing did not own or directly control. Those suppliers would invest their own money, assume a portion of the development risk, and recover their investment through future production revenues. Boeing was therefore not merely dispersing manufacturing geographically. It was transferring substantial portions of the aircraft's detailed design, capital requirements, and program risk beyond the boundaries of the company itself.

Many industry observers — and particularly many veteran Boeing insiders — were stunned. They understood that outsourcing a greater share of an aircraft's components could reduce costs and expand access to specialized expertise. But some capabilities were considered sacrosanct for an airplane manufacturer, especially one with Boeing's history. Chief among them was the design and manufacture of the wing.

Over decades, Boeing had developed an unrivaled depth of experience in creating wing planforms optimized for specific aircraft and missions. A wing's planform encompassed its span; chord, or the distance from leading edge to trailing edge; taper from root to tip; sweep angle; aspect ratio; and total surface area. Those geometric characteristics, together with the wing's airfoil sections, twist, thickness distribution, and structural design, determined much of an airplane's aerodynamic performance, including lift, drag, fuel efficiency, stall behavior, stability, and handling qualities.

Designing an effective wing involved far more than selecting a set of dimensions. It required balancing aerodynamic efficiency, structural strength, weight, manufacturability, fuel capacity, and performance across a wide range of speeds and operating conditions. Boeing had accumulated generations of institutional knowledge in making those tradeoffs.

The company was so accomplished at wing design that some engineers described the discipline as bordering on a black art — part science, part experience, and part intuition refined through decades of testing and flight. Boeing was widely regarded as one of the industry's foremost authorities in the field. To many insiders, surrendering substantial responsibility for such a defining capability was not simply another outsourcing decision. It risked giving away a core element of what made Boeing Boeing.

What Boeing was proposing, in the eyes of many veteran insiders, amounted to giving away a large portion of its institutional inheritance — nearly a century of accumulated knowledge about how to design, engineer, and manufacture wings. Much of that expertise had never existed solely in drawings, formulas, or technical manuals. It resided in the experience of Boeing's engineers and production workers — in the judgments, techniques, and lessons refined across generations of aircraft programs.

Transferring substantial responsibility for wing design and construction to an outside company risked transferring far more than work. It meant teaching an independent supplier how Boeing performed one of the most technically demanding and strategically important tasks in commercial aircraft development. Once that knowledge had migrated beyond the company, it could not easily be reclaimed.

To many critics, Boeing appeared to be documenting the means of its own demise. It was pursuing a permanent solution to a short-term financial problem, sacrificing a core capability in order to reduce the immediate cost and risk of launching one airplane. The savings might appear attractive on a balance sheet, but the long-term consequences could prove detrimental and irreversible.

For Stonecipher, the logic was straightforward. By transferring responsibility for the design and manufacture of major aircraft sections to outside partners, Boeing could launch a new commercial-airplane program while committing far less of its own capital. The suppliers would finance much of the development work, assume a share of the technical and financial risk, and recover their investments through future production revenues.

Stonecipher pitched to the board that the arrangement would align the suppliers' interests with Boeing's. Because their own money was at risk, the partners would have every incentive to control costs, solve problems quickly, and deliver their sections on schedule. In theory, they would approach the program not as ordinary vendors paid to complete assigned work, but as investors whose returns depended upon the airplane's success.

There was one more inducement presented to the board — speed. Stonecipher pressed for an extraordinarily aggressive schedule, and Boeing committed to flying the new aircraft sometime in 2007. That meant Boeing and its partners would have roughly three years to complete the detailed design of an unprecedented composite airliner, develop entirely new manufacturing techniques, establish a global production network, certify thousands of new systems and components, and assemble the first flightworthy aircraft.

It was a breathtaking timetable. Boeing was attempting not merely to develop a new airplane, but to reinvent the way commercial aircraft were designed, financed, manufactured, and assembled — and to do all of it simultaneously. Yet the promise of limited investment, shared risk, and rapid entry into service proved irresistible. The board took the carrot and swallowed it whole.

The board authorized Boeing to begin formally offering the new jetliner to airlines. By then, the aircraft had shed its developmental 7E7 designation and acquired its permanent name: the 787 Dreamliner. The name was intended to capture the promise Boeing was making to both passengers and airlines. For travelers, the aircraft would offer a more comfortable cabin, larger windows, higher humidity, and a lower cabin altitude. For airlines, it promised substantially lower fuel consumption, maintenance costs, and overall lower operating expenses. In Boeing's marketing vision, it would be a "dream" for passengers to fly aboard and a "dream" for airlines to operate.

Boeing needed a substantial launch order before committing formally to the new program, and it knew which airline to approach first. At the top of the list was Japan's All Nippon Airways.

ANA's route network was particularly well suited to the aircraft's core value proposition. The airline operated numerous high-frequency domestic and regional routes on which widebody capacity and passenger comfort were desirable, but demand did not always justify an aircraft as large as the 777, which the airline had in its fleet. It also required an efficient long-range airplane capable of serving international routes that were too thin to support the largest widebodies. The 787 had been conceived precisely for that broad middle of the market. It promised to combine the range and comfort of a widebody aircraft with the operating economics needed to serve medium-density routes profitably.

On April 26, 2004, ANA's board approved an order for 50 aircraft — the largest launch order for a new Boeing airliner up to that time — and Boeing's board formally launched the 787 program that same day. ANA's initial order reflected both requirements, comprising 30 short-range 787-3s and 20 long-range 787-8s. The airline intended the new aircraft eventually to replace much of its fleet of 767s, along with a small number of Airbus A321s.

ANA provided Boeing with a highly credible launch partner. It was one of the manufacturer's most loyal and technically sophisticated customers, operating large fleets of 767s and 777s. Its pilots, maintenance personnel, and management were already familiar with Boeing's aircraft, training practices, and support systems. That familiarity reduced some of the commercial and operational uncertainty inherent in introducing such a technologically ambitious airplane.

Just as important, ANA's order offered real-world validation of Boeing's market strategy. The airline demonstrated that a major international carrier was prepared to invest heavily in a smaller, highly efficient widebody that could replace older aircraft, increase frequency on established routes, and open markets that could not profitably support a larger jet. ANA did not merely provide the order that launched the 787; it helped validate the entire business case upon which the airplane had been conceived.

ANA was also willing to accept the development risk associated with an all-new aircraft in exchange for early access, favorable launch-customer pricing, and an opportunity to influence aspects of the airplane's design. As the first customer, the airline could help shape the 787 around its operational requirements while securing an early position in the production queue.

For Boeing, winning ANA carried significance far beyond the size of the order. ANA was widely regarded as a conservative, technically sophisticated, and quality-conscious operator with exacting standards for reliability and performance. If an airline of

ANA's reputation was willing to become the first customer for such an ambitious aircraft, the decision would lend immediate credibility to the 787's underlying concept.

ANA's commitment served as a powerful signal to the broader market. It suggested that the airplane's promised economics and operating capabilities were sufficiently compelling to justify the risks of introducing an entirely new design. That endorsement would make it easier for other airlines to conclude that the 787 was not merely an intriguing Boeing proposal, but a commercially credible aircraft worthy of serious consideration.

ANA was not Boeing's only consideration in securing a blue-chip launch customer. The Japanese market as a whole figured prominently in the company's strategy for the 787. Japan had long been one of Boeing's strongest markets and, for many years, had been dominated almost entirely by Boeing aircraft. With Airbus making gradual inroads among customers Boeing had historically regarded as its own, the company was determined to preserve that position.

Under the new risk-sharing business model being introduced with the 787, Boeing offered Japanese industry responsibility for some of the aircraft's most valuable and technically sensitive structures — effectively, portions of the company's crown jewels. A consortium consisting of Mitsubishi Heavy Industries, Kawasaki Heavy Industries, and Fuji Heavy Industries (later renamed Subaru Corporation), received major design and manufacturing assignments.

Mitsubishi Heavy Industries would manufacture the composite wing box and major wing structures. Kawasaki Heavy Industries would produce portions of the forward fuselage, while Fuji Heavy Industries would design and manufacture the center wing box — the massive structural assembly through which the wings attach to the fuselage and through which many of the aircraft's principal flight loads are transferred. These companies were not newcomers to Boeing programs. All three had worked with Boeing on earlier aircraft, including the 767 and 777, and had accumulated significant experience manufacturing major structures and composite components.

They also possessed the financial strength Boeing required of its new risk-sharing partners. Each company was large enough to invest heavily in engineering, tooling, facilities, and production capacity while accepting the risk that those investments would be recovered only if the 787 sold in sufficient numbers. By assigning such important work to Japanese industry, Boeing was doing more than outsourcing major structures. It was binding some of Japan's most powerful industrial companies — and, indirectly, the Japanese airline market — to the long-term success of the 787.

When Boeing outsourced the design and manufacture of these major structures, it was seeking more than a reduction in its own financial exposure. The company also intended to use the supply chain as a tool for securing the 787's commercial success. The wings, in particular, could not be assigned to just any supplier. Boeing wanted them entrusted to industrial partners whose participation would send a powerful message to the market. By awarding such a prestigious and technologically critical role to Japanese companies, Boeing was signaling that Japan — and, by extension, the broader Asia-Pacific region — stood at the center of the 787's commercial strategy.

That calculation reflected where Boeing believed the future growth of aviation would occur. Passenger traffic across Asia, particularly around the Pacific Rim, was expanding rapidly and was expected to account for a substantial share of worldwide aviation growth over the following two decades. The region's airlines would require hundreds of new aircraft to serve expanding domestic markets, connect rapidly growing cities, and open new long-range routes. Boeing's leadership viewed Asia not merely as another important market, but as one of the principal engines of commercial aviation's future.

By giving Japanese industry responsibility for the 787's wings and other critical structures, Boeing created a powerful alignment of interests. The success of the airplane would benefit not only Boeing and its airline customers, but also some of Japan's largest industrial companies and the extensive network of suppliers supporting them. That industrial commitment could, in turn, strengthen the 787's position with Japanese airlines and reinforce Boeing's broader presence throughout the region.

Other major sections of the airplane were distributed among additional risk-sharing partners. Alenia Aeronautica of Italy, now part of Leonardo, was assigned responsibility for sections of the center and rear fuselage as well as the horizontal stabilizer. Dallas-based Vought Aircraft Industries was responsible for manufacturing major aft-fuselage sections at its North Charleston plant that was a joint-venture with Global Aeronautica, a subsidiary of Alenia. Others would be responsible for the nose section and other components of the aircraft. Taken together, these assignments reflected the global character of Boeing's new production model — the 787 would be designed and built not primarily within Boeing's own factories, but by a worldwide network of independent industrial partners whose financial fortunes were tied to the aircraft's success.

The transformation to this new way of manufacturing required extensive vetting, engineering reviews, financial assessments, and, most important, complex contract negotiations. Boeing had to determine whether each company possessed the technical capability, management discipline, production capacity, and financial strength to design and manufacture major aircraft structures while investing substantial amounts of its own capital. Yet much of this was completed within only six to twelve months. That was an extraordinarily compressed period in which to evaluate prospective partners and negotiate agreements governing intellectual property, design authority, certification responsibilities, production schedules, quality standards, financial exposure, and revenue sharing over what could become a multi-decade program.

At the time, the combined investment required from Boeing and its partners was estimated to exceed $8 billion, but most industry observers felt that was a very conservative estimate. Each participant therefore had to place a substantial ante on the table merely to secure a seat in the program. For Boeing, the attraction was obvious: much of the financial burden would be shifted beyond its own balance sheet. The danger was equally apparent. The company was committing the success of its most important commercial-aircraft program in decades to a global network of partners selected, evaluated, and contractually bound together on an exceptionally aggressive timetable.

This new approach to building the next generation of commercial jet transport would upend supply chains that Boeing had refined over decades. Long-established flows of parts, materials, engineering data, and production work — many of which had operated at a predictable rhythm across successive aircraft programs — would now be broken apart and redistributed among partners scattered around the world.

Entirely new supply chains would have to be created, qualified, and synchronized. Composite structures, advanced electrical systems, new engines, specialized software, and unprecedented levels of systems integration would all have to come together according to a tightly controlled schedule. Timing would therefore become critical. A delay or quality problem at any major partner could prevent Boeing from completing the aircraft, regardless of how well the rest of the production network performed. Because the large sections were supposed to arrive substantially finished, problems originating deep within a partner's own network might not become visible until they reached Boeing's final assembly line.

The result would be a supply chain of unprecedented complexity in commercial aviation. Boeing was not merely introducing a radically different airplane. It was simultaneously creating a radically different industrial system to design, manufacture, transport, and assemble it. The technical novelty of the 787 and the novelty of its production model would therefore become inseparable risks.

Even the transportation of the 787's major structures would have to be reinvented. Boeing's principal partners were scattered around the world, yet they would be producing components of unprecedented size. Each wing would measure nearly 100 feet, or about 30 meters, in length, while some fuselage sections would exceed 18 feet, or 5.7 meters, in width and 19 feet, or 5.9 meters, in height. Once completed, these enormous structures somehow had to be moved quickly and reliably to Boeing's final assembly plant.

Transporting them by sea was impractical. Ocean shipment would take too long and would still require each component to be moved overland from the supplier's factory to a port, loaded aboard a ship, unloaded at the destination port, and transported overland once again to final assembly. The specialized infrastructure, handling equipment, storage facilities, and scheduling required would have added both cost and delay to a production system intended to operate with minimal inventory.

Boeing decided to move the major sections by air. The company acquired used 747-400 passenger aircraft and converted them into enormous freighters known as Dreamlifters. Their fuselages were dramatically enlarged to create cavernous cargo holds capable of carrying complete wings and major fuselage sections. The rear fuselage swung open laterally, allowing the oversized structures to be loaded and unloaded through the back of the aircraft.

Boeing initially planned to operate three Dreamlifters. Each would circulate through the global production network, collecting completed structures from supplier-partners and transporting them to final assembly. One flight might deliver a fuselage section from Italy or Japan, while another would retrieve other major assemblies from a different partner. The aircraft would function as the connective tissue of Boeing's new worldwide production system.

The entire operation was intended to follow a just-in-time manufacturing model. Rather than maintaining large inventories of expensive structures or subassemblies near the assembly line, Boeing expected components to arrive only when they were required. Ideally, a completed section would be unloaded from a Dreamlifter and moved directly onto the production floor just in time for it to be mated up with other structures.

The financial logic was compelling. Inventory sitting in storage represented cash tied up in assets that were not yet generating revenue. By minimizing work in process and reducing storage requirements, Boeing could preserve liquidity, shorten production cycles, and lower the amount of capital required to build each aircraft.

If the system worked as envisioned, final assembly would become remarkably fast. Boeing workers would join the major sections, connect the electrical, hydraulic, plumbing, and control systems crossing the structural interfaces, complete testing, and prepare the aircraft for delivery. Boeing claimed that once the supplier and distribution network had matured and production was operating smoothly, a 787 could potentially move through final assembly in as little as three days.

The company also projected an eventual production rate approaching sixteen aircraft per month — an extraordinary pace for a widebody airliner. Such a rate promised to shorten delivery backlogs and allow airlines to receive newly ordered aircraft far more quickly than had been possible under traditional widebody production systems.

But the entire concept depended upon flawless synchronization. Just-in-time production worked only if every major section arrived complete, properly engineered, fully inspected, and precisely on schedule. A single missing component, unfinished system, or dimensional mismatch could halt final assembly and transform Boeing's lean production system into an expensive bottleneck.

***

The effort to maximize outsourcing on the 787 was only one part of Harry Stonecipher's broader campaign to change how Boeing operated. After being brought out of retirement following Condit's abrupt resignation, Stonecipher undertook a top-to-bottom review of the company's organization, operations, and capital assets.

He understood that he had a relatively narrow window in which to remake Boeing according to his vision. Stonecipher believed he knew how to make the company more disciplined, more efficient, and, above all, more effective at generating returns for shareholders.

His model drew heavily from the approach he had followed at Sundstrand and McDonnell Douglas — concentrate the company on the activities it considered most valuable, impose strict financial discipline, and shed operations that failed to earn an adequate return on the capital invested in them. If Boeing was going to remain in the commercial airplane business, Stonecipher believed it had to find a leaner and less capital-intensive way to build aircraft. That meant reducing the company's ownership of factories, equipment, and manufacturing operations wherever outside suppliers could perform the work more economically.

Becoming lean also required disposing of assets that management considered underutilized, financially burdensome, or insufficiently profitable. Boeing owned facilities and operations scattered across the country and around the world, many of which required substantial continuing investment. Stonecipher's team evaluated them according to a relatively straightforward financial standard: if an asset could not generate cash flow comfortably exceeding the cost of the capital required to own, maintain, and operate it, Boeing should consider selling it.

Among the first major operations subjected to that scrutiny was the company's Wichita division, commonly known as Boeing Wichita. It was not a peripheral property but one of Boeing's largest and most important manufacturing centers. Nevertheless, under Stonecipher's emerging philosophy, even a strategic industrial asset was expendable if its work could be transferred outside the company and the capital tied up in it could be put to more profitable use.

Wichita, Kansas, known as the "Air Capital of the World," had been deeply rooted in aviation since the early 1920s. As American aviation expanded rapidly after World War I, numerous aircraft manufacturers established operations there. The region's steady winds provided favorable conditions for testing new airplanes, while Wichita's location near the geographic center of the continental United States made it a logical stopping point for early transcontinental mail and passenger services. Its extensive rail connections also made it relatively easy to transport raw materials, engines, and other components into the city.

By 1928, Wichita was home to roughly sixteen aircraft manufacturers producing approximately 120 airplanes each week — about one-quarter of all aircraft then being built in the United States. Its importance grew even further during World War II, when the city became one of the central pillars of the nation's aircraft-manufacturing industry.

Boeing had established its presence in Wichita in the 1930s through its acquisition of the Stearman Aircraft Company, which produced some of the era's most widely used civilian and military training aircraft. When the United States entered World War II, Boeing rapidly expanded the Wichita operation and turned it into one of the principal production centers for the B-29 Superfortress. Boeing maintained the facility after the war, and Wichita subsequently played a major role in producing the B-47 Stratojet, B-52 Stratofortress, and KC-135 Stratotanker. As commercial aviation expanded during the 1960s, the plant increasingly manufactured major subassemblies for Boeing's jetliners.

Beginning in 1966, constrained by limited factory space and labor availability in the Seattle area, Boeing assigned Wichita responsibility for manufacturing the complete fuselage of the new 737. Finished fuselages were loaded onto specially designed railcars and transported to Boeing's Renton plant in Washington, where the aircraft underwent final assembly.

Over the following decades, Wichita became extraordinarily proficient at producing the 737 fuselage. During the program's early years, the plant manufactured approximately four to six fuselages per month. By 2004, output had risen to roughly twenty-four per month. The increase reflected not simply greater efficiency, but decades of accumulated experience, production-line expansion, process refinement, automation, workforce specialization, and continuous improvement. Wichita had evolved into one of the most mature and productive manufacturing operations within Boeing's commercial-aircraft system.

Despite the efficiencies Wichita had achieved, the operation required enormous amounts of working capital. Supporting high production rates meant maintaining substantial inventories of raw materials, partially completed structures, tooling, and other production assets. The facilities also demanded continuing investment to maintain equipment, modernize production lines, and preserve the capacity needed to support Boeing's commercial aircraft programs.

From Stonecipher's perspective, however, Wichita was producing work that had become increasingly commoditized. After manufacturing thousands of fuselages and subassemblies, the plant was turning out products that management regarded as commodities — important to the airplane, certainly, but not necessarily activities Boeing itself had to own. If an outside company could manufacture the same structures at lower cost while assuming the capital investment and inventory burden, what value did Boeing gain by keeping the work in-house?

In April 2004, Boeing announced that it was studying the sale of its commercial-aircraft operations in Wichita, Kansas, along with facilities in Tulsa and McAlester, Oklahoma. In February 2005, the company reached an agreement to sell those operations to an investment group led by Onex Corporation for approximately $1.2 billion. The transaction closed on June 16, 2005, and the former Boeing facilities emerged as an independent company named Spirit AeroSystems.

The decision fit perfectly with Stonecipher's asset-light philosophy. Boeing would continue purchasing 737 fuselages and other major structures from the same factories and, in many cases, from many of the same workers. The difference was that Boeing would no longer own the plants, carry their inventories, finance their modernization, or employ their workforce. Those financial and operational responsibilities would now belong to Spirit.

Like several other decisions made during the Condit-Stonecipher era, the divestiture would eventually come back to haunt Boeing. The company had separated itself from one of its most experienced and strategically important manufacturing operations, only to become heavily dependent upon that operation as an outside supplier. Problems involving cost, production stability, quality, and coordination would increasingly expose the risks of placing such a critical link in the commercial aircraft production system beyond Boeing's direct control.

Boeing was disposing of other operations during the same period. In 2004, Boeing Capital sold its commercial-finance business to General Electric Capital, although Boeing retained Boeing Capital itself and continued using it to support aircraft financing and leasing. In 2005, Boeing also agreed to sell Rocketdyne Propulsion & Power to Pratt & Whitney. Boeing had acquired Rocketdyne through its purchase of Rockwell's aerospace and defense businesses less than a decade earlier.

Each divestiture was presented as part of an effort to create a more focused, leaner, and financially disciplined company. Assets that consumed capital without producing sufficiently attractive returns would be sold, while Boeing concentrated on the activities its leadership considered most strategically valuable. The underlying theory was that shedding factories, financing portfolios, and specialized manufacturing businesses would make Boeing more competitive. The longer-term question was whether the company was becoming leaner — or gradually disposing of the industrial capabilities upon which its competitiveness depended.

There was one more strategic objective Stonecipher wanted addressed quickly: limiting Airbus' ability to obtain government launch aid for a new aircraft that could compete with the 787.

The dispute had its roots in a 1992 bilateral agreement between the United States and the European Community governing government support for large commercial aircraft. The agreement did not eliminate subsidies altogether. Instead, it prohibited certain forms of production support and permitted governments to finance as much as one-third of a new aircraft program's development costs, subject to repayment and other conditions.

Airbus' launch of the A380 intensified Boeing's objections to that arrangement. The governments of France, Germany, Britain, and Spain had committed substantial reimbursable launch investment to the program, and Boeing maintained that the financing distorted the market by allowing Airbus to pursue an aircraft whose commercial prospects might not otherwise have justified the enormous development expense.

Boeing's initial reaction to the A380 had been relatively restrained. Many within the company believed Airbus was committing billions of dollars and much of its engineering capacity to an aircraft serving a limited market. If the A380 proved to be the commercial boondoggle Boeing expected, it would weaken Airbus financially and perhaps undermine European governments' willingness to support another ambitious program.

The impending arrival of the 787 changed that calculation. Boeing understood that if the new aircraft delivered its promised reductions in fuel consumption and operating costs, it could force Airbus to respond much sooner than planned. Airbus might then seek another round of government financing to develop a direct competitor even while it remained heavily committed to the A380.

Therefore, Stonecipher supported a more confrontational approach. Boeing pressed the U.S. government to challenge the European support system and prevent Airbus from using launch aid to finance its response to the 787. The objective was not merely to punish Airbus for the A380. It was to raise the financial and political cost of launching whatever aircraft came next.

On October 6, 2004, the United States formally requested consultations at the World Trade Organization over European support for Airbus. At the same time, Washington announced that it was terminating its participation in the 1992 bilateral agreement. The American complaint alleged that launch aid and other measures provided to Airbus constituted prohibited or actionable subsidies under WTO rules.

The European Community responded that same day with its own WTO complaint against the United States. It alleged that Boeing had benefited from extensive federal, state, and local support, including tax incentives, NASA and Department of Defense research programs, and other government measures. The dispute thus immediately expanded from a challenge to Airbus launch aid into a broader examination of how both manufacturers had benefited from public support.

Boeing and the U.S. government may have calculated that Airbus would hesitate to request launch aid for another aircraft so soon after the A380. Doing so would reinforce Boeing's contention that Airbus could not launch a major commercial aircraft program without government financing. Nevertheless, the WTO proceedings did not prevent European governments from supporting Airbus' eventual response to the 787, the A350 XWB.

What followed became one of the longest and most expensive trade disputes in WTO history. The WTO ultimately found that both Airbus and Boeing had received subsidies that violated international trade rules or caused adverse effects to the other side. The litigation led to authorization for billions of dollars in retaliatory tariffs by both the United States and the European Union.

The dispute was not truly "settled" after a dozen years. It continued for roughly sixteen years before the United States and European Union agreed in 2021 to suspend their retaliatory tariffs for five years while pursuing a negotiated framework governing support for large commercial aircraft. By then, however, Boeing had long since launched the 787 — and now faced the far more immediate challenge of delivering the revolutionary airplane it had promised.

Accelerating Towards the Precipice

When Airbus launched the A380 in late 2000, the program's development schedule called for the aircraft to make its first flight by mid-2004 and enter commercial service in early 2006. But mid-2004 came and went, and Airbus was eventually forced to acknowledge what had already become obvious: the program had encountered serious technical problems and was falling behind schedule.

The aircraft was not rolled out for public view until January 2005, during a lavish ceremony attended by several European heads of state. Four months later, it finally made its first flight, nearly a year later than originally planned. Shortly afterward, Airbus delivered more troubling news to the airlines that had ordered the aircraft. The program was significantly behind schedule, and entry into commercial service would be delayed until late 2007, approximately 18 months later than originally planned.

One of the principal causes of the delay was a mismatch in the computer-aided design software being used across Airbus' European operations. The software, known as Computer-Aided Three-Dimensional Interactive Application, or CATIA, was being used in two different versions by Airbus' French, German, Spanish, and British partners. France and the United Kingdom were using the most current version, while Germany and Spain were working with an older one.

Although the two versions were broadly compatible, the older version lacked some of the newer system's precision and design capabilities. As a result, engineering data developed at one facility did not always translate accurately when incorporated into work being performed elsewhere. The consequences became particularly severe during the manufacture and installation of the A380's vast electrical system.

When thousands of miles of wiring began to be installed in the prototype aircraft, engineers discovered that many of the wire bundles did not fit properly within the spaces allocated for them.

Instead of routing closely along bulkheads, frames, and other structural members as intended, some bundles had to be stretched more directly through the aircraft. In many cases, this caused them to interfere with other systems and components being installed nearby.

CATIA had been intended to allow Airbus to accomplish with the A380 what Boeing had achieved with the 777 — design and build a highly complex commercial aircraft digitally, without first constructing a complete full-scale physical mock-up. But the incompatible software versions undermined that effort. The digital model did not always correspond precisely with the aircraft being assembled on the factory floor.

Much of the wiring had to be re-fabricated and installed by hand, an extraordinarily labor-intensive process that contributed to the delay in the aircraft's first flight. Airbus then had to redesign the wiring system and develop a more efficient method of manufacturing and installing it on the production line. That work caused the A380's entry into commercial service to slip even further.

Save for its enormous size, the A380 was a relatively conventional airliner when compared with Boeing's 777 or Airbus' own A330 and A340, yet the program still encountered major difficulties. What, then, did that suggest about the 787 Dreamliner? Although the 787 may have appeared conventional from the outside, almost nothing about it was conventional. Its structure, systems, materials, supply chain, and manufacturing strategy represented a radical departure from the way Boeing had designed and built its previous commercial aircraft.

The warning bells should have been sounding in Chicago and Seattle, as well as at the facilities of Boeing's strategic partners around the world. Instead, there appears to have been a measure of cautious satisfaction within Boeing as Airbus struggled with the

A380. At last, its European rival seemed to have been humbled after biting off more than it could chew.

But if Boeing's leadership took any comfort in Airbus' difficulties or believed that the competitive pendulum had finally begun to swing back in Boeing's favor, those perceptions would prove premature.

***

On February 23, 2005, a member of Boeing's board of directors and the company's legal and ethics officers received a collection of emails from an anonymous source. Many of the messages were sexually explicit and documented an affair between Harry Stonecipher, who was married, and Debra Peabody, a female Boeing vice president.

After a brief investigation, Stonecipher admitted that the relationship had been consensual. Although investigators found no evidence of favoritism, harassment, misuse of company resources, or harm to Boeing's business, the board concluded that his conduct had violated the company's code of ethical conduct and asked for his resignation.

It was the same code that Stonecipher himself had imposed companywide and demanded strict adherence to as part of his effort to restore Boeing's reputation following the scandals involving the Air Force tanker lease and the company's launch-services programs.

The irony was difficult to overlook. Stonecipher had been brought out of retirement to lead Boeing out of a succession of scandals and restore its credibility. He had insisted that the highest standards of ethical conduct were necessary to accomplish that task, had imposed those standards with uncompromising intensity, and had ultimately fallen because he failed to live by them himself.

More troubling than the affair alone was Peabody's position within Boeing. She was a vice president who oversaw operations for the office of Boeing's chief Washington lobbyist. Although there was no evidence that the affair involved government business, her proximity to Boeing's political and lobbying operations gave the scandal an added dimension that the company could ill afford. Stonecipher's short tenure as Boeing's chief executive had begun under the shadow of the federal government, and it ended with that shadow still looming over him.

Harry Stonecipher submitted his resignation and left Boeing on March 7, 2005.

The search for a new chief executive began immediately, but in the meantime Boeing once again found itself at the center of a headline-grabbing scandal. The only consolation was that, on this occasion, the company acted quickly and decisively. Although the board and Wall Street had viewed Stonecipher as the executive most capable of restoring Boeing's financial and operational performance, particularly from a shareholder's perspective, there was little hesitation in removing him once the affair became known.

Boeing's stock barely moved following the announcement of Stonecipher's resignation, suggesting that investors remained confident the company could continue moving forward without him.

However, the manner in which the affair was exposed should have raised concerns within the boardroom. Someone inside or close to Boeing had learned of Stonecipher's relationship with Peabody, obtained copies of their email correspondence, and deliberately forwarded the evidence to the company's directors and ethics officials. The episode suggested that serious resentment toward Boeing's senior leadership had taken hold within the company.

That resentment had been building since Philip Condit's elevation to the chief executive's office in 1996. It intensified after Boeing acquired McDonnell Douglas, moved its headquarters out of the Seattle area and into Chicago, and installed Stonecipher and other former McDonnell Douglas executives in influential leadership positions. To many longtime Boeing employees, the company had not merely acquired McDonnell Douglas; McDonnell Douglas had gradually taken control of Boeing.

By the time Stonecipher became chief executive, that resentment had hardened into open disdain. Contemporary reports suggested that word of the affair had circulated inside the company before someone obtained the emails and sent them to Boeing's leadership. Whether motivated by ethical concern, personal animosity, or a desire to remove Stonecipher, the source had found the evidence needed to bring him down.

The episode exposed the widening divide between Boeing's executive leadership and its rank-and-file employees. It should have served as a warning to the board as it began searching for Stonecipher's replacement. Instead, the old adage that some things never change appeared to remain firmly in place.

Among the leading candidates were Alan Mulally and James Albaugh, both of whom had also been considered following Philip Condit's resignation. A third candidate was James McNerney, a member of Boeing's board who had likewise been considered during the previous succession process. By considering Mulally and Albaugh alongside McNerney, the board appeared to be weighing both internal and external candidates.

Of the three, Mulally was still widely regarded by Boeing employees and many industry observers as the executive most capable of restoring the company's traditional engineering-centered culture and repairing its strained relationship with its workforce. He had spent his career inside Boeing, had led the development of the 777, and remained closely identified with the company's commercial-airplane heritage. The board, however, had other priorities.

On June 30, 2005, following what it described as an exhaustive search, Boeing announced that James McNerney would become its next chairman, president, and chief executive officer. If Boeing's employees had hoped that Stonecipher's departure would bring a return to the Boeing of old, they were sorely disappointed.

McNerney had spent most of his career at General Electric under the tutelage of Jack Welch and was cut from much the same managerial cloth as Stonecipher. He was not an engineer. His career had instead been defined by a steady progression through operational, international, and high-stakes executive assignments, with an emphasis on financial performance, organizational change, and the management of complex global businesses. Those qualities were likely among the principal reasons the board favored him over the other candidates.

Any hope among the rank and file that Stonecipher's sudden departure would produce a change in direction was about to be extinguished. McNerney's elevation to the chief executive's office would not reverse the course Boeing had taken since acquiring McDonnell Douglas eight years earlier. It would accelerate it.

In the meantime, Boeing had a revolutionary new airliner to build, and the company's third chief executive in less than two years was about to have his hands full.

***

Besides the Stonecipher scandal, his resignation, and McNerney's appointment as chief executive, 2005 would prove to be a pivotal year for Boeing.

The price of oil continued to rise, driven largely by increasing demand from an accelerating global economy. Much of that growth was coming from China, whose appetite for energy had soared as the country rapidly emerged as the world's factory floor. Few industries felt the increase in oil prices more acutely than the airlines, whose operating costs were particularly sensitive to the price of fuel.

With oil prices expected to remain high, fuel efficiency assumed even greater importance in airline fleet planning. Carriers began taking a serious look at the 787 Dreamliner and the operating efficiencies Boeing promised it would deliver. By the end of 2005, airlines had ordered 235 additional 787s, bringing the program's total order book to nearly 300 aircraft. There was no longer much doubt that Boeing had once again correctly anticipated the market's demand for a new kind of airliner.

Airbus watched the orders pile up but publicly dismissed them as part of the customary surge that often followed the launch of a new aircraft. It argued that the 787 would not necessarily be able to compete effectively with an upgraded version of its own A330, which Airbus planned to make more fuel-efficient by tweaking the design.

Airbus also contended that because the 787 differed so substantially from conventional airliners, airlines would face significant costs integrating it into their fleets. Those additional expenses, Airbus suggested, could offset some of the fuel-efficiency benefits the Dreamliner promised.

Even so, Airbus could not ignore what the 787 offered. At first, it characterized the Dreamliner as little more than an updated version of the A330, hoping to emphasize the advantages its existing aircraft already possessed. One Airbus executive went so far as to deride the 787 as a "Chinese copy" of the A330, an especially inflammatory remark given the importance of the rapidly expanding Chinese market to Airbus and its hopes for the A380.

Once the 787 was formally launched, Airbus shifted into high gear. It proposed a substantially modified version of the A330 incorporating a new composite wing, upgraded systems, and the same new generation of engines being developed for the 787. Airbus claimed that the improved A330 could deliver performance that, while not exactly comparable to the 787, with a lower acquisition cost it could negate any benefits the 787 would bring to an airline's fleet. And if an airline already had Airbus planes it its fleet then the integration would be far easier and less costly. The initial market reaction was tepid.

At the same time, Boeing had not overlooked Airbus' own early surge of orders for the A380. By the end of 2004, Airbus had accumulated 122 orders for the giant aircraft, most of them placed shortly after the program's launch. Orders had since slowed to a trickle. That pattern was not necessarily unusual for a newly launched airliner, but the technical delays affecting the A380 undoubtedly gave airlines still sitting on the fence reason to hesitate.

Boeing continued to analyze the potential demand for very large aircraft, and its conclusions remained far below Airbus' projections. Airbus forecast a market for approximately 1,500 to 1,700 very large airliners over the next 20 years. Boeing believed the actual demand would be far lower than that amount.

Boeing's analysis concluded, among other things, that the 777 would continue to erode the lower end of the very-large-aircraft market, leaving only the highest-capacity routes for an aircraft as enormous as the A380. It also showed that most airlines interested in the A380 were seeking to replace aging 747s on routes too large for the 777 to serve effectively.

Of the airlines that had ordered the A380, only two would ultimately acquire more than a dozen. Emirates ordered the lion's share and would eventually account for nearly half of all A380s produced because it constructed much of its route network around the aircraft. Most other carriers used the A380 merely to augment existing networks that had heavy-traveled long-distance routes within them.

Singapore Airlines would ultimately order 24 A380s, while the other dozen airlines that operated the aircraft ordered between three and 14 apiece. In practical terms, most carriers that acquired the A380 were introducing what amounted to an orphan fleet into their operations. Supporting a very small number of aircraft —

especially airplanes of such extraordinary size and complexity — would add substantially to maintenance, training, inventory, and operating costs.

If most A380 customers were acquiring the aircraft to replace 747s on routes that required a 747-sized airplane or something larger, Boeing faced an obvious question: What if it introduced a new and modestly improved 747?

For an airline already operating the 747, an updated version would be far easier to integrate into its fleet than the all-new A380. It would also be more economical than introducing an entirely new aircraft type in very limited numbers. An updated 747 could also fill the capacity gap between the largest twin-engine aircraft and the A380. With the A380 optimized for approximately 500 passengers or more, the next rung down remained occupied almost entirely by the 747.

As one of his first actions after returning as chief executive in 2004, Stonecipher authorized Boeing's engineers to study such an aircraft. Initially known as the 747 Advanced, the proposed model would feature a stretched fuselage capable of carrying approximately 34 additional passengers, modified wings for improved efficiency, and selected technologies being developed for the 787 Dreamliner.

Boeing's internal analysts estimated that the updated 747 could be brought to market for approximately $2 billion, far less than the estimated $12 billion required to develop an entirely new aircraft. There was also another market the new 747 could serve: air cargo.

The 747, whose origins could be traced to the Air Force's CX heavy-transport competition four decades earlier, carried the DNA of a freighter in its very design. It had been conceived during the same era as the supersonic transport and was expected by some planners to be relegated primarily to cargo service once fleets of SSTs filled the skies.

That future never arrived, but the 747's freight-carrying capabilities remained unmatched. With the global economy booming, air cargo traffic — measured in freight tonne-kilometers, the industry standard combining weight carried with distance flown — was expected to double over the following decade.

In November 2005, in one of the first major program decisions made under McNerney's leadership, Boeing formally launched the 747-8. Two versions would be offered: the 747-8 Freighter and the passenger-carrying 747-8 Intercontinental, commonly referred to as the 747-8I.

Cargolux and Nippon Cargo Airlines almost immediately placed orders for a combined 18 freighters, getting the program underway. It would be another year before Boeing secured the first order for the passenger version.

By the end of 2005, Boeing had racked up a combined total of 1,029 net commercial aircraft orders across its product line for that year, which included 257 orders for the 787 Dreamliner. It was a record year that shattered every previous company sales mark. There was only one problem: Airbus exceeded Boeing's total by approximately two dozen aircraft. The difference was not large, but it allowed Airbus to retain its claim as the commercial-airliner industry's sales leader.

The following year, 2006, would prove to be another banner year for both Boeing and Airbus as they continued riding the crest of the booming global economy. But storm warnings were beginning to appear.

By the middle of the year, reports began circulating throughout the industry that the 787 program was encountering serious development problems. Boeing's supplier-partners were struggling to perfect the processes required to manufacture the aircraft's enormous composite wings and fuselage barrels. Components that met the required structural specifications were also arriving substantially overweight.

Any increase in structural weight would directly affect the aircraft's promised fuel consumption. In an era of rising fuel prices, fuel burn had become one of the most important factors in an airline's purchasing decision. Even a small increase could have a measurable effect on an airline's operating costs and bottom line.

Because of the growing weight problem, entire subassemblies — and in some cases individual parts — had to be redesigned. One consequence was the decision to use substantially more titanium throughout the aircraft.

Although the 787's wings and fuselage barrels were constructed primarily from carbon-fiber-reinforced polymer, or CFRP, they still contained thousands of other components that had originally been intended to be manufactured from aluminum and traditional metal alloys. Even though each fuselage barrel was produced largely as a single composite section, it still required internal framing and structural supports to maintain its shape and carry loads.

Many of those internal structures were themselves composite, but others, including portions of the cabin-floor framework, were made from conventional metals. The wings likewise contained numerous metallic fittings, brackets, fasteners, and subassemblies.

As weight increased, Boeing and its suppliers began substituting titanium for more of those parts. Titanium was exceptionally strong and relatively light for a metal, but it was also expensive and difficult to machine. The change created the additional challenge of finding suppliers — or expanding the capabilities of existing ones — that could fabricate titanium components in the required quantities and at the necessary production rate.

It was becoming increasingly difficult to believe that the 787 would meet its planned first-flight date in 2007. In Chicago, however, Boeing's leadership maintained that such problems were normal for a new-aircraft program and that the Dreamliner remained on schedule to meet its major milestones. Wall Street appeared to accept those assurances, as reflected in Boeing's rising share price, but many aviation-industry experts were becoming doubtful. With 787 orders continuing to arrive, blue skies still appeared to lie ahead. But clouds were beginning to gather.

In the months leading up to the 2006 Farnborough International Airshow, airlines approached by Airbus about its proposed A330 upgrade responded with little enthusiasm. Carriers that had already reviewed Boeing's projected performance figures for the 787 told Airbus bluntly that their proposed improvements would not give the A330 the capabilities necessary to compete effectively with the Dreamliner. Airbus clearly had to do something or risk surrendering a large portion of the lucrative widebody market to Boeing.

At Farnborough in July, Airbus stunned the commercial aviation industry by announcing that it would develop an all-new, clean-sheet airliner constructed primarily from composite materials and intended to compete directly with the 787. The new aircraft would be designated the A350 XWB.

The letters stood for "extra widebody," emphasizing that the A350's fuselage would be approximately five inches wider than the 787's. Five inches may not have seemed significant to the average passenger, but to an airline it could make a substantial difference in cabin configuration and passenger comfort.

The 787 had originally been promoted as comfortably accommodating eight-abreast seating in economy class in a two-four-two arrangement. That configuration would provide passengers with generous shoulder and elbow room while still allowing airlines to operate the aircraft profitably.

When airlines examined the 787's cabin dimensions more closely, however, they determined that the aircraft could accommodate nine-abreast seating in a three-three-three arrangement. The tradeoff was narrower seats, reduced shoulder room, and slightly narrower aisles.

However, the additional seat in each row could dramatically increase passenger capacity, revenue, and profit margins. As a result, nearly every airline that ordered the 787 selected a nine-abreast economy configuration.

Airbus took careful notice. When it unveiled the A350 XWB, it immediately emphasized the advantage of its slightly wider fuselage. Its message to airlines was straightforward: They could install nine-abreast economy seating without subjecting passengers to the same degree of crowding found in a similarly configured 787.

What truly shocked the industry at Farnborough was not merely Airbus' announcement of the new aircraft. Airbus also revealed that it had secured a letter of intent from Singapore Airlines, one of the world's most respected carriers, even though Singapore had signed a letter of intent for 20 Boeing 787s only the previous month.

The announcement was striking because the 787 was already in development and its production system was beginning to take shape, while the A350 XWB remained little more than a concept on paper.

Airbus had again demonstrated its marketing savvy. Singapore's letter of intent was essentially conditional. If Airbus produced an aircraft capable of delivering the promised performance, Singapore Airlines intended to order it. The airline had little to lose by signing such an agreement, while Airbus gained the endorsement of a prestigious global carrier at the very moment it introduced the new program.

Airbus projected that the A350 XWB would enter commercial service by mid-2012. It would arrive after the 787, but Airbus hoped that its wider fuselage and other features would slow the Dreamliner's momentum and prevent Boeing from dominating the next generation of long-range widebody aircraft. As stunned as the aviation industry was by the announcement, no one was more surprised than Boeing's employees and senior leadership.

As though pouring salt into a fresh wound, Airbus also stated openly that it intended to seek launch aid from its European sponsor governments. Its A380 program, whose development cost had ballooned beyond $20 billion amid continuing technical problems, had still not entered commercial service, yet Airbus was already preparing to launch another entirely new airliner.

Boeing's executive suite and the United States government protested loudly. Had Airbus been an ordinary commercial enterprise, Boeing argued, it would have faced enormous difficulty obtaining the financing necessary to launch another major aircraft program while still carrying the escalating financial burden of the A380.

The stakes in the continuing World Trade Organization dispute over government support for Airbus could scarcely have been higher. Unfortunately for Boeing, the matter would not be resolved for years. In the meantime, there was little the company could do except accept the new competitive reality.

The 787 was no longer alone in the arena, and Boeing could no longer assume that the Dreamliner would remain the uncontested efficiency champion of the widebody market.

***

Despite mounting reports that the 787 was encountering serious development problems, Boeing continued assuring the aviation industry and Wall Street that the program remained on schedule. With each reassurance, however, more people began scratching their heads. The first major components for the aircraft did not begin arriving in Seattle until late April 2007.

When Boeing launched the program, it said the first aircraft would be rolled out for public viewing sometime in mid-2007. By the beginning of 2007, the company had committed itself to a specific date: July 8, 2007 — or 7-8-7. That left Boeing little more than two months to assemble the first aircraft after the major sections began arriving in Everett.

In the meantime, Boeing pulled out all the stops in planning the rollout ceremony. It was intended to be a public spectacle unmatched in the commercial aviation industry. Airbus' A380 rollout in 2005 had been attended by several European heads of state, but Boeing expected a far larger audience for the Dreamliner's unveiling, both in person and around the world. The event would be broadcast live and would include participation from the industrial partners responsible for building the aircraft's major sections. A well-known public figure would be selected to serve as master of ceremonies. But back in Everett, problems appeared as soon as the first major components arrived.

Discreetly captured photographs began circulating online showing the forward fuselage section, comprising the nose and cockpit and manufactured by Spirit AeroSystems in Wichita, being joined to a center-fuselage section built by Kawasaki Heavy Industries. One photograph revealed a large gap where the two sections were supposed to meet. The misalignment was so substantial that someone was able to slip their open hand through it.

Other fit problems emerged as well. Workers had to use shims and make various adjustments to join components that were supposed to arrive ready for final assembly. In at least one instance, part of the edge of a composite section reportedly had to be ground away so that it would fit against the adjoining structure.

As those reports spread through aviation circles, Boeing continued assuring investors and the public that the program was proceeding well. When the photographs exposed the extent of the fit problems, Boeing maintained that the sections had been manufactured correctly. The problem, the company explained, was that they had been shipped without sufficient internal support structures to hold their shape. During transportation, the large composite barrels had flexed or warped slightly. Boeing said future shipments would include temporary internal bracing to ensure that the sections retained their proper shape and could be joined more easily after arriving in Everett.

The explanation did little to ease concerns that Boeing would miss the July 8 rollout date. Many observers expected the company to announce a postponement at any moment. No such announcement came.

On July 8, 2007, Boeing staged its extravagant rollout ceremony at its Everett factory at Paine Field. Tom Brokaw, the iconic American television-news anchor who had recently retired from network broadcasting, served as master of ceremonies. Nearly 15,000 employees, airline executives, suppliers, dignitaries, and members of the media attended in person. The event was broadcast live by satellite and webcast around the world, allowing millions more to watch.

At the appointed moment, the enormous hangar doors opened and revealed the first 787 Dreamliner, freshly painted in Boeing's new corporate livery. The thousands gathered at Paine Field were awed by the sight of the aircraft. Boeing appeared to have delivered exactly what it had promised, and precisely on schedule.

The Dreamliner seemed to validate Boeing's "new ways of working" philosophy. Major sections of the aircraft had been manufactured at locations around the world, transported to Everett, and assembled into a graceful new airliner in a remarkably short period of time.

But to some of the guests who were able to examine the aircraft more closely, something did not seem right. When they looked through the cockpit windows from ground level, they could see virtually nothing inside. Normally, even from below, some part of an aircraft's cockpit interior would be visible — the overhead panel, instrument panel, pilot seats, or seat backs. In the case of the 787, the cockpit appeared to be an empty shell.

That raised an obvious question: Was the rest of the aircraft an empty shell as well?

When an aircraft manufacturer rolled out a new prototype, the airplane was ordinarily in a substantially completed state. Its first flight would typically follow within a few months, after its electrical, hydraulic, flight-control, avionics, and other systems had undergone extensive ground testing. That had been the pattern for earlier Boeing prototypes and for new aircraft produced by other manufacturers, including Airbus. But the 787 on display did not appear to be in such a condition.

Observers began noticing other peculiarities. Several panels and small service doors on the fuselage appeared to show a wood grain beneath their fresh coat of paint. Why would an advanced composite airliner have wooden access panels? Large bolts with protruding hexagonal heads could also be seen securing some of the aircraft's sections together. For an airplane expected to derive part of its efficiency from a smooth, aerodynamically clean exterior, those bolts could not possibly be part of the finished design. They were not.

It did not take long for rumors to spread that the aircraft had been hastily assembled as little more than a display shell so Boeing could meet its highly publicized 7-8-7 rollout date. The rumors were substantially correct.

Although Boeing acknowledged that the aircraft was far from flight-ready, it continued insisting that the first flight would occur within a few months. Few outside the company believed it.

On September 5, Boeing announced that the first flight would be delayed by approximately three months. Then, on October 10, it announced another delay, pushing the flight to the end of March 2008 and postponing the first airline deliveries until late 2008 — approximately a year later than originally planned.

Additional delay announcements soon followed in rapid succession. It had become apparent that Boeing's "new ways of working" were not working as intended. Reports circulated that components and subassemblies for subsequent aircraft were arriving in Everett incomplete, improperly documented, or not arriving at all. Boeing and its suppliers had underestimated the difficulty of coordinating a global production system in which major partners were expected not merely to manufacture sections of the airplane but to design, integrate, test, and certify much of the work themselves before delivery.

The software for the aircraft's flight-control system was also proving more difficult to complete than expected. The 787 would rely upon a deeply integrated fly-by-wire architecture, and engineers were taking longer than planned to identify and resolve problems within its software. The aircraft's unusually extensive electrical systems were creating additional complications that designers had not fully anticipated.

The first 787 would not finally take to the air until December 15, 2009 — more than two years after its originally scheduled first flight. Only then could the formal flight-test and certification program begin.

Boeing had intended the 787's certification campaign to be among the shortest ever attempted for an all-new commercial airliner. The company estimated that it would take approximately nine months and involve the first six 787s produced, including the original prototype.

Under Boeing's original plan, the six test aircraft would eventually be refurbished and delivered to launch customers at substantial discounts after the certification program was completed. But the airplanes accumulated so many problems — including excessive weight, incomplete systems, nonstandard components, and extensive structural modifications — that none proved suitable for normal airline service. Each differed too greatly from the standardized production aircraft that followed. In industry parlance, they had become "bastards" — nonuniform airplanes that would be costly and difficult for an airline to maintain and operate within an otherwise common fleet.

None of the six aircraft entered regular airline service. Some were eventually donated to museums or educational institutions. Others were retained or repurposed by Boeing for continued testing, while one was dismantled entirely.

***

On October 26, 2011, more than three years after the 787's originally scheduled entry into service, the aircraft completed its first revenue-generating passenger flight. Operated by launch customer All Nippon Airways, the flight departed Tokyo's Narita International Airport and landed in Hong Kong.

It was an enormously important milestone for Boeing, but also a painful one. As company leaders looked back over the program's troubled development, the financial and organizational damage left in its wake was becoming increasingly apparent.

Boeing had originally estimated that developing the 787 would cost approximately $8 billion, with the company funding about half and its risk-sharing partners absorbing much of the remainder. By the time the aircraft entered commercial service in late 2011, however, some estimates placed the program's total development cost at nearly $36 billion, with Boeing responsible for approximately $32 billion of that amount.

Under the original business plan, analysts estimated that Boeing could break even after selling between 300 and 400 aircraft. The company publicly projected that it could sell as many as 2,000 Dreamliners during an anticipated 20-year production run, suggesting that the program had the potential to become highly profitable.

But with approximately $32 billion now embedded in the program, some analysts estimated that Boeing might have to deliver 1,600 aircraft or more merely to break even. Others began to question whether the 787 would ever generate a meaningful profit over its lifetime.

If Harry Stonecipher had believed he had the formula for bringing an all-new, clean-sheet airliner to market inexpensively, the experiment had failed spectacularly. It had left an enormous hole not only in Boeing's finances, but also in the company's confidence in its own ability to manage a major development program. Its airline customers' confidence in the company would also take a huge hit.

The 787 would prove superlative in many respects, but those superlatives encompassed the bad as well as the good. And the bad was not over. In 2009, largely because of the continuing delays, airlines canceled orders for 76 aircraft. Boeing was struggling mightily with its global supply chain and with persistent production problems. Nearly every Dreamliner moving through the assembly line required extensive traveled work — tasks that should have been completed earlier by suppliers but instead had to be performed later, often in Everett.

Those supply-chain and assembly failures delayed customer payments and widened the gap in Boeing's revenue projections. They also exposed the company to contractual penalties for failing to deliver aircraft on schedule, while inflicting further damage on its reputation.

Even after more than two years of commercial service, Boeing had delivered barely 50 Dreamliners when the program suffered another headline-grabbing setback.

On January 7, 2013, a Japan Airlines 787 experienced a serious battery overheating event and fire while parked and unoccupied at Boston's Logan International Airport. Firefighters required nearly 40 minutes to extinguish the blaze, which caused damage in the aircraft's aft electrical-equipment area and to surrounding structure.

The fire originated in one of the 787's lithium-ion battery systems. Like batteries used on other commercial aircraft, it supplied electrical power for critical functions when other sources were unavailable and supported aircraft operations on the ground. Boeing had selected lithium-ion technology because it offered substantial weight savings over the nickel-cadmium batteries traditionally installed in airliners. Lithium-ion batteries, however, can be vulnerable to a phenomenon known as thermal runaway. When a cell within the battery becomes unstable, it can begin generating heat faster than that heat can dissipate. The process can spread from one cell to another, producing smoke, extreme temperatures, and fire.

That was what occurred aboard the Japan Airlines aircraft in Boston. As investigators from the Federal Aviation Administration, the National Transportation Safety Board, Boeing, and other organizations began examining the damaged aircraft, another battery incident occurred just nine days later.

On January 16, 2013, an All Nippon Airways 787 made an emergency landing at Takamatsu Airport in Japan after the flight crew received a battery warning and detected smoke. The passengers and crew evacuated on the runway using emergency slides. Investigators discovered that one of the aircraft's lithium-ion batteries had overheated in another thermal-runaway event.

Two apparently similar battery failures involving the same aircraft type within nine days rattled the aviation world.

That same day, shortly after learning of the ANA incident, the FAA issued an emergency airworthiness directive grounding all 787s operated by United States carriers. Aviation authorities around the world quickly followed with their own grounding orders.

It was the first complete grounding of a commercial-airliner type by the FAA since the McDonnell Douglas DC-10 was grounded following the American Airlines Flight 191 crash in Chicago 34 years earlier. Many observers — particularly those who believed Boeing had lost its way after acquiring McDonnell Douglas — could not overlook the comparison.

The DC-10 had been a must-do program intended to preserve McDonnell Douglas' standing in the commercial-aircraft industry. It had been developed under intense financial pressure and rushed into service amid tight budgetary constraints. Design weaknesses contributed to several serious accidents, a temporary grounding, and lasting damage to the aircraft's reputation. Combined with disappointing sales and high development costs, those problems helped make the program a financial disappointment for the company.

The 787 appeared to be following a disturbingly similar path and the ghost of McDonnell Douglas seemed to be haunting Boeing. The Dreamliner had already cost Boeing tens of billions of dollars in overruns, and the grounding would add to that burden for every day the aircraft remained unavailable to its airline customers.

One of the most troubling aspects of the episode was that investigators never established a single definitive initiating cause for both battery failures. They identified thermal runaway as the process that produced the heat, smoke, and fire, but determining precisely what first triggered the chain reaction proved far more difficult.

Possible contributing factors included internal short circuits, manufacturing defects, contamination, charging abnormalities, and deficiencies in the battery's design and testing. Boeing increased oversight of GS Yuasa, the Japanese company that manufactured the batteries, and introduced changes intended to reduce the likelihood and consequences of another failure. But investigators did not identify one conclusive root cause that fully explained both incidents.

Rather than wait indefinitely for such a determination, Boeing proposed a containment system designed to prevent a future battery fire from threatening the aircraft. The batteries would be enclosed in strengthened stainless-steel cases capable of containing heat, fire, and debris. Ventilation tubes would carry smoke and gases directly outside the aircraft. Boeing also modified the batteries, their chargers, and associated monitoring systems.

The FAA approved the redesigned system and authorized the 787 to return to service in April 2013. The formal grounding directive was lifted on April 25, after which airlines began modifying their aircraft with the Boeing-supplied containment and ventilation system.

The battery episode was another major setback for the 787 program. It shook the confidence of the aviation industry, airline customers, and passengers. Yet because the worldwide Dreamliner fleet was still relatively small, the disruption was not widely felt by the traveling public, and the story gradually receded from the headlines.

Nevertheless, the incidents raised an unsettling question: Had Boeing attempted to incorporate too many cutting-edge technologies into one aircraft without adequately understanding how those technologies would interact? The question also invited comparisons with the Airbus A350 XWB.

Airbus maintained a professional posture throughout the 787 grounding and publicly expressed sympathy for its competitor. At the same time, it did not hesitate to distinguish the A350 from the Dreamliner. Although the Airbus aircraft incorporated many of the same broad innovations, the company emphasized that it was taking a more measured approach to introducing new technology.

One frequently cited example involved the aircraft's batteries. Airbus initially planned to use lithium-ion batteries in the A350 but reconsidered that decision during the 787 grounding and elected to introduce the aircraft with traditional nickel-cadmium units. Although heavier, nickel-cadmium batteries had a long service history and were less susceptible to the type of thermal-runaway event that had grounded the Dreamliner.

Boeing had chosen lithium-ion batteries largely because of their reduced weight. Each of the 787's two battery units weighed approximately 62 pounds, compared with roughly 107 pounds for an equivalent nickel-cadmium installation. But the containment cases, venting equipment, and associated modifications added considerable weight, eroding much of the original advantage.

Airbus also emphasized that it would not rush the A350 into service, an unmistakable criticism of Boeing's belief that it could bring such a revolutionary aircraft to market on an exceptionally compressed schedule.

When Airbus announced the A350 XWB in 2006, its development plan anticipated a first flight around 2012 and entry into service during 2013, allowing approximately 15 months for flight testing and certification. That represented a development cycle of roughly six years, compared with Boeing's original ambition of developing the 787 in approximately three years.

Although the A350 would also be constructed largely from composite materials, Airbus adopted a somewhat more conventional manufacturing method. Instead of forming much of the fuselage as complete cylindrical barrels, as Boeing had done with the 787, Airbus fabricated large composite fuselage panels and attached them to an underlying structural framework.

The approach was broadly similar to the traditional method used to construct metal aircraft, in which aluminum skin panels were fastened to frames and stringers. Airbus believed the panel method would allow the composite material to be laid up and cured with greater precision. It would also make it easier to replace an individual panel if the aircraft suffered ground damage or if access were required for future structural modifications.

The 787's repeated delays, production failures, and battery grounding gave airlines ample reason to pause and consider the Airbus alternative. Many ultimately concluded that the A350 represented the safer choice.

From the launch of each program through the end of 2013, Boeing accumulated approximately 1,030 gross orders for the 787, compared with roughly 812 for the A350, although the Dreamliner had enjoyed a head start of approximately two years.

The more revealing comparison came during the five-year period beginning in 2008, when the 787's delays and production problems became most severe. From 2008 through 2013, the A350 secured approximately 518 net orders, compared with roughly 213 for the 787. Boeing had clearly surrendered the initiative.

The company had led the world into the age of commercial jet travel, the era of the widebody airliner, the expansion of long-range point-to-point service, and now the pursuit of unprecedented fuel efficiency. This time, however, Boeing had stumbled after taking the first step and had allowed Airbus to seize much of the momentum.

***

James McNerney was not fazed by the 787's setbacks. His years in senior corporate leadership had taught him to remain disciplined and focused despite the noise surrounding a troubled program. One of his principal mandates upon becoming Boeing's chief executive was to see the Dreamliner's production and delivery issues resolved, and he was not about to fail in that regard.

But the 787 was not the only source of turmoil demanding his attention. In September 2005, only a few months after McNerney assumed the chief executive's office, more than 18,000 Boeing machinists went on strike. The dispute centered on pensions, retiree medical coverage, health-care costs, job security, and Boeing's increasing reliance on outside suppliers. The walkout halted commercial-aircraft production in the Puget Sound region and elsewhere for nearly four weeks before the two sides reached a new contract.

Three years later, in September 2008, approximately 27,000 machinists walked off the job again. The 57-day strike involved many of the same concerns, including outsourcing, job security, wages, pensions, and health-care benefits.

The timing could scarcely have been worse. The 787 program was already badly behind schedule and struggling with incomplete components, traveled work, and a deeply troubled global supply chain. The strike halted work across Boeing's commercial-aircraft product line and compounded the delays affecting the Dreamliner. It also cost the company billions of dollars in deferred revenue, disrupted production, and contractual penalties. After the second strike, McNerney became increasingly determined to reduce Boeing's vulnerability to organized labor and the production disruptions a strike could cause.

By 2009, with the 787's first flight finally appearing to be within reach, Boeing was considering how to increase production. Only delivered aircraft generated the bulk of the cash Boeing desperately needed from the program, and the company had to find a way to move airplanes through final assembly and into customers' hands more quickly.

Despite cancellations and repeated delays, the Dreamliner still had a formidable backlog. Even after the problems in its supply chain and production system were mitigated, the Everett assembly line would have a practical limit on how many aircraft it could produce. So, Boeing announced that it would establish a second 787 final-assembly line.

Washington state and local officials offered substantial tax and economic incentives to persuade Boeing to place the line in the Puget Sound region. But the smoke from the previous year's machinists' strike was still hanging heavily over McNerney and the board. To many employees and labor leaders, Boeing appeared determined to demonstrate that the unions could no longer hold the company's production system hostage without consequences.

Boeing had already begun establishing a significant presence in South Carolina. In 2008, the company acquired Vought Aircraft

Industries' 50 percent interest in Global Aeronautica, the North Charleston joint venture responsible for integrating major 787 fuselage sections. Then, in July 2009, Boeing agreed to purchase Vought's adjacent South Carolina operations.

Vought's North Charleston plant manufactured and assembled the 787's aft-fuselage sections. The facility had encountered severe production and financial problems that were contributing to the Dreamliner's delays. Boeing maintained that the program could no longer afford to wait for Vought to correct those deficiencies and that acquiring the operation would give Boeing greater control over a critical portion of the 787 supply chain.

That explanation was entirely plausible. The outsourcing strategy had left Boeing without sufficient visibility into or control over work being performed by some of its most important partners, and Vought had become one of the weakest links in the chain.

But industry observers and Boeing's unions suspected that the acquisition served another purpose as well. Only a few months later, in October 2009, Boeing announced that it had selected North Charleston as the location for the second 787 final-assembly line.

Many employees believed Boeing had intended all along to establish the second line outside Washington state. To them, the decision reflected McNerney's increasingly hard-line approach toward organized labor and amounted to retaliation for the 2008 strike. South Carolina was a right-to-work state with comparatively little union influence, giving Boeing a source of commercial-aircraft production that would not automatically shut down during a strike by its Puget Sound machinists.

The International Association of Machinists filed an unfair-labor-practice charge with the National Labor Relations Board. In April 2011, the NLRB's acting general counsel issued a formal complaint alleging that Boeing had chosen South Carolina partly to retaliate against unionized employees for their past strikes and to discourage future labor activity.

Boeing denied the allegation. It argued that no union work had been transferred from Washington because the company was adding production capacity rather than relocating an existing assembly line. The company also maintained that employment in the Puget Sound region had continued to grow despite the South Carolina expansion.

The legal dispute became intensely political, with critics accusing the NLRB of attempting to prevent a company from expanding in a right-to-work state. It ended in December 2011 after Boeing and the machinists reached a new four-year labor agreement that included a commitment to build the forthcoming new version of the 737 in Renton, Washington. The union asked that the case be withdrawn, and the NLRB complaint was dismissed without a ruling on its merits.

The dispute nevertheless established an unmistakable demarcation line between Boeing's management and much of its unionized workforce. Relations between labor and management had grown increasingly toxic since the McDonnell Douglas acquisition. Boeing's workforce had been heavily unionized for decades, and the company's labor history was by no means conflict-free. Employees had staged several major walkouts during the postwar period, including a 140-day strike in 1948, the longest in Boeing's history.

Yet before the merger, there had generally remained a sense of family and common purpose within Boeing. Management and labor frequently disagreed, sometimes bitterly, but both sides shared a deep attachment to the company and its products. That bond helped foster engineering excellence, craftsmanship, and extraordinary employee loyalty.

However, the McDonnell Douglas merger introduced a new volatility into that relationship. Its leadership culture placed greater emphasis on cost reduction, financial performance, efficiency, and shareholder returns. Many longtime Boeing employees came to believe that management no longer regarded them as members of a common enterprise but as costs to be controlled, reduced, outsourced, or moved elsewhere. Beyond any doubt, an us-against-them mentality had taken hold on both sides.

Good corporate leadership must be measured across a broader spectrum than efficiency, earnings, and shareholder value alone. A chief executive represents the entire organization and must balance the interests of investors, customers, employees, suppliers, and the communities in which the company operates.

Disagreement within such a large organization is inevitable. One of a leader's most important responsibilities, however, is to foster cohesion and common purpose by understanding the legitimate concerns of those on opposing sides. That had not happened during McDonnell Douglas' final years as an independent company. It did not happen when former McDonnell Douglas executives assumed influential positions within Boeing, and it did not appear to be happening under James McNerney.

McNerney adopted an increasingly confrontational posture toward Boeing's unions. His attitude toward the workforce appeared to reveal itself in 2014, when he was asked during an employee meeting whether he intended to retire upon reaching Boeing's customary chief-executive retirement age of 65. McNerney joked that employees would be "cowering" rather than celebrating because the board had extended his tenure.

He later apologized, describing the remark as a joke that had fallen flat. But to employees who already believed Boeing's leadership viewed them with contempt, the comment seemed to confirm their worst suspicions. Whatever humor McNerney had intended was lost amid the bitterness that had come to define labor-management relations.

A chief executive is accountable to a corporation's board of directors, even when — as with McNerney and several previous Boeing chief executives — he also serves as its chairman.

McNerney's mandate following Stonecipher's abrupt and ignominious departure was to continue moving Boeing through its post-merger transformation and restore its dominance in the commercial-airliner market.

When the 787 program went off the rails, his immediate priority became getting it back under control. In that limited but important respect, McNerney ultimately succeeded. Under his leadership, Boeing restored momentum to the Dreamliner despite a seemingly endless procession of delays, cost overruns, production failures, and headline-grabbing incidents.

The problems did not end when the aircraft entered service or when the battery grounding was lifted. Extensive traveled work continued to burden aircraft moving through the assembly lines. In 2014, Boeing temporarily halted some deliveries after hairline cracks were discovered in wing components manufactured by Mitsubishi Heavy Industries. Quality-control deficiencies continued to emerge throughout the supply chain, and software problems periodically affected the airplane after it entered service.

In July 2013, an Ethiopian Airlines 787 parked at London's Heathrow Airport was badly damaged by a fire traced to the aircraft's emergency-locator transmitter. Investigators found that wires associated with the transmitter's lithium-metal battery had become trapped beneath its cover, creating a short circuit and fire. The incident was unrelated to the 787's main lithium-ion battery system, but it generated another round of alarming headlines for an aircraft already struggling to overcome questions about its safety and reliability.

Despite those continuing difficulties, production gradually increased as Boeing established a steadier rhythm at the Everett and North Charleston final-assembly plants. The problems never disappeared completely, but McNerney succeeded in moving the program beyond the near-paralysis of its early years and suppressing many of its most immediate crises.

The stock market reflected that progress. When McNerney became chief executive in 2005, Boeing shares ended the year trading at approximately $70. The stock continued rising before losing much of its value during the 2008 financial crisis, when the worldwide economic downturn coincided with repeated 787 delays and weakening confidence in the program.

However, by the end of 2013, after the Dreamliner had returned to service and production was accelerating, Boeing's stock was trading above $130. McNerney had clearly won the confidence and admiration of Wall Street. Within the leadership culture that had taken hold at Boeing, that appeared to be the measure that mattered most.

***

As the global economy recovered and orders across Boeing's commercial-aircraft product line accelerated, the company began rebuilding its cash reserves. Under McNerney, Boeing then embarked upon a buying spree of sorts — not the traditional kind involving factories, equipment, or acquisitions, but a massive campaign to purchase its own shares.

The practice is known as a stock repurchase, or stock buyback. A company uses its cash to buy shares from investors, ordinarily through transactions on the open market. The purchased shares are then retired or held as treasury stock, reducing the number of shares outstanding.

Boeing had repurchased shares before McNerney arrived, but under his leadership the company pursued buybacks on a far grander scale. Boeing suspended the practice during the financial crisis and the most difficult years of the 787's development, but resumed it in 2013 as commercial aircraft deliveries and cash generation improved. The company spent approximately $6 billion repurchasing shares in 2014 alone and another $6.75 billion in 2015, although McNerney stepped down as chief executive midway through that year. By the end of his tenure, Boeing had spent well over $15 billion buying back its own shares, with substantially larger authorizations already approved for the years ahead.

For much of the 20th century, open-market stock repurchases were viewed with suspicion because regulators feared they could be used to manipulate share prices. That changed significantly in 1982, when the Securities and Exchange Commission adopted Rule 10b-18. The rule did not simply legalize all repurchases or exempt them from securities laws; rather, it created a voluntary safe harbor protecting companies from certain market-manipulation claims when their purchases complied with limits governing timing, price, volume, and the manner in which the transactions were conducted.

Even after Rule 10b-18, stock buybacks remained controversial. Repurchasing shares can be a legitimate means of returning excess capital to shareholders, particularly when management believes the stock is undervalued and the company has no more productive use for the cash. Reducing the number of outstanding shares can also increase earnings per share because the company's earnings are divided among fewer shares.

That same mechanism, however, can make a company's financial performance appear stronger even when its underlying earnings have not increased. A repurchase may also support or elevate the market price of the remaining shares by increasing demand and reducing supply.

The most immediate beneficiaries are the shareholders who sell their stock or retain shares that may appreciate in value. But senior executives can benefit as well when a substantial portion of their compensation is tied to earnings per share, total shareholder return, stock options, restricted shares, or other equity-based performance measures.

McNerney's compensation, like that of other senior Boeing executives, included significant incentives connected directly or indirectly to Boeing's financial and stock-market performance. His reported compensation was valued at approximately $13.8 million in 2006, his first full year as chief executive. A later correction to Boeing's compensation disclosures placed the accounting value of his total 2006 package above $20 million because of the way certain pension and deferred-compensation benefits were reported. By 2014, his final full year as chief executive, his reported total compensation had risen to nearly $29 million.

There was nothing illegal about Boeing's repurchases. The more important question was whether they represented the wisest use of the company's capital.

The 787 had left behind tens of billions of dollars in development costs, a damaged supplier network, recurring production problems, and an exhausted workforce. Boeing could have used more of its available cash to strengthen the Dreamliner's supply chain, modernize its production facilities, expand engineering resources, or improve quality control. It could have invested more heavily in research and development or begun laying the technological groundwork for the next generation of commercial aircraft.

The company's unions viewed the buybacks with particular resentment. Boeing was demanding concessions from its workers, seeking to limit pension obligations, shifting work to nonunion locations, and insisting that it needed greater cost competitiveness. At the same time, it was committing billions of dollars to purchasing its own stock. To employees who had endured strikes, outsourcing, layoffs, benefit disputes, and the production turmoil surrounding the 787, the contrast appeared to be a slap in the face.

That did not mean every dollar devoted to repurchases could simply have been redirected to wages, factories, or research with an equivalent return. Nor did buying back shares automatically prevent Boeing from investing in those areas. The company was generating substantial cash and could pursue more than one objective at a time.

But capital allocation is ultimately a declaration of corporate priorities. Boeing's leadership was choosing to return an increasing amount of cash to shareholders rather than retain all of it for long-term investment in the company's products, people, infrastructure, and technical capabilities.

Repurchasing shares could produce visible results relatively quickly. Fewer outstanding shares generally improved earnings per share, while the buying activity could support the company's stock price. Investments in engineering, workforce development, manufacturing systems, and new-aircraft technology might not produce measurable returns for years and could reduce reported earnings in the meantime.

Boeing's leadership was therefore choosing between two different concepts of value. One emphasized immediate and readily measurable returns to shareholders. The other required a longer view and an acceptance that investments in engineering capability, product development, manufacturing resilience, and employee relations might not reward investors until years later. Under McNerney, Boeing increasingly favored the first.

And the more than $15 billion spent repurchasing shares during his tenure would pale in comparison with the amount the company would devote to buybacks in the years that followed. Boeing's stock price would eventually rise to extraordinary heights, but the company's growing fixation on shareholder returns would come at a cost.

In time, it would appear that Boeing had begun losing sight of the fundamental purpose for which the company existed — designing and building safe, technologically advanced aircraft.

Teetering on the Precipice

Despite all the challenges that had plagued the 787, there was no denying that the airplane was going to be a game changer for the airline industry. That raised an obvious question: Could the benefits of Boeing's Yellowstone technologies be brought to other segments of the commercial aircraft market?

Of the three Yellowstone concepts Boeing had studied during the 1990s — Y1, Y2, and Y3 — Y2 was the first to be developed, ultimately becoming the 787. Boeing had prioritized that segment because sales of the 767 had stalled in the face of the superior operating economics offered by Airbus' A330 and A340 families. Boeing had attempted to extend the 767's commercial life during the 1990s by introducing a stretched version, the 767-400ER, but the model achieved only limited market penetration. Just 37 passenger versions were produced, along with one additional aircraft that was placed into service as a VIP transport. Only two airlines, Delta and Continental, operated the passenger version.

There was little reason for Boeing to pursue Y3 at the time because it covered the same general market as the 777, which was proving extremely popular with airlines and still had a long commercial life ahead of it. But could Y1 bring the same revolutionary economics to the smaller aircraft market served by the 737?

Despite the 787's delays and teething problems, airlines were showing intense interest in its promised capabilities. Many began asking whether comparable improvements in operating economics could be brought to the short-haul market. In 2008, CFM International, a joint venture between General Electric and Safran, and Pratt & Whitney announced separate programs to develop a new generation of engines that could potentially be installed on the 737 NG and A320. Both engines held out the promise of reducing fuel burn by as much as 15 percent compared with the engines then powering the two airplanes.

Airbus was already publicly acknowledging that it was studying the feasibility of mounting the new engines on its existing A320 platform. Because the A320 was a more modern design, Airbus maintained that accommodating the larger engines would require relatively little engineering effort. An A320 fitted with the new powerplants could deliver operating-efficiency improvements approaching 15 percent without requiring Airbus to invest in an entirely new airplane.

Boeing could theoretically offer a similar upgrade to the 737, but its options were more limited. Because the 737 sat lower to the ground on its relatively short landing gear, there was considerably less space between the underside of its wings and the pavement than there was beneath the wings of the A320. Boeing's analysis indicated that CFM's new engine could be installed on the 737 only if it were modified with a smaller-diameter fan. Pratt & Whitney's new engine, however, was not considered a feasible option.

Pratt & Whitney's engine, formally designated the PW1000G, incorporated a geared-fan architecture that allowed the large fan at the front of the engine to rotate more slowly than the turbine components behind it. The gearbox allowed the turbine section within the engine's core to operate at the higher rotational speed at which it was most efficient, while the much larger fan could turn more slowly at a speed better suited for moving bypass air. It was a revolutionary design that promised significant improvements in fuel efficiency and noise reduction. But the complex planetary gearbox and large fan substantially increased the diameter of the front of the engine, requiring more ground clearance than the 737 could provide.

The result was that a re-engined 737 would be limited to a single engine option, while the A320 could be offered with engines from both CFM and Pratt & Whitney. Although the two engines promised broadly comparable fuel-efficiency improvements of approximately 15 percent, Pratt & Whitney's geared-fan design offered another important advantage: It was expected to be nearly 50 percent quieter than the CFM engine. That was a significant benefit for airlines operating from airports subject to strict noise-abatement requirements because of the communities surrounding them.

By offering airlines a choice between two new-generation engines, Airbus would possess a clear competitive advantage over any comparably re-engined 737. It was becoming increasingly apparent that Boeing's 1965 decision to position the 737 close to the ground — partly because the company was eager to secure Lufthansa as the launch customer — was beginning to hobble the airplane more than four decades later.

Boeing's engineers had already been studying whether the technologies developed under Yellowstone could be applied to a new short-to-medium-haul airplane, but the emerging engine programs caused them to redouble their efforts. Unfortunately, their analysis repeatedly came up short. Their calculations indicated that the lightweight materials and advanced engineering that gave the 787 such impressive operating economics would produce only a modest benefit when applied to a smaller airframe. At best, an airframe incorporating the 787's technologies might deliver an efficiency improvement of only 2 to 3 percent.

Combined with the new engines under development, an all-new short-to-medium-haul airplane might therefore deliver an overall efficiency improvement of approximately 17 to 18 percent compared with the existing generation of aircraft. But roughly 15 percentage points of that improvement would come from the engines alone.

Although an additional 2 or 3 percent improvement was not insignificant, it was insufficient to justify the cost of developing an entirely new airplane. A clean-sheet design would require Boeing to invest billions of dollars. Meanwhile, the 787's mounting production and engineering problems were adding billions more to the tens of billions of dollars already spent bringing that airplane to market. Within Boeing's C-suite, there was little appetite for launching another enormously expensive development program.

Moreover, the small additional efficiency advantage of a clean-sheet Boeing design could easily be neutralized if Airbus simply reduced the selling price of its re-engined A320. With a relatively modest investment in aerodynamic refinements or wing improvements, Airbus might also be able to close much of the remaining 2-to-3-percent gap between an updated A320 and any new airplane Boeing could develop.

The relatively limited benefit of applying Yellowstone's advanced technologies and composite materials to a smaller airplane resulted largely from the shorter routes typically flown by aircraft in the 737's category. On a typical long-range flight, a 787 might spend 20 to 30 minutes climbing to its cruising altitude. That takeoff and climb could consume up to 10 percent of the aircraft's total fuel load depending on takeoff weight, cruise altitude, temperatures, winds, and routing. The airplane would then spend many hours at cruise speed and altitude, where its advanced aerodynamics, lighter composite structure, and other technologies delivered their greatest benefits.

A 737 operating a short-haul route followed a very different flight profile. It could consume as much as 20 percent of its fuel during takeoff and climb, then remain at cruising altitude for only one or two hours before beginning its descent. Because the airplane spent far less time in the cruise phase, the economic benefits provided by Yellowstone's advanced materials and engineering were greatly diminished.

There was therefore little economic justification for bringing Yellowstone's Y1 concept to market as a replacement for the 737 — not yet, at least. But airlines were becoming increasingly anxious to obtain the benefits offered by the new engines being developed by CFM and Pratt & Whitney, and the 737 was not looking like the ideal aircraft on which to install them. With its more modern design and, most important, its taller landing gear and greater ground clearance, the A320 appeared ideally suited for the new generation of powerplants.

In Chicago, a palpable fear was developing that the A320, which had already captured slightly more than half of the short-to-medium-haul market, might seize an overwhelming share of it if Airbus launched a version equipped with the new engines.

To forestall a potential stampede toward Airbus, Boeing undertook a somewhat clandestine sales campaign aimed at persuading airlines to delay ordering any re-engined A320 until Boeing could bring the benefits of Y1 to the same market segment. But airlines were eager to acquire aircraft offering fuel-efficiency improvements of as much as 15 percent. The best Boeing could tell them was that its clean-sheet replacement for the 737 was at least five years away, and possibly longer.

That was how long Boeing believed its engineers would need merely to refine a new design capable of delivering the approximately 25-percent efficiency improvement the company considered necessary to justify an all-new airplane. Even then, the aircraft would still require years of detailed engineering, testing, certification, and production preparation before entering airline service. In practical terms, a new Boeing design was still a decade away.

To hedge its bets, Boeing directed its engineers to study the existing 737 and determine what modifications would be required to accommodate the new engine being developed by CFM. The standard engine CFM planned to offer was the LEAP-1A. But CFM, determined to preserve its position as the exclusive engine supplier for the 737, proposed a modified version with a smaller-diameter fan to compensate for the Boeing airplane's limited ground clearance.

That modified engine became the LEAP-1B. Its fan diameter was almost one foot smaller than that of the LEAP-1A, and it consequently produced less thrust. The smaller diameter, however, would create less drag and allow the 737 to cruise slightly faster than the A320. But because the 737 was somewhat heavier than the A320, its overall fuel efficiency would be slightly lower. Its takeoff performance would also be less favorable.

Another race to produce the best airplane in a major market segment was taking shape. But for once, one of the two competitors was entering the contest at a clear disadvantage — even though both companies would be using variations of the same engine technology.

Assuming Boeing chose to proceed with a re-engined 737, it would be constrained by decisions embedded in an airframe that had first been designed during the 1960s. Yet Boeing was still desperately searching for a way to gain the upper hand over Airbus. The company believed the best way to win the emerging contest was to develop an entirely new airplane incorporating Yellowstone's technologies in much the same way the 787 had.

The problem was that Boeing first had to determine how such an airplane could deliver the 25-percent efficiency improvement necessary to make it another industry-changing aircraft. And it had to do so before Airbus changed the market.

***

On December 1, 2010, Airbus formally launched the A320neo. The suffix stood for "new engine option," reflecting Airbus' plan to offer the airplane with either CFM International's LEAP-1A (LEAP was an acronym for Leading Edge Aviation Propulsion) or Pratt & Whitney's PW1000G geared-turbofan engine.

Boeing's initial public response was one of resistance. Company leaders maintained that re-engining the 737 would provide only marginal gains compared with the benefits that might be achieved through an entirely new airplane developed under the Yellowstone program. Jim McNerney emphasized the advantage of waiting for emerging technologies to mature so Boeing could introduce a clean-sheet design that, when combined with the new generation of engines, might leapfrog the A320neo rather than merely match it.

But in December 2010, with the 787 program mired in delays amid massive cost overruns, Boeing lacked both the financial resources and the engineering capacity to launch another all-new airplane. Its production lines were also humming, turning out 737s at a healthy rate against a large and profitable backlog. Boeing was understandably reluctant to disrupt or prematurely replace one of its most important cash-generating products.

McNerney had ample reason to hope that airlines would hold off ordering the A320neo. He publicly indicated that Boeing was content to watch the market's reaction before committing itself to either a re-engined 737 or an entirely new design. It did not take long for the market to force him to reveal his hand.

Within several months of launching the A320neo, Airbus had accumulated nearly 1,000 orders and commitments for the refreshed airplane. Pressure mounted on Boeing's engineers to produce a breakthrough in their work on a 737 replacement. Then, in May 2011, Boeing's executive offices in Chicago were rocked by news that American Airlines was preparing to place a massive order for Airbus A320 and A320neo aircraft.

During discussions between Boeing sales representatives and American's senior executives, the airline confirmed that it was considering an order for more than 200 Airbus airplanes, with options for hundreds more. American was moving urgently to replace its aging fleet of MD-80s and 757s. It operated one of the oldest fleets among the major U.S. carriers, and the high fuel consumption and maintenance requirements of those older airplanes were driving up its operating costs. American needed to modernize its fleet to remain competitive with airlines already operating newer and more efficient aircraft.

The news was a devastating blow to Boeing. Although the company had been required to dissolve its exclusive supplier agreement with American in 1997 to satisfy European regulators during the acquisition of McDonnell Douglas, American had remained one of Boeing's most loyal customers.

Until then, many A320neo orders had come from airlines in Airbus' own European backyard, Asian start-up carriers, and aircraft-leasing companies that intended to place the airplanes with airlines as demand arose. An order for hundreds of A320s from one of the world's largest airlines — and an American carrier at that — would reverberate throughout the industry. It would provide an extraordinary endorsement of the A320neo and risk establishing the A320 family as the obvious choice for airlines purchasing aircraft in that market segment.

Boeing scrambled to get in front of American's leadership and persuade the airline to delay its decision long enough for Boeing to present an alternative.

More than any other event, American Airlines' planned order for hundreds of A320 and A320neo aircraft forced Boeing to give up its immediate pursuit of Yellowstone's Y1 replacement for the 737. Instead, the company would modify and upgrade the 737 for an unprecedented fourth major generation in order to remain competitive.

The order came down from Chicago for Boeing's engineers to determine as quickly as possible how the 737 could be adapted to accept the new generation of engines — specifically, the only one that could realistically be made available for the airplane, CFM International's LEAP-1B, which would be developed specifically for Boeing.

In what many industry observers regarded as a shrewd move, American agreed to divide its planned purchase between the two manufacturers. By placing the companies in direct competition for one of the largest aircraft orders in history, American was virtually assured of obtaining highly favorable pricing from both.

The airline also needed new airplanes as quickly as possible. Airbus could not deliver enough existing A320s soon enough to satisfy American's immediate fleet-renewal needs. Ordering Boeing's current 737 NG alongside the A320 would allow American to begin replacing its fuel-hungry and maintenance-intensive older aircraft without waiting for the re-engined models to enter service.

Although the A320 and 737 competed in broadly the same market, there were sufficient differences in their operating characteristics for American to assign each airplane to the routes and missions for which it was best suited. American also already operated a substantial fleet of 737s. Ordering Boeing's re-engined version would allow the airline to preserve a degree of commonality with its existing flight crews, maintenance personnel, facilities, and support infrastructure rather than bearing the full cost of converting its narrow-body fleet entirely to Airbus aircraft.

On July 20, 2011, American Airlines announced an order for 260 Airbus A320-family aircraft. The purchase consisted of 130 A320neo aircraft and 130 examples of the existing A320 series, which Airbus was beginning to distinguish as the A320ceo — the "current engine option."

At the same time, American announced an order for 200 Boeing 737s — 100 of the proposed re-engined version and 100 existing 737 NG aircraft.

Both manufacturers were expected to begin delivering their current-generation airplanes to American in 2013. Airbus planned to begin deliveries of the A320neo in 2016, while Boeing's re-engined 737 was expected to follow in 2017.

The following month, Boeing's board authorized the launch of the new 737 derivative, and the company announced that it would be called the 737 MAX. During discussions over how to distinguish the airplane from previous generations, Boeing chose

"MAX" to convey the maximum advantage — or, according to other contemporary descriptions, the maximum benefits — the new model would offer. Some would jokingly whisper that the MAX denoted this was the maximum that Boeing could ever extract from the 1960s-designed 737.

At its formal launch, Boeing announced three primary versions: the 737 MAX 7, MAX 8, and MAX 9, with each successive number identifying a larger-capacity aircraft. The three models would correspond broadly to the 737 NG's -700, -800, and -900 series. Boeing would later introduce an even larger version, designated the 737 MAX 10.

From the outset, many industry observers questioned whether Boeing was accepting an enormous risk by modifying the 737 yet again to compete with the A320neo. Boeing would have to make substantial changes to the airplane to accommodate the LEAP-1B. Airbus, by comparison, needed to make relatively modest modifications to the existing A320 airframe.

Airbus estimated that development of the A320neo would cost approximately $1.3 billion. Boeing initially projected that development of the 737 MAX would cost between $2 billion and $3 billion. By the time the first of the series, the MAX 8, entered commercial service, that figure would exceed $4 billion but would still be accumulating costs as the company worked to bring the rest of the series up to certification standards.

Even CFM's decision to modify the LEAP-1A specifically for the 737 would not make the conversion a straightforward engine replacement. The LEAP-1B's nacelle would still be approximately 12 inches larger in outside diameter than that of the CFM56-7 engines powering the 737 NG. Even more consequentially, the new engine would weigh nearly 1,000 pounds — approximately 450 kilograms — more.

Boeing would need considerable engineering ingenuity to overcome limitations embedded in an airplane whose basic architecture had been established more than four decades earlier, when the company had designed the original 737 in part around the operating requirements of launch customer Lufthansa.

The MAX's wing structure would have to be strengthened to absorb the additional weight and stresses created by the larger engines, a challenge Airbus also faced with the A320neo. But Boeing confronted an additional and far more difficult problem. Because of the 737's short landing gear and limited ground clearance, the LEAP-1B could not simply be suspended beneath the wing in the same location as the engines on the 737 NG. The engine would have to be repositioned.

***

One of the 737's distinctive design features was that it did not have doors that completely enclosed its main landing-gear wheels after they were retracted. When the airplane was originally conceived during the 1960s, Boeing's engineers sought to keep it mechanically simple, lightweight, inexpensive, and easy to service at smaller airports.

The main landing gear retracted inward into wells beneath the fuselage. Once retracted, the outer faces of the tires remained exposed to the airstream, lying nearly flush with the underside of the airplane. Aerodynamic hubcaps and seals reduced drag and helped prevent water and debris from entering the wheel wells. Eliminating full-sized main-gear doors also eliminated their actuators, hinges, linkages, hydraulic plumbing, structural supports, weight, and maintenance requirements. Although partially exposed wheels were not unprecedented in aircraft design, they were unusual on a commercial jet airliner.

The arrangement was closely tied to the 737's original configuration. Its engines were mounted relatively close to the fuselage, which reduced the yawing moment produced if one engine failed. The main landing gear was therefore positioned relatively close to the fuselage as well. Its short legs helped place the cabin, cargo holds, engines, and service points near the ground — an important feature for Lufthansa and other airlines expected to operate the airplane from airports with limited ground-support equipment.

By the time Boeing began developing the MAX, those decisions had become deeply embedded in the airplane's structure. Moving the main landing gear farther outboard or replacing it with significantly longer gear would not have been a minor modification. It would have required extensive redesign of the wing, center wing box, wheel wells, supporting structure, hydraulic systems, and surrounding portions of the fuselage. Such an undertaking would have sharply increased the program's cost, complexity, development time, and certification burden. Boeing instead concentrated on changing the installation of the engines themselves.

To obtain sufficient ground clearance for the larger LEAP-1B engines, Boeing mounted them farther forward and higher relative to the wing than the CFM56 engines on the 737 NG. A newly designed pylon and strut allowed the engine nacelles to be positioned closer to the level of the wing's upper surface. Boeing also lengthened the nose landing gear by eight inches, which raised the front of the airplane while preserving ground clearance comparable to that of the 737 NG.

These changes gave the LEAP-1B nacelles sufficient clearance from the pavement during taxiing, takeoff, landing, and crosswind operations while reducing the possibility of foreign-object ingestion. But repositioning the engines produced aerodynamic consequences.

The principal concern was not simply that the LEAP-1B generated more thrust than the engine it replaced. Nor was it merely the normal tendency of an airplane's pitch attitude to change when thrust was added. The more consequential difference emerged when the MAX was flown manually, with its flaps retracted, at elevated angles of attack.

Because the LEAP-1B nacelles were larger and positioned farther forward, airflow around them could generate additional aerodynamic lift as the airplane's angle of attack increased. Since that lift acted ahead of the airplane's center of gravity, it created an additional nose-up pitching moment.

Under ordinary operating conditions, the effect was modest. At higher angles of attack and certain airspeeds, however, it became more pronounced. As the airplane approached a stall, the additional nacelle lift could cause the force required to continue pulling back on the control column to increase less progressively than regulators required. Meaning on the 737 NG the force required to continue pulling back would increase noticeably, whereas on the MAX it would increase modestly.

This distinction was critical. The MAX was not inherently uncontrollable, nor would its nose suddenly rise whenever the pilots advanced the throttles. The concern was that during particular high-angle-of-attack maneuvers, the airplane's control forces and pitch response could differ from those of the 737 NG. A pilot continuing to pull aft on the control column might not receive the steadily increasing physical resistance expected as the airplane approached a stall.

A wing stalls when its critical angle of attack is exceeded — the angle between the wing's chord line and the relative airflow — not merely when the airplane falls below a particular airspeed. Airspeed, aircraft weight, load factor, bank angle, and maneuvering all influence when that critical angle may be reached, but it is excessive angle of attack that causes airflow over the wing to separate and lift to deteriorate.

Boeing's engineers investigated aerodynamic means of correcting the MAX's altered handling characteristics. Among the possibilities considered were changes to the wings and the placement of vortex generators — small aerodynamic surfaces used to energize airflow and delay separation. Boeing incorporated aerodynamic refinements, but those changes alone were insufficient to produce the control-force characteristics required throughout the airplane's certification envelope.

The most straightforward operational solution would have been to recognize that the MAX possessed handling characteristics that differed from those of the 737 NG and train pilots accordingly. Pilots could have been taught how the airplane behaved at elevated angles of attack and how its pitch response differed from that of earlier 737s.

But such training would have undermined one of the central commercial promises of the MAX program: that airlines could place pilots already qualified on the 737 NG into the new airplane with very minimal additional training.

Therefore, Boeing sought another solution — one that would make the MAX feel to a pilot more like the 737 NG. That solution would be software.

***

When Airbus launched the A320neo, one of its strongest selling points was that the airplane would deliver much of the economic benefit of an entirely new design — most notably, a promised fuel-burn improvement of approximately 15 percent — without the disruption and complexity normally associated with introducing a new aircraft type.

Those gains would come from an airplane that airlines already knew how to operate. Airbus was mating new engines to the existing A320 platform while preserving the cockpit layout, systems architecture, maintenance infrastructure, ground-support equipment, and operating procedures of the original aircraft family. Pilots already qualified on the A320 would not have to undergo extensive transition training to fly the neo.

In the past, performance improvements of that magnitude usually accompanied the introduction of a new model. Airlines would have to invest in new simulators, pilot-training programs, maintenance procedures, spare-parts inventories, manuals, and support equipment. The beauty of the A320neo was that it promised much of the economic benefit of a new airplane without forcing airlines to treat it as one.

When Boeing scrambled to counter the A320neo's sales momentum — and, in particular, to prevent American Airlines from awarding Airbus its entire narrow-body order — the company found itself in a difficult position. The MAX would be at a serious sales disadvantage if it could not fit seamlessly into an airline's existing 737 fleet. That was especially important to American, which already operated a substantial number of 737 NGs.

It was even more important to carriers such as Southwest Airlines in the United States and Ryanair in Europe, both of which had built their business models around fleets composed entirely of 737s. Each operated hundreds of the airplanes and derived enormous financial and operational benefits from fleet commonality.

Those airlines would naturally be interested in an aircraft offering double-digit improvements in fuel efficiency. But if introducing the MAX required simulator training, extensive new maintenance programs, or major changes to their operating infrastructure, much of the airplane's commonality advantage would disappear. At that point, they could more credibly consider the A320neo — or at least use the possibility of ordering it to place greater pricing pressure on Boeing.

To preserve the 737's market position, Boeing had little choice but to offer much the same promise of seamless integration that Airbus was making for the A320neo. When the company formally launched the MAX, it emphasized the airplane's commonality with the 737 NG and maintained that pilots already qualified on the NG would require only limited difference-training.

Many industry observers understood that fulfilling that promise would not be easy. The MAX would incorporate new engines, revised displays, aerodynamic changes, and other modifications that distinguished it from the NG. Boeing would have to ensure that those changes did not require airlines to treat the MAX as a substantially different airplane. The airlines understood that as well.

To reinforce its promise, Boeing included an extraordinary provision in its sales agreement with Southwest Airlines. If regulators required Southwest's pilots to undergo simulator training before transitioning from the 737 NG to the MAX, Boeing could be required to provide the airline with a rebate of $1 million for each affected airplane it delivered.

With Southwest ordering hundreds of MAX aircraft, Boeing's potential exposure amounted to hundreds of millions of dollars. The provision gave the company a powerful financial incentive to ensure that regulators approved a training program that did not require flight-simulator instruction.

The stakes extended beyond Southwest. Simulator training would have increased the cost of introducing the MAX for every airline already operating the 737 NG. It would also have weakened Boeing's claim that the MAX could be incorporated into existing 737 fleets with minimal disruption.

Therefore, Boeing faced two simultaneous demands. The MAX had to deliver the fuel savings necessary to compete with the A320neo, and it had to preserve enough continuity with the 737 NG to avoid costly transition training. The aerodynamic differences created by the MAX's new engine installation threatened that promise.

Boeing's engineers had already concluded that the airplane's altered handling characteristics could not be corrected sufficiently through aerodynamic changes alone. Training pilots to recognize and manage those differences remained an option, but doing so would have undermined one of the MAX program's central commercial commitments.

The company turned to a solution intended to make the MAX behave more like the 737 NG without requiring pilots to learn substantially different handling techniques.

That solution was the Maneuvering Characteristics Augmentation System — MCAS.

***

The Maneuvering Characteristics Augmentation System, or MCAS, was a new flight-control law Boeing incorporated into the 737 MAX's Speed Trim System. It resided within the airplane's flight-control computers, not its flight-management system.

MCAS was not originally conceived as a stall-prevention system in the conventional sense. Its purpose was to alter the control forces a pilot felt while manually maneuvering the MAX at elevated angles of attack. Because the larger LEAP-1B nacelles produced additional lift ahead of the airplane's center of gravity in those conditions, the force required to continue pulling back on the control column did not increase as progressively as certification standards required.

MCAS was intended to correct that characteristic by automatically commanding nose-down stabilizer trim. By doing so, it increased the amount of back-pressure a pilot would have to apply to continue raising the airplane's nose, making the MAX's handling characteristics feel more like those of the 737 NG.

The original system operated only when the airplane was being flown manually, the autopilot was disengaged, the flaps were fully retracted, and the sensed angle of attack exceeded a threshold that varied with airspeed. MCAS was therefore not intended to activate during takeoff or landing while the flaps remained extended.

On paper, the concept appeared straightforward. MCAS would operate only in an unusual portion of the flight envelope that airline pilots were unlikely to encounter during normal operations. However, if the airplane reached that condition, the system would quietly adjust the stabilizer and restore the control-force characteristics expected of a 737.

That expectation contributed to Boeing's decision not to describe MCAS in the flight manuals or training materials provided to most MAX pilots. Boeing maintained that pilots did not need to know about a system that operated only in the background and whose effects, if it malfunctioned, could supposedly be addressed through procedures pilots already knew.

The company was also under intense pressure to preserve the promise that pilots qualified on the 737 NG could transition to the MAX without simulator training. Adding a new flight-control system to the training curriculum risked inviting questions from regulators and airlines about whether the MAX's handling characteristics differed enough from those of the NG to require more extensive instruction.

MCAS was therefore designed to perform its function without requiring any deliberate pilot action. It did not move the control column forward as though another pilot were pushing it. Instead, it used the airplane's electric stabilizer-trim system to reposition the horizontal stabilizer in a more nose-down direction.

The pilots could nevertheless receive physical and audible indications that the stabilizer was moving. The trim wheels alongside the center pedestal inside the cockpit rotated when the electric stabilizer trim operated, producing a recognizable sound and motion. But because pilots had not been told that MCAS existed, they had no reason to associate an unexpected stabilizer movement with a system responding to angle-of-attack data.

That distinction would become critically important. A pilot confronting an unexplained nose-down movement would have to diagnose what the airplane was doing without knowing that a newly installed automated system might be commanding it.

MCAS contained several design vulnerabilities that would ultimately combine with one another in catastrophic fashion.

The first was its use of the horizontal stabilizer rather than the elevator. The elevator is the pilot's primary means of controlling pitch through movement of the control column. The stabilizer is primarily used to trim away sustained control forces. But the stabilizer has considerably more long-term pitching authority than the elevator and, if moved far enough in the nose-down direction, can make it extremely difficult for a pilot to hold the airplane's nose up using the control column alone.

When MCAS activated, it changed the stabilizer position without moving the pilots' control columns. A pilot pulling back could therefore feel the required force increasing without immediately understanding that the stabilizer itself was moving against him.

The pilots could counter an MCAS command by using their electric trim switches. Doing so stopped the MCAS command and moved the stabilizer back toward a nose-up position. But if the triggering angle-of-attack signal remained above the programmed threshold, the system could reset and command another nose-down movement several seconds later.

The second major vulnerability was the amount of stabilizer authority eventually given to MCAS.

In its earlier design, MCAS was limited to approximately 0.6 degrees of incremental nose-down stabilizer movement during an activation. That amount was considered sufficient for the higher-speed maneuvering condition for which the system had initially been developed.

During flight testing, however, Boeing discovered that the MAX also failed to produce the desired control-force progression during certain lower-speed, elevated-angle-of-attack maneuvers. Because the stabilizer was less aerodynamically effective at lower speeds, the original 0.6-degree authority was inadequate.

Boeing therefore expanded MCAS' maximum authority at low Mach numbers to approximately 2.5 degrees of incremental stabilizer movement — more than four times the original amount. The system could command that movement over approximately nine seconds.

This was not just an increase from 0.6 degrees per second to 2.5 degrees per second. It was an increase in the total stabilizer movement MCAS could command during a single activation.

That distinction did not make the change any less consequential. A system originally assessed with substantially less authority could now move the stabilizer far enough to create powerful nose-down forces. More importantly, the safety analyses provided to parts of the FAA certification process did not fully account for the effects of the system's expanded authority and repeated operation.

Boeing's assumptions rested heavily on the expectation that pilots would recognize unexpected stabilizer movement as runaway trim and respond promptly. Every 737 pilot was trained to address runaway stabilizer trim by holding the control column firmly, disengaging the automatic pilot if necessary, and moving the stabilizer-trim cutout switches to stop electrical power from driving the stabilizer.

But MCAS did not necessarily resemble the continuous movement pilots traditionally associated with runaway trim. It operated intermittently. It commanded nose-down trim, stopped, allowed the pilots to counter it with electric nose-up trim, and then could activate again if the triggering condition remained present. That behavior could make the problem appear to have been corrected, only for the nose-down command to return seconds later.

The third major vulnerability was the source of the information MCAS used to determine whether the airplane had reached an elevated angle of attack.

The 737 MAX had two angle-of-attack sensors, one on each side of the forward fuselage. But the way MCAS was designed originally, it did not compare their readings before activating. Instead, it relied on only one sensor during each flight.

The airplane had two flight-control computers, although only one served as the master during any particular flight. The master computer normally alternated from one flight to the next. When the left flight-control computer was active, MCAS used the left, captain's-side angle-of-attack sensor. When the right computer was active, it used the right, first-officer's-side sensor.

In either case, MCAS accepted the reading from that single sensor without checking it against the sensor on the opposite side. The system's fundamental vulnerability was, therefore, that it depended on only one sensor at a time. The consequences of the MCAS' design could be profound.

A faulty reading from the active angle-of-attack sensor could convince MCAS that the airplane was approaching a dangerously high angle of attack even when it was flying at an entirely normal attitude. Once the flaps were fully retracted and if the autopilot was disengaged, a false high-angle-of-attack signal satisfied the conditions for MCAS activation. The system would then command nose-down stabilizer movement even though the airplane was not actually approaching a stall.

If the pilots countered the movement with electric nose-up trim, MCAS could reset. As long as the defective sensor continued reporting an excessive angle of attack, the system could activate again and command another nose-down movement.

Boeing's safety assumptions treated an erroneous MCAS command as something pilots would recognize and correct within seconds. What those assumptions failed to consider adequately was the confusion created when a system the pilots did not know existed responded repeatedly to a sensor failure while numerous other warnings competed for their attention.

MCAS had been conceived as an unobtrusive means of making the MAX handle like the 737 NG. But Boeing had given that unseen system substantial control authority, allowed it to rely on a single sensor, permitted it to reactivate repeatedly, and withheld its existence from the pilots expected to overcome it if it failed.

***

Dennis Muilenburg was born in 1964 and grew up on a farm near Sioux Center, Iowa. From an early age, he was fascinated by the airplanes he watched passing high overhead and displayed a natural curiosity about how mechanical things worked. That combination of fascination and aptitude eventually led him to Iowa State University, where he earned a bachelor's degree in aerospace engineering in 1986. He later earned a master's degree in aeronautics and astronautics from the University of Washington.

While still attending Iowa State, Muilenburg secured an internship with Boeing in 1985. After completing his undergraduate degree, he joined the company full-time and began working within its defense operations.

Muilenburg's first assignment involved structural analysis, but he steadily advanced through a succession of engineering, program-management, and executive positions. One of the defining assignments of his early career was his work on Boeing's X-32 entry in the competition to develop the Joint Strike Fighter. Muilenburg held several engineering and management positions on the program, eventually serving as director of weapon systems.

Boeing lost the competition in October 2001 when the Department of Defense selected Lockheed Martin's X-35, which became the F-35. But the defeat did little to impede Muilenburg's rise through the company. Like many of the Boeing executives who had played prominent roles in the embarrassingly unsuccessful program, he continued to receive increasingly important assignments.

In 2009, Muilenburg was named president and chief executive officer of Boeing Defense, Space & Security, giving him responsibility for the company's sprawling portfolio of military aircraft, weapons, satellites, spacecraft, and related services. The position placed him in charge of one of Boeing's largest business units and gave him responsibility not only for enormous profit-and-loss accounts, but also for maintaining the company's critical relationships with the Pentagon, Congress, and other government customers.

It was during this period that Muilenburg attracted the attention of Boeing chairman and chief executive officer James McNerney. McNerney valued Muilenburg's engineering background, familiarity with complex government programs, and reputation for operational discipline. The two developed a close professional relationship, and Muilenburg emerged as one of the leading candidates to succeed him.

In December 2013, McNerney elevated Muilenburg to vice chairman, president, and chief operating officer of Boeing. The promotion placed him alongside McNerney in overseeing the company's day-to-day operations and relationships with its major stakeholders. It was also an unmistakable signal that Muilenburg was being prepared for the chief executive's office.

On July 1, 2015, McNerney relinquished the CEO position after a decade at Boeing's helm, and Muilenburg succeeded him as president and chief executive officer. McNerney remained chairman for several additional months before Muilenburg assumed that position as well in March 2016.

By the time McNerney stepped aside as CEO, he had led Boeing through one of the most turbulent periods in its history. His tenure began in the aftermath of the contracting and ethics scandals that had brought down Philip Condit and Harry Stonecipher. It subsequently encompassed the disastrous development and industrialization of the 787, including years of delays, massive cost overruns, supplier failures, and production problems.

By 2015, however, those scandals appeared to have receded into the company's past, and the 787 program finally seemed to be gaining momentum. Boeing was producing approximately ten Dreamliners a month between its assembly operations in the Seattle area and North Charleston, South Carolina. Plans were underway to increase the rate to twelve airplanes a month in 2016 and eventually to fourteen.

The 737 production system was also operating at a relentless pace. By mid-2015, Boeing was producing 42 airplanes a month in Renton and preparing to increase that rate to 47 in 2017 and 52 in 2018. Demand for both the 737 NG and the forthcoming 737 MAX was being driven by expanding global air travel, relatively strong airline finances, and an enormous worldwide appetite for more fuel-efficient single-aisle airplanes.

Boeing had begun assembling the first MAX on a new production line in Renton so its introduction would not initially disrupt the lines already producing the 737 NG. Over time, the MAX would gradually displace the NG until the newer model accounted for the entire 737 production system.

The demand appeared to validate Boeing's decision to re-engine the 737. By the time Muilenburg became CEO, the MAX had accumulated close to 3,000 firm orders even though the first airplane had not yet flown.

Airbus nevertheless retained a numerical advantage with the A320neo family. More importantly, it was establishing a formidable position at the upper end of the single-aisle market with the A321neo. The airplane was the re-engined successor to the stretched A321 and could be powered by either the CFM LEAP-1A or Pratt & Whitney PW1100G engine.

The A321neo was giving Airbus a strong presence in a segment once dominated by Boeing's 757. But the final 757 had rolled off Boeing's production line in 2004 after the company decided that the market no longer justified keeping the program alive.

That decision was becoming increasingly consequential. As air travel expanded, airlines were searching for an airplane capable of carrying substantial passenger loads over routes that fell between those traditionally served by smaller narrow-body airplanes and larger twin-aisle aircraft.

Airbus now occupied that portion of the market largely by itself, and the decision under Philip Condit and Harry Stonecipher to terminate the 757 program was coming back to haunt Boeing. If an airline wanted a newly manufactured airplane approaching the 757's combination of capacity and range, the A321neo was becoming the only practical choice. It was a striking reversal from the position Boeing had enjoyed when the 757 was still in production and Airbus lacked a directly comparable alternative.

In 2015, however, Boeing did not appear excessively concerned about the gap. The company was studying larger MAX derivatives, although it had not yet committed to what would eventually become the 737 MAX 10. That model would not be formally launched until 2017.

Although the MAX 10 would carry a roughly comparable number of passengers in certain configurations, its takeoff performance and range would not fully match those of the A321neo. Boeing's decision to remain within the architectural limits of the 737 would constrain how far the airplane could be stretched and how much additional performance could be extracted from it.

For the moment, though, the global market appeared large enough to sustain both manufacturers. Orders were flowing into Boeing and Airbus faster than either company could build airplanes. McNerney could therefore leave the CEO's office confident that Boeing had made the correct decision when it approved the MAX rather than undertaking the far more expensive and risky development of an entirely new airplane.

McNerney's departure on July 1, 2015, was accompanied by relatively little fanfare. His tenure had been characterized by financial discipline, aggressive cost control, and a relentless emphasis on improving margins, reducing risk, and generating returns for shareholders. Contemporary assessments credited his steady leadership with guiding Boeing through one of the most difficult periods in its history.

But his tenure had also completed the cultural transformation begun under Condit and Stonecipher. The cost-conscious, financially driven management philosophy that entered Boeing with the McDonnell Douglas merger was no longer merely competing with the company's traditional engineering culture. It was now firmly embedded throughout the enterprise.

Perhaps that was one reason McNerney regarded Muilenburg as an ideal successor. McNerney came from the Jack Welch school of corporate management, with its intense focus on cost reduction, financial performance, and shareholder value. Nearly two decades after the McDonnell Douglas merger began reshaping Boeing, McNerney had put the finishing touches on the company's transformation into an organization operating according to that model.

Yet Muilenburg was not another executive recruited from General Electric or another outside corporation. He was a lifelong Boeing employee and an aerospace engineer who had spent his entire career working on some of the company's most technically complex programs.

Dennis Muilenburg projected an almost boyish quality. An avid cyclist and fitness enthusiast, he carried himself more like an athlete than a corporate executive and was known to ride his bicycle for 25 miles or more several times a week.

Nothing about his appearance immediately suggested that he occupied one of the most powerful positions in American industry. Placed in a room full of strangers, he might have been mistaken for a senior manager or divisional vice president rather than the chief executive of one of the largest and most influential corporations in the world.

Muilenburg was also relatively soft-spoken for the leader of a Fortune 50 company. During public appearances, he sometimes seemed less commanding than one might expect from a corporate chief executive, occasionally displaying a tentative, almost apprehensive expression while answering questions.

At only 51 years old when he became CEO, Muilenburg appeared to have a long tenure ahead of him. If Boeing continued prospering under his leadership, he could remain in the position for many years and potentially become one of the company's longest-serving chief executives since Bill Allen. That was, of course, assuming everything went well.

Despite being an engineer by education and profession, and despite having spent his entire 30-year career with Boeing, Muilenburg had also learned the importance the modern company placed on shareholder value. Most of his career had unfolded during the cultural transformation initiated by Condit and Stonecipher and accelerated under McNerney. By the time Muilenburg reached the executive suite, cash generation, margin growth, financial performance, and returns to shareholders had become central measures of managerial success.

If McNerney had been the architect of Boeing's renewed emphasis on stock repurchases, Muilenburg would become its most enthusiastic executor. McNerney oversaw approximately $15 billion in stock repurchases during his decade as CEO. Under Muilenburg, Boeing dramatically accelerated the program, spending nearly $30 billion repurchasing its own shares in only a few years. Whatever expectations may have existed that

Muilenburg's engineering background would cause him to govern the company differently from McNerney, his embrace of the buyback program suggested otherwise.

After three decades inside a corporation increasingly shaped by the management philosophy associated with Jack Welch and General Electric, Muilenburg had plainly absorbed its central tenets. Some detractors inside the company whispered he had drunk the proverbial Kool-Aid.

There can be legitimate reasons for a corporation to repurchase its own stock. Management may believe the shares are undervalued. Repurchases can offset dilution caused by employee stock-compensation programs, provide flexibility in returning excess cash to investors, stabilize the share price during periods of market volatility, or offer a more tax-efficient alternative to increasing dividends. Buybacks can also be justified when a mature company is generating more cash than it can productively reinvest.

But Boeing was not lacking opportunities for productive investment. The global economy was expanding, air travel was growing rapidly, and airlines were ordering airplanes in unprecedented numbers. Airbus was challenging Boeing across virtually every segment of commercial aviation and had already established a significant advantage with the A320neo family, particularly at the upper end of the single-aisle market with the A321neo.

Boeing was also not an asset-light technology company. It was one of the most complex industrial enterprises in the world. It depended on capital-intensive manufacturing, vast factories, highly skilled workforces, complicated global supply chains, expensive testing and certification programs, and products that could take the better part of a decade and tens of billions of dollars to bring from conception into commercial service.

Compared with the enormous repurchase programs undertaken by technology companies such as Apple, Microsoft, and

Google's parent company, Alphabet, Boeing's approximately $45 billion in buybacks under McNerney and Muilenburg was not the largest in absolute terms. But that comparison obscured a critical distinction.

The technology giants were generally high-margin, asset-light companies that generated far more cash than was required to maintain their physical operations. Boeing was a capital-intensive manufacturer operating on much narrower margins while confronting mounting competitive, technological, and industrial challenges.

That was why Boeing's stock-buyback program raised so many eyebrows. The concern was not simply that the company was returning money to shareholders. It was the amount of cash consumed over such a relatively short period while Airbus was steadily eroding Boeing's competitive position.

Even a fraction of the approximately $45 billion spent repurchasing shares could have been reinvested in Boeing's engineering workforce, production facilities, supplier network, quality-control systems, or the development of an entirely new single-aisle airplane capable of competing directly with the A321neo and whatever Airbus developed next. Instead, tens of billions of dollars left the company to purchase its own shares in the open market.

The repurchases reduced the number of shares outstanding, increased earnings per share, supported Boeing's stock price, and rewarded investors — including senior executives whose compensation was tied heavily to financial performance and share value.

But they did not design a new airplane. They did not strengthen a production line. And they did not restore the competitive advantage Boeing was steadily surrendering to Airbus.

At Boeing's headquarters in Chicago and within its boardrooms, momentum was building behind the belief that the company's worst days were finally behind it. More than two decades after the acquisition of McDonnell Douglas, the strategy that followed from that merger appeared to finally be paying off. Confidence was high that Boeing's existing product line was good enough for the foreseeable future and that the correct path forward was to continue reducing costs, improving margins, and increasing shareholder value.

For the moment, at least, few seemed inclined to question that conclusion.

Over the Precipice

On the morning of October 29, 2018, Lion Air Flight 610 taxied onto Runway 25L at Soekarno–Hatta International Airport in Jakarta, Indonesia. Its destination was Pangkal Pinang, roughly 90 minutes away. On board were two pilots, six flight attendants, and 181 passengers. The aircraft was a Boeing 737 MAX 8 that had been delivered new to the airline only two months earlier, in August.

Cleared for takeoff, Flight 610 accelerated down the runway as the first light of dawn broke across the horizon. Its wheels lifted from the pavement at 6:20 a.m. local time. Thirteen minutes later, the aircraft crashed into the Java Sea about 18 miles offshore. Witnesses on a nearby oil platform reported seeing it strike the water at high speed in a steep nose-down attitude. All 189 people on board died instantly. The disaster would claim a 190th victim several days later when a professional diver died during the search-and-recovery operation.

In the immediate aftermath, it emerged that only minutes after takeoff the pilots had requested permission to return to the airport because of "flight-control problems." The aircraft's two flight recorders — the cockpit voice recorder and the flight data recorder, which captured hundreds of operating parameters — were eventually recovered from the seafloor. What they revealed to investigators was a harrowing sequence of events compressed into the flight's final minutes.

The first sign of trouble appeared during rotation. As the nose lifted from the runway, the captain's control-column stick shaker activated, warning of an impending stall, while the first officer's did not. Investigators would later recognize this asymmetry as evidence that the two angle-of-attack sensors — one mounted on each side of the aircraft — were providing conflicting information. The pilots, however, received no angle-of-attack disagree alert.

The erroneous data from the left-side AOA sensor, which was responsible for triggering the captain's control-column stick shaker, also affected other systems. Shortly after liftoff, the pilots' primary flight displays began showing airspeed and altitude-disagree warnings. The captain's indicated airspeed initially differed from the first officer's by about 10 knots, while his altitude indication at times differed by as much as 250 feet.

Those warnings appeared prominently on both pilots' primary flight displays at one of the most demanding moments of flight, just after the aircraft had become airborne. At the same time that the captain's stick shaker produced a loud, continuous rattling as it vigorously vibrated his control column, the first officer's column remained still. The conflicting indications created a stressful and deeply confusing cockpit environment precisely when both pilots most needed reliable information.

The pilots initially climbed with the flaps extended, a configuration that inhibited MCAS. Once the flaps were fully retracted, however, the system became active. Relying on the erroneous reading from the captain's angle-of-attack sensor, MCAS interpreted the aircraft as approaching an excessive angle of attack and commanded nose-down stabilizer trim.

The pilots knew nothing of the system's existence. Already contending with a continuous stick shaker and conflicting airspeed and altitude indications, they now faced repeated, uncommanded nose-down trim movements at the very moment they were trying to stabilize the climb and determine what was wrong.

The first officer began working through emergency checklists and eventually contacted air traffic control to request an immediate return to the airport. Meanwhile, the captain, confronted with continuous stall warnings through the stick shaker, kept the throttles advanced to preserve airspeed against what he believed might be an impending stall.

Each time MCAS activated, it drove the horizontal stabilizer farther toward nose-down trim. The captain countered by pulling back on the control column and using the electric trim switches on the yoke to command nose-up trim. His inputs temporarily stopped and reversed the movement, but after a brief pause MCAS activated again and resumed trimming nose-down.

The struggle continued through repeated cycles. Each MCAS command moved the stabilizer farther toward nose-down, while each pilot trim input provided only temporary relief. As the aircraft accelerated, the aerodynamic forces working against the pilots grew more severe.

By the final moments, the stabilizer had been driven to an extreme nose-down position, and the captain could no longer pull hard enough on the control column to arrest the descent. The elevator alone could not overcome the stabilizer's powerful nose-down command.

Flight 610 struck the Java Sea at high speed. The aircraft's actual angle of attack had never justified MCAS intervention — the system had been responding to a single failed sensor. To the pilots, who had never been told that MCAS existed, it must have seemed as though an unseen force had taken control of the airplane.

When word of the crash reached Boeing's headquarters in Chicago on the evening of October 28 — Jakarta was 12 hours ahead — the initial assumption was that some external event unrelated to the aircraft, such as a bomb, might have brought it down. Over the next 24 hours, however, reports emerged that the crew had experienced flight-control problems and had requested permission to return to the airport. Among some Boeing engineers in Seattle, suspicion began to turn toward a system few outside the company even knew existed: MCAS.

***

When Boeing launched the 737 Next Generation series in 1993, its engineers redesigned the flight deck around six large liquid-crystal displays, bringing the 737 closer to contemporary cockpit standards and making it more competitive with the Airbus A320. The digital displays also made it possible to present pilots with information that had not been available on earlier 737 instrument panels.

Among the features developed for the NG were two related but distinct angle-of-attack displays. The first was an Angle-of-Attack Disagree alert, which would appear when the aircraft's two angle-of-attack sensors differed by more than a specified amount for a sustained period. Such a disagreement could indicate that one of the sensors had failed, been damaged, or fallen out of calibration.

The distinction was important because each side of the aircraft had its own angle-of-attack sensor. Among other functions, the sensor on each side could activate the stick shaker on the corresponding pilot's control column if it indicated that the aircraft was approaching a stall. If the two sensors reported substantially different values, the Angle-of-Attack Disagree alert provided the pilots with an additional indication that one of the readings might be unreliable.

The second feature was an angle-of-attack indicator. This displayed the actual reading from the sensor serving that pilot's instruments on a scale incorporated into the primary flight display. It allowed the pilot to observe the indicated angle of attack in real time and compare it with other flight information. Unlike the disagreement alert, however, the indicator was considered supplemental information rather than a feature necessary for the safe operation of the aircraft.

Boeing therefore treated the two features differently. The Angle-of-Attack Disagree alert was intended to be standard equipment, while the angle-of-attack indicator was offered as a customer-selected option. Although Boeing never publicly released a formal price list for the feature, the optional indicator was widely reported to cost approximately $80,000 per aircraft.

On an airplane costing tens of millions of dollars, another $80,000 might appear insignificant. Airlines, however, commonly operate on narrow profit margins, and relatively small charges become substantial when multiplied across fleets containing dozens or hundreds of aircraft. Because the angle-of-attack indicator was not required for certification or routine flight operations, most airlines declined to purchase it. Only about 20 percent of MAX customers ultimately selected the option.

Both features were carried forward when Boeing developed the 737 MAX. Maintaining them in substantially the same form supported Boeing's broader objective of preserving cockpit commonality between the NG and the MAX. Pilots transitioning from one model to the other would encounter familiar indications and would not require instruction on a newly designed alerting philosophy.

The MAX nevertheless received an extensively redesigned display system. Its four large primary display units were brighter, larger, and capable of presenting more information than those installed in the NG. Rockwell Collins was responsible for supplying the display system, including the display processing computers and much of the associated software. Rockwell Collins was descended from the portion of Rockwell International that Boeing had not acquired in 1996, when Boeing purchased Rockwell's aerospace and defense businesses.

As development of the MAX began, Boeing supplied Rockwell Collins with the design requirements for the new cockpit displays, including the requirement that the Angle-of-Attack Disagree alert remain a standard, standalone feature while the angle-of-attack indicator remained optional.

The problem began with an error in the requirements documentation. In 2014, a comment in one of Boeing's documents incorrectly suggested that the Angle-of-Attack Disagree alert should be linked to the optional angle-of-attack indicator. Rockwell Collins questioned the apparent linkage, and Boeing clarified later that year that the alert was supposed to operate independently. The supplier corrected the software accordingly.

In 2015, however, Rockwell Collins was addressing an unrelated problem involving the way the disagreement alert appeared on the cockpit displays. During that work, the supplier inadvertently restored the link between the standard alert and the optional indicator, relying on the erroneous comment that had remained in Boeing's requirements document. Boeing subsequently tested and approved the revised software but failed to detect that the two features had once again been linked.

The result was that the Angle-of-Attack Disagree alert would function only on aircraft whose customers had purchased the optional angle-of-attack indicator. This meant that on the roughly 80 percent of MAX aircraft that lacked the optional indicator, the supposedly standard disagreement alert was inoperative.

The FAA certified the 737 MAX 8 in March 2017, and Boeing began delivering aircraft to customers two months later. In August, approximately three months after the MAX entered commercial service, Boeing engineers discovered that the display-system software did not conform to the company's design requirements. Aircraft had already been delivered without a functioning Angle-of-Attack Disagree alert, and additional aircraft with the same defect were moving through production.

Boeing documented the defect in a formal problem report and notified Rockwell Collins. In September 2017, Boeing reviewed the matter under its internal procedures and classified it as a non-safety issue. The reasoning was that the alert provided supplemental awareness, did not require an immediate pilot response, and was not required by FAA safety regulations. Boeing decided that the software could be corrected during a future display-system update planned for 2020, when the MAX 10 was then expected to enter service.

The determination was technically defensible within Boeing's internal classification system, but it overlooked the broader operational consequences. The Angle-of-Attack Disagree alert appeared in Boeing's flight-crew documentation, and pilots and airlines therefore had reason to believe it was functioning. A crew encountering conflicting airspeed, altitude, stick-shaker, or other indications could have used an Angle-of-Attack Disagree message as another clue that one of the aircraft's angle-of-attack sensors was producing erroneous data.

That omission became considerably more significant on the MAX because angle-of-attack data no longer merely supported cockpit indications and stall warnings. A single angle-of-attack sensor also supplied the data that could activate the Maneuvering Characteristics Augmentation System. If that sensor produced a falsely high reading after the flaps were retracted, MCAS could repeatedly command nose-down stabilizer movement even though the aircraft was not approaching a stall.

Boeing considered informing operators about the inoperative alerts. Internal discussions included proposals for an operations-manual bulletin or a less prominent fleet communication advising customers that they might not have the alert they expected. Ultimately, neither notice was issued. Boeing continued delivering MAX aircraft, and the flight manuals supplied to airlines continued to describe the Angle-of-Attack Disagree alert without clearly explaining that it would not function unless the optional angle-of-attack indicator had been purchased.

Because Boeing had classified the defect as non-safety-related, the company was not required under its established procedures to notify the FAA immediately. The FAA therefore remained unaware of the problem throughout the following year.

On October 29, 2018, Lion Air Flight 610 crashed after erroneous data from a damaged and improperly calibrated angle-of-attack sensor activated the captain's stick shaker and repeatedly triggered MCAS. The aircraft had not been purchased with the optional angle-of-attack indicator. Consequently, its supposedly standard Angle-of-Attack Disagree alert was also inoperative.

Only after the crash did Boeing inform the FAA and its airline customers of the defect. In December 2018, Boeing convened a formal Safety Review Board to reconsider the matter. The board upheld the earlier determination that the missing alert did not constitute a safety-of-flight issue, and the FAA later concurred with that classification.

The absence of the alert was not identified as a direct cause of the MAX accident. Nevertheless, Boeing had knowingly delivered aircraft that did not conform to their approved design and had allowed pilots to believe that a potentially useful warning was available when it was not.

What had begun as an error in a requirements document, compounded by a supplier's software change and Boeing's failure to detect it during testing, became another example of how seemingly minor technical and organizational failures could align throughout the MAX program.

***

The pilots of Lion Air Flight 610 had more than 10,000 hours of flight time between them, much of it on the 737. Because Boeing had required no simulator training for pilots transitioning from the 737 NG to the MAX, they were expected to encounter essentially the same cockpit indications described for the earlier model — including an Angle-of-Attack Disagree alert when the readings from the aircraft's two sensors differed. But no such warning appeared. Lion Air, like most MAX operators, had not purchased the optional angle-of-attack indicator, and because of the software defect, the supposedly standard Angle-of-Attack Disagree alert would not function without it.

Had the alert appeared, it might have helped the crew recognize more quickly that the captain's stick shaker and some of the flight information displayed on his side of the cockpit were being generated by an unreliable angle-of-attack sensor reading. The alert would not have identified MCAS or explained the repeated nose-down trim commands, but it would have provided another clue that the aircraft was receiving conflicting information from its AOA sensors.

As investigators examined the aircraft's maintenance history, they discovered a remarkable sequence of events involving its flight from Denpasar to Jakarta the previous evening. During that flight, the aircraft had experienced nearly the same malfunction that would confront the crew of Flight 610 the following morning. The difference was that a third Lion Air pilot happened to be riding in the cockpit jump seat.

The additional pilot was traveling as an extra crew member rather than as a fare-paying passenger to position himself for a later assignment. Airline pilots traveling in this manner are commonly referred to as deadheading. When available, the cockpit jump seat allows an authorized pilot to travel without occupying a passenger seat in the cabin.

Shortly after the aircraft departed Denpasar, the captain's stick shaker activated and remained on. After the flaps were fully retracted, erroneous data from the captain's AOA sensor began triggering repeated MCAS nose-down stabilizer commands. The crew repeatedly countered the automatic commands with electric nose-up trim, only to have MCAS activate again several seconds later.

Amid the confusion, the pilots eventually recognized the repeated uncommanded stabilizer movement and moved the stabilizer-trim cutout switches to CUTOUT, disabling the automatic trim system. The third pilot riding in the jump seat assisted the operating crew as they diagnosed and responded to the malfunction.

Once the switches were moved to CUTOUT, the repeated nose-down commands stopped, and the crew continued safely to Jakarta while controlling the stabilizer with the manual trim wheel. The presence of a third pilot — someone able to observe the developing situation without simultaneously flying the aircraft — may have been decisive.

The chain of events had begun before that flight departed Denpasar. After the aircraft experienced recurring airspeed and altitude indication problems on earlier flights, Lion Air mechanics replaced the angle-of-attack sensor on the captain's side. The replacement was a repaired unit supplied by Xtra Aerospace, a Florida company authorized to overhaul and repair aircraft components.

During its earlier repair, however, the sensor had been improperly calibrated. It produced readings approximately 21 degrees higher than the vane's actual position. The calibration error was not detected before Xtra Aerospace released the unit for return to service. Consequently, when the sensor was later installed on the Lion Air aircraft, it was already un-airworthy.

Lion Air's maintenance procedures required mechanics to conduct an installation test after replacing the sensor. Properly performed, that test should have revealed the approximately 21-degree discrepancy between the captain's and first officer's angle-of-attack readings. The mechanic recorded that the test had been completed successfully, but investigators concluded that the mis-calibration was not identified during the test even though the mechanic signed off on the paperwork.

After the aircraft reached Jakarta following the previous evening's flight, the captain reported problems involving the aircraft's airspeed and altitude indications. However, he did not clearly document the continuous stick shaker or the repeated automatic nose-down trim commands, nor did he report that the crew had moved the stabilizer-trim cutout switches to CUTOUT during the flight and returned them to their normal position after landing.

Without that critical information, the Jakarta maintenance crew focused on the reported indication problems, performed additional checks, and returned the aircraft to service. The next morning, Flight 610 departed with the same mis-calibrated angle-of-attack sensor still installed.

***

Because the crash occurred in Indonesian territory and involved an Indonesian airline, Indonesia's National Transportation Safety Committee — known by its Indonesian initials, KNKT — led the investigation under international aviation protocols. Since the aircraft had been designed, manufactured, and certified in the United States, the U.S. National Transportation Safety Board participated as the accredited representative of the state of design and manufacture. Boeing, as the aircraft's designer and manufacturer, served as a technical adviser to the NTSB.

With the flight data already pointing toward an automatic flight-control response triggered by erroneous angle-of-attack information, Boeing faced the possibility that its newest commercial airplane — and a system about which its pilots had not been informed — might be implicated in the deaths of 189 people.

On November 6, only eight days after Flight 610 went down, Boeing issued an Operations Manual Bulletin to MAX operators worldwide. The bulletin warned that erroneous angle-of-attack data could cause repeated nose-down stabilizer-trim commands and directed crews to follow the existing runaway-stabilizer procedure. It did not identify MCAS by name or explain that a new flight-control law was producing the commands.

The following day, November 7, the FAA issued Emergency Airworthiness Directive 2018-23-51, requiring MAX operators to revise their flight manuals and instruct pilots how to respond if erroneous angle-of-attack data caused repeated nose-down stabilizer movement. Like Boeing's bulletin, the directive described the symptoms and prescribed the runaway-stabilizer response, but it did not mention MCAS.

The directive warned pilots that erroneous angle-of-attack data could produce several indications, including continuous or intermittent stick-shaker activation, airspeed and altitude-disagree alerts, and increasing nose-down control forces. If the automatic nose-down trim continued, pilots were instructed to move the stabilizer-trim cutout switches to CUTOUT and control the stabilizer with the manual trim wheel.

The omission of MCAS was critical. The bulletin and directive told pilots how to respond to the symptoms, but not what system was causing them, why that system could be activated by a single failed angle-of-attack sensor, or why it could reactivate after a pilot used electric trim to counter its commands. Pilots were therefore being asked to recognize an unfamiliar failure mode without being told that the MAX contained a new flight-control law capable of repeatedly moving the stabilizer without their command.

Pilots who read the documents immediately began asking questions. A malfunctioning angle-of-attack sensor on an earlier 737 might activate a stick shaker or produce conflicting flight information, but it did not cause the stabilizer to move repeatedly in the nose-down direction. Why, then, would erroneous angle-of-attack data on the MAX require pilots to use the stabilizer-trim cutout switches? Was there something fundamentally different about the MAX that Boeing had not disclosed?

Pilots at major MAX operators, including American Airlines and Southwest Airlines, raised concerns with their companies and unions. Representatives of the Allied Pilots Association, which represented American's pilots, and the Southwest Airlines Pilots Association pressed Boeing for an explanation. The Air Line Pilots Association, which represented pilots at several other airlines, also sought additional information.

As the questions mounted, Boeing began explaining MCAS directly to airlines and pilot groups. For the first time, many MAX pilots learned that the Maneuvering Characteristics Augmentation System was a flight-control law capable of commanding nose-down stabilizer movement without pilot input when it received a sufficiently high angle-of-attack reading.

The revelation was alarming not merely because the system existed, but because it appeared nowhere in the pilots' flight manuals or differences-training materials. Boeing had introduced a system capable of moving the horizontal stabilizer — a primary means of controlling the aircraft's pitch — without ensuring that the pilots responsible for flying the airplane understood what the system did, when it could activate, or how it would behave after they countered it.

Throughout aviation history, there has been a fundamental expectation, some would even say a cardinal rule, that if an aircraft is going to do something without a pilot's command, the pilot should understand what the aircraft may do, why it may do it, and under what circumstances it would do it. It was now becoming apparent to the aviation community that Boeing had violated that expectation. In doing so, the company had damaged the trust of the pilots who flew its airplanes.

On November 12, the first major media reports began revealing that the MAX contained a flight-control system that had not been disclosed to the pilots flying the airplane. The reports described MCAS as capable of automatically commanding nose-down stabilizer movement and raised the possibility that it had played a central role in the crash of Lion Air Flight 610.

The emerging reports provoked intense criticism from pilots, aviation experts, and the news media. Boeing responded by emphasizing that the MAX remained safe, that the company had complied with applicable regulatory requirements, and that the proper response to uncommanded stabilizer movement was already contained in the established runaway-stabilizer procedure.

Rather than immediately confronting the implications of having withheld information about MCAS from pilots, Boeing's public position rested heavily on the argument that it had stayed within regulatory requirements and that existing procedures should have been sufficient to control the airplane.

***

In response to the growing and sometimes explosive media coverage, Boeing suddenly found itself on the defensive. Its initial position was that the Operations Manual Bulletin and the existing runaway-stabilizer checklist adequately addressed the symptoms pilots might encounter if erroneous angle-of-attack data caused uncommanded stabilizer movement. At the same time, the company frequently declined to address questions about the accident directly, referring inquiries to the Indonesian authorities leading the investigation.

On November 19, only a week after the first major reports about the previously undisclosed MCAS began to appear, Dennis Muilenburg issued an internal message to Boeing employees. Addressing reports that the system had been omitted from the pilots' operating manuals, he wrote, "That's simply untrue. The relevant function is described in the Flight Crew Operations Manual, and we routinely engage with customers about how to operate our airplanes safely."

The statement was, at best, deeply misleading. The Flight Crew Operations Manual described the established runaway-stabilizer procedure, but it did not identify MCAS by name or explain that the system could respond to a single erroneous angle-of-attack reading by repeatedly commanding nose-down stabilizer movement. Pilots may have been given a procedure that Boeing believed would arrest the consequences of an MCAS malfunction, but they had not been told that the system existed, what activated it, or why it could resume operating after they countered it with electric trim.

The distinction between describing a response to uncommanded stabilizer movement and describing the system that caused it was not a trivial one. Nevertheless, Muilenburg publicly treated the two as though they were interchangeable. Rather than acknowledge that Boeing had withheld consequential information from MAX pilots, he defended the company's actions by arguing that the relevant pilot response was already contained in the manuals.

The statement also came at the beginning of a major investigation involving one of Boeing's most important products. A more prudent response would have been to acknowledge the existence of MCAS, explain that the system was under investigation, and refrain from making categorical claims until investigators had established precisely how it had contributed to the accident. Instead, Boeing, and its chief executive officer, adopted a defensive posture before all the evidence had been assembled.

At the same time, the company continued pursuing a software correction for the inoperative Angle-of-Attach Disagree alert. Boeing pressed Rockwell Collins to accelerate an update to the primary-flight-display software so that the supposedly standard alert would function independently of the optional angle-of-attack indicator. Correcting that defect was necessary, but it did not address the more fundamental questions raised by the accident: why MCAS had been given such extensive control authority, why it depended on data from only one angle-of-attack sensor at a time, why pilots had not been told about it, and how those decisions had survived Boeing's engineering and safety-review processes.

A proper root-cause analysis would have examined not merely the defective sensor or the pilots' response, but whether Boeing's established design, certification, communication, and safety-assessment procedures had been followed. Such an analysis would eventually expose failures extending far beyond a single software function. Information had been fragmented among different engineering and certification groups, assumptions about pilot behavior had gone largely unchallenged, and consequential changes to MCAS had not been communicated adequately to everyone responsible for evaluating the system.

Many of the procedures governing Boeing's certification work existed within the FAA's Organization Designation Authorization program. Because the FAA did not possess enough personnel to examine every detail of every new aircraft design, it delegated specified certification functions to qualified employees of aircraft manufacturers. At Boeing, those employees were known as ODA Unit Members.

The ODA system did not allow Boeing simply to certify an aircraft however it wished. The FAA approved the organization's delegated authority, established the limits under which it operated, and retained legal responsibility for issuing the aircraft's type certificate. ODA Unit Members were expected to perform their delegated work independently and to represent the FAA's safety interests when making findings of compliance.

They nevertheless remained Boeing employees. Their salaries, careers, performance evaluations, and daily working relationships were tied to the same corporation whose products they were evaluating. That arrangement created an inherent tension — an ODA Unit Member was expected to act independently on the FAA's behalf while working inside an organization facing intense commercial pressure to complete the MAX on schedule and preserve its competitive advantages.

The FAA had determined in 2012 that the 737 MAX qualified for certification as an amended version of the existing 737 type design and that much of the certification work could be managed through Boeing's ODA. As the program progressed, the FAA delegated an increasing share of the compliance findings to Boeing. This allowed the agency to concentrate its limited resources on matters it considered most critical, but it also meant that the FAA's understanding of the aircraft depended heavily on the accuracy and completeness of the information Boeing provided.

Separate from the ODA was Boeing's internal safety-review structure. Among its components was the Safety Review Board, which evaluated certain technical problems and determined whether they constituted safety-of-flight issues. The SRB did not exercise delegated FAA certification authority, nor was it an FAA-approved substitute for regulatory oversight. It was an internal Boeing body that considered information developed through the company's engineering and safety processes.

After Boeing discovered in 2017 that the Angle-of-Attack Disagree alert would not function unless an airline had purchased the optional angle-of-attack indicator, the company classified the defect as a non-safety issue. Boeing concluded that the alert provided supplemental information and was not required for the safe operation of the aircraft. That determination allowed the company to postpone correcting the defect until a later software update and meant that the FAA was not immediately informed.

The decision reflected the fragmented manner in which hazards associated with the MAX were being evaluated. Considered in isolation, an inoperative disagreement alert might appear relatively minor. But the same AOA sensor that could generate unreliable cockpit indications also supplied the data that activated MCAS. Boeing's review did not adequately consider how the absent alert might affect a pilot attempting to understand several simultaneous and apparently unrelated warnings while the stabilizer repeatedly moved in the nose-down direction.

Another breakdown occurred in the communication between Boeing's flight technical personnel and the FAA's Aircraft Evaluation Group, which was responsible for determining the training required for pilots transitioning to the MAX. During simulator work in 2016, a Boeing flight technical pilot encountered MCAS behavior that revealed the system had greater operational authority than he had previously understood. This was because of the previous flight testing that revealed MCAS might not have sufficient authority moving the stabilizer at a rate of 0.6 degrees per second, which necessitated the modification that allowed it to move instead at a much faster 2.5 degrees per second.

The pilot communicated internally about the unexpected behavior, but Boeing did not ensure that the FAA's Aircraft Evaluation Group received a complete explanation of how MCAS had evolved. That omission was consequential because the AEG was responsible for determining the minimum training required for pilots transitioning from the 737 NG to the MAX. Boeing had made the absence of simulator training one of the MAX program's most important commercial promises, and disclosing that MCAS could now operate through nearly the aircraft's entire speed range, with far greater authority than originally understood, risked forcing the FAA to reconsider whether the system constituted a significant enough difference to require additional training.

Rather than disclose the full extent of the change, Boeing's flight technical organization continued pressing the AEG to remove references to MCAS from the MAX's flight-standardization report. The justification presented to the FAA was that MCAS operated outside the aircraft's normal operating envelope and would therefore be effectively transparent to pilots during routine airline service.

But the flight technical pilots knew by then that the system's operating range and authority had been expanded considerably. By withholding that information, they allowed the AEG to evaluate MCAS according to an obsolete and incomplete description. The FAA agreed to remove the system from the final report, preserving Boeing's desired training determination while ensuring that the pilots who would fly the MAX entered service without being taught that MCAS existed, what could activate it, or how it could behave following an erroneous sensor input.

This was not the result of a single employee secretly certifying his own work. It was a broader organizational failure involving engineering, flight technical personnel, certification specialists, managers, and regulators. Information known in one portion of Boeing did not necessarily reach another, and no one with a complete understanding of the system appears to have forced the company to reconsider its assumptions before the aircraft entered service.

Among those assumptions was the belief that a properly trained 737 pilot would quickly recognize uncommanded stabilizer movement, perform the immediate runaway-stabilizer actions that pilots were required to know from memory, and move the stabilizer-trim cutout switches to CUTOUT.

That expectation became a central defense of the MAX after Flight 610. Boeing maintained that, regardless of what had initiated the trim movement, the existing procedure should have allowed the crew to retain control of the aircraft.

That position reduced the accident to a question of pilot response while avoiding the design decisions that had created the emergency. It did not adequately account for the continuous stick shaker, conflicting airspeed and altitude information, repeated cockpit warnings, heavy control forces, and the fact that MCAS resumed operating after the pilots used electric trim to oppose it. Nor did it account for the pilots' complete lack of knowledge that such a system existed.

In the immediate aftermath of Flight 610, Boeing had enough information to know that MCAS had repeatedly commanded nose-down stabilizer movement in response to erroneous data from the captain's angle-of-attack sensor. A company still governed by an engineering-first culture might have treated that discovery as evidence of a systemic design failure requiring immediate and comprehensive examination. It might have asked why a single sensor could activate the system, why the system could operate repeatedly, why the pilots had not been informed about it, and why the hazard assessments had not anticipated the cockpit conditions encountered by the crew.

Instead, Boeing concentrated on defending the aircraft's certification, emphasizing the existing runaway-stabilizer procedure, and correcting the separate Angle-of-Attack Disagree software defect. Its public posture suggested that the airplane remained fundamentally sound and that adherence to established pilot procedures would prevent a recurrence.

The company's priorities had changed considerably from those that had guided Boeing through much of its history. Under Philip Condit, Harry Stonecipher, and James McNerney, financial performance, cost control, schedule discipline, and shareholder returns had assumed increasing importance. Engineering remained essential to the company's products, but engineers no longer possessed the institutional authority they had once exercised.

Those priorities shaped Boeing's reaction to Flight 610. Acknowledging that MCAS itself contained fundamental design deficiencies would have threatened the MAX program, raised questions about its certification, potentially required additional pilot training, and jeopardized one of the company's most profitable products. Emphasizing pilot adherence to established procedures was far less disruptive. And, of course, it involved minimal cost.

Boeing continued telling its airline customers that pilots should treat repeated uncommanded stabilizer movement as a runaway-stabilizer condition. If the movement continued, they were to hold the control column firmly, disengage the automatic trim system by moving the stabilizer-trim cutout switches to CUTOUT, and control the stabilizer manually.

The company also continued portraying MCAS as a limited handling-quality feature rather than as a system central to routine aircraft control. Some defenders compared it with the flight-envelope protections incorporated into Airbus aircraft, which could prevent pilots from commanding maneuvers that would place the aircraft in an unsafe condition.

That comparison was deeply flawed. Airbus' flight-control protections were central, documented features of its fly-by-wire design philosophy. Pilots were taught how those protections worked, under what circumstances they operated, and how aircraft behavior changed if portions of the system became unavailable. MCAS, by contrast, had been deliberately omitted from MAX pilot manuals and transition training. The people flying the aircraft could not account for the behavior of a system they did not know existed.

As the weeks passed and Flight 610 gradually receded from the world's headlines, a belief took hold within Boeing that the accident might have been an isolated event. MCAS may have responded to erroneous sensor data, but the company continued to maintain that a properly trained crew should have controlled the resulting stabilizer movement by following the established runaway-stabilizer procedure.

That belief allowed Boeing to frame the accident primarily as a failure of pilot proficiency rather than as evidence of either inadequate training or a fundamental design problem. The distinction was consequential. If the pilots had lacked sufficient training, Boeing might have been forced to acknowledge that MAX crews required additional instruction or simulator time, undermining one of the program's most important promises to airline customers. If the aircraft itself contained a fundamental design flaw, Boeing would have faced even more serious questions about the MAX's certification and continued airworthiness. By describing the accident instead as a matter of proficiency, the company placed responsibility on the pilots for failing to apply procedures they were already expected to know.

Boeing's defensive position protected the MAX in the short term. It also prevented the company from fully confronting what the crash of Flight 610 had revealed — a single faulty sensor had activated an undocumented system that repeatedly moved a primary flight-control surface, and Boeing's assumptions about how pilots would respond had proved disastrously incomplete.

***

On the morning of March 10, 2019, Ethiopian Airlines Flight 302 taxied onto Runway 07R at Bole International Airport in Addis Ababa, Ethiopia. On board were 149 passengers, two pilots, five cabin crew members, and one in-flight security officer — a total of 157 people. Its destination was Nairobi, Kenya, roughly two hours away.

At 8:38 a.m. local time, the captain, who was the pilot flying, advanced the throttles and began the takeoff roll. Within seconds of liftoff, the left angle-of-attack sensor began transmitting a value that differed from the right sensor by nearly 60 degrees. The captain's stick shaker activated almost immediately and continued for the remainder of the flight. Conflicting airspeed and altitude information appeared on the pilots' displays, accompanied by multiple cockpit warnings.

The crew engaged the autopilot in an effort to reduce their workload while they attempted to understand the conflicting indications. It remained engaged for approximately thirty-three seconds. After the flaps were fully retracted, however, the autopilot disconnected and the aircraft began pitching downward. The captain pulled back on his control column and used electric stabilizer trim to raise the nose.

With the flaps retracted and the left angle-of-attack sensor continuing to transmit a falsely high reading, MCAS became active and commanded its first nose-down stabilizer movement. The captain countered with nose-up electric trim, but approximately five seconds after he released the trim switch, MCAS activated again. The system repeatedly drove the stabilizer toward nose-down while the captain pulled back on the control column and applied electric trim in the opposite direction.

The captain reported flight-control problems to air traffic control and requested permission to maintain his heading while the crew tried to trouble-shoot their aircraft. He later requested clearance to return to the airport. Meanwhile, the aircraft continued accelerating as the pilots struggled against the increasingly heavy control forces.

Recognizing the repeated uncommanded stabilizer movement, the crew discussed the runaway-stabilizer procedure and moved both stabilizer-trim cutout switches to CUTOUT. The automatic trim commands stopped because electrical power had been removed from the stabilizer-trim motor. But by then, MCAS had already moved the stabilizer far toward nose-down. Although the pilots could still control the elevators by pulling on their control columns, the stabilizer remained badly mis-trimmed and continued exerting a powerful nose-down force.

The crew's only remaining means of changing the stabilizer position was the manual trim wheel mounted beside the pilots on the center pedestal. The first officer attempted to turn it, but the aerodynamic load on the stabilizer made the wheel extraordinarily difficult to move. The aircraft was already traveling at high speed, and the throttles had remained near takeoff power. As its speed continued increasing, so did the aerodynamic forces acting on the horizontal stabilizer.

The trim wheel was not mechanically incapable of moving the stabilizer, but under the conditions confronting the crew it could not be turned effectively or fast enough because it was directly linked to the stabilizer, which meant the crew was fighting the forces pushing against it.

Relieving the load might have required the pilots to momentarily reduce the force they were applying to the control columns, allowing the nose to drop enough to unload the stabilizer. At their low altitude and already excessive speed, however, deliberately allowing the aircraft to pitch farther downward would have appeared both dangerous and counterintuitive. The technique for dealing with a severely loaded manual trim system was also no longer emphasized in contemporary 737 training materials.

The crew faced an impossible dilemma. Pulling back on the control columns kept the aircraft from descending more steeply, but the resulting aerodynamic load made the manual trim wheel nearly impossible to turn. Releasing the control columns might have reduced that load, but it would also have allowed the nose to fall dramatically while the aircraft was close to the ground and accelerating rapidly.

Unable to correct the stabilizer position manually, the pilots restored electrical power to the trim system so they could once again use the yoke-mounted switches. The captain immediately commanded nose-up electric trim, and for several seconds the stabilizer moved in the direction the crew needed. But restoring electrical power also re-enabled MCAS.

Receiving the same false high-AOA signal, MCAS activated again approximately five seconds after the captain released the electric-trim switch. It commanded another sustained nose-down stabilizer movement, erasing much of the nose-up trim the captain had just applied. The crew had escaped one trap only by reopening the door to the system that had created it.

The pilots pulled back on their control columns with tremendous force, but the elevators could no longer overcome the nose-down moment produced by the mis-trimmed stabilizer. The aircraft pitched steeply downward and rapidly lost altitude. The captain transmitted a final call to air traffic control before Flight 302 struck the ground at high speed near the town of Bishoftu, approximately six minutes after takeoff.

All 157 people on board were killed instantly. The impact created a crater approximately 33 feet deep and 92 feet wide. Much of the aircraft was fragmented and driven deep into the earth, while smaller pieces of wreckage and human remains were scattered around the site. Recovery teams spent months excavating and painstakingly examining the crater, searching for aircraft components, personal effects, and the fragmented remains of those who had been aboard.

***

As news of Flight 302 spread, regulators, airlines, passengers, and Boeing confronted the same urgent question: Did the 737 MAX contain an inherent design flaw? Not since the de Havilland Comet had a newly introduced commercial jetliner suffered two catastrophic crashes in such rapid succession.

Before Flight 302 went down, its captain had informed air traffic control that the aircraft was experiencing flight-control problems and requested permission to return to Addis Ababa. The circumstances bore disturbing similarities to those of Lion Air Flight 610.

Both aircraft were nearly new 737 MAX 8s, both encountered serious control problems shortly after takeoff, and both crashed within minutes. Although investigators had not yet established that the two accidents shared the same cause, aviation authorities immediately began reconsidering whether the MAX posed an unacceptable risk to the flying public. The People's Republic of China acted first — and wasted no time doing so.

Over the previous decade, China had invested billions of dollars in developing a commercial aerospace industry capable of competing with Boeing and Airbus. Its effort echoed the government-backed support that had helped establish Airbus in Europe during the late 1960s and 1970s, although China lacked Europe's long aviation history, established aerospace infrastructure, and reputation for adherence to Western industrial standards. Its challenge was not merely to build a viable airliner, but to convince the world that a Chinese-manufactured aircraft could meet the same standards of safety, reliability, and quality as its Western competitors.

China also needed its aircraft to obtain certification from the FAA and the European Aviation Safety Agency if it hoped to sell them widely outside its domestic and allied markets. At the same time, it sought to elevate its own regulator, the Civil Aviation Administration of China, or CAAC, into an internationally respected authority capable of standing alongside the FAA and EASA.

At the time of the two MAX crashes, China was pressing to bring the Comac C919 — a state-backed competitor to the Boeing 737 and Airbus A320 families — to market. The aircraft had completed its first flight in 2017 and was undergoing flight testing and certification, with an expected entry into service of 2020. However, its eventual entry into commercial service would be delayed until 2023.

The crashes of Flights 610 and 302 handed China two significant strategic opportunities. The first was commercial. Worldwide demand for new narrow-body aircraft was already exceeding the production capacity of Boeing and Airbus. If investigators discovered that the MAX contained a fundamental design flaw and the aircraft remained grounded for an extended period, the C919 might be able to capture orders that otherwise would have gone to Boeing — perhaps eventually including orders from Western airlines.

The second opportunity was institutional. By moving quickly and authoritatively against the MAX, particularly after evidence had already connected MCAS to the Lion Air crash, the CAAC could demonstrate that it was prepared to exercise independent regulatory judgment rather than wait for guidance from the FAA. A decisive response would elevate the Chinese regulator's international stature and help position it alongside the FAA and EASA as one of the world's leading civil-aviation authorities.

On March 11, less than twenty-four hours after Flight 302 went down, the CAAC ordered Chinese airlines to suspend commercial operations of the 737 MAX 8. China then operated the world's largest national MAX fleet, with ninety-six aircraft, so its decision immediately removed a substantial portion of the worldwide fleet from service.

The symbolism was as important as the number of airplanes involved. For decades, the global aviation community had generally looked to the FAA for leadership on questions involving American-designed aircraft. This time, China had acted first, effectively declaring that it was unwilling to rely on the judgment of the regulator that had certified the MAX.

Other regulators and airlines across Asia quickly followed. Indonesia grounded the MAX, as did Singapore, Australia, Malaysia, and several other countries. Airlines elsewhere voluntarily suspended operations as public concern intensified. On March 12, the EASA suspended all MAX operations within Europe and prohibited the aircraft from entering European airspace.

Despite the mounting pressure and the obvious similarities between the two accidents, the FAA held fast. On March 11, it issued what it called a Continued Airworthiness Notification to the International Community. The notice stated that the FAA was participating in the investigation of Flight 302 but that, based on the information then available, it had identified no systemic performance issue that justified grounding the MAX.

Because of the FAA's global stature and its position as the regulator responsible for certifying the aircraft, the notification was intended to reassure the international aviation community that the MAX remained airworthy. Instead, it had little effect. Regulators around the world were grounding the aircraft, airlines were suspending its use, and passengers were refusing to fly on it. The FAA's declaration increasingly sounded like a weather forecaster predicting clear skies while torrential rain was producing widespread flooding. Almost no one appeared willing to accept the agency's assurances.

Why did the FAA maintain that position when much of the international regulatory community was moving in the opposite direction?

One factor was that the agency had not yet obtained evidence conclusively linking the Ethiopian crash to the same flight-control system involved in the Lion Air accident. The flight-data and cockpit-voice recorders, recovered the day after the crash, had not yet been analyzed and the FAA maintained that a grounding order needed to be supported by evidence rather than by mere similarities between the two accidents.

But the FAA was also under intense pressure from Boeing. In the immediate aftermath of Flight 302, Dennis Muilenburg and other senior Boeing executives began lobbying U.S. authorities to resist grounding the aircraft. Muilenburg contacted officials at the highest levels of the federal government, including the White House, and personally argued that the MAX remained safe. Boeing executives also pressed the FAA, maintaining that established pilot procedures were sufficient to control any uncommanded stabilizer movement and that no evidence justified removing the aircraft from service.

The campaign reflected the enormous stakes confronting Boeing. The MAX was the company's best-selling product, with thousands of aircraft on order and billions of dollars in future revenue dependent on its continued operation. A worldwide grounding would not merely disrupt airlines and delay deliveries; it would signal that regulators no longer trusted the safety of Boeing's newest commercial airplane.

The lobbying pressure by Boeing and its chief executive officer proved unable to hold back the growing international consensus. By March 13, the United States and Canada were virtually alone among major aviation markets in continuing to permit MAX operations. New satellite tracking data indicated that Flight 302 had followed a pattern of vertical movement resembling that of Lion Air Flight 610. Investigators had also recovered evidence from the Ethiopian crash site suggesting that the aircraft's stabilizer had been positioned substantially nose-down.

At approximately 3:00 p.m. Eastern Time on March 13, the FAA reversed course and issued an emergency order grounding every 737 MAX under its jurisdiction. The order prohibited further operation of the aircraft and required those still airborne to land at their destinations before being removed from service.

At the time the order was issued, approximately thirty MAX aircraft were in the air over the United States, all operated by Southwest Airlines or American Airlines. As each one landed, it was taken out of service. Within hours, every 737 MAX in the world was grounded.

The FAA — the agency that had certified the aircraft and long served as the world's preeminent aviation regulator — had been one of the last major authorities to act. China, seeking both an opening for its own commercial aircraft industry and greater standing for its aviation regulator, had acted first.

***

Following the FAA's grounding order, Dennis Muilenburg and Boeing faced a crisis unlike any in the company's more than century-long history. All 371 MAX aircraft then operating around the world were removed from service, disrupting thousands of flights each week and leaving the future of Boeing's most important commercial product suddenly uncertain.

Boeing first had to help airline customers preserve and maintain aircraft that could no longer carry passengers. Some airplanes remained parked at their home airports, while others had to be ferried — many with flaps extended to avoid MCAS activation — without passengers to locations where they could be stored and maintained under approved procedures. Engines, hydraulic systems, flight controls, fuel systems, batteries, tires, and other components all required periodic attention to prevent deterioration during an indefinite grounding.

The company also had to confront what to do with a production system that was building 737s at a rate of fifty-two aircraft per month. Airplanes already moving through the Renton factory could not be delivered to customers, yet abruptly stopping production would have inflicted enormous disruption on Boeing's employees and its worldwide supplier network. Every unfinished aircraft represented tens of millions of dollars tied up in inventory, and every completed aircraft required parking space, preservation work, inspections, and eventual preparation for delivery.

Dennis Muilenburg projected confidence and insisted that the MCAS software update begun after the Lion Air crash would be accelerated. But his carefully controlled public statements failed to reassure regulators, airline customers, pilots, or the traveling public. Boeing's credibility had been badly damaged, and assurances from the same leadership that had minimized the significance of MCAS after the first crash no longer carried much weight.

Among those increasingly troubled by Muilenburg's statements were officials at the FAA. On April 4, barely three weeks after Flight 302 went down, Muilenburg said Boeing was nearing completion of the MCAS software enhancement and expected to submit the final certification package to the FAA within weeks. His public statements created the impression that the principal obstacle to returning the MAX to service was the completion and approval of a relatively straightforward software update.

The reality was considerably more complicated. The FAA had to understand the relationship between the two accidents, evaluate the redesigned MCAS, examine its dependence on angle-of-attack data, assess the broader flight-control architecture, review pilot procedures, determine new training requirements, and conduct extensive ground, simulator, and flight testing. Other aviation regulators, whose confidence in the FAA had been shaken, also intended to conduct their own reviews rather than automatically accept the FAA's conclusions.

The FAA did not need to await the publication of the final accident reports before permitting the MAX to return to service. Accident investigations commonly take many months or years, and regulators can act once they possess enough evidence to identify an unsafe condition and mandate an effective correction. But the agency did need to be satisfied that Boeing had identified every material failure exposed by the accidents and that the proposed changes addressed more than the most obvious flaw in MCAS.

Boeing's early public forecasts therefore appeared premature. Investigators were still examining wreckage, analyzing flight-recorder data, reconstructing the accident sequences, and evaluating the actions of the crews. Although it had become apparent that MCAS had played a central role in both crashes, regulators could not yet assume that changing a few lines of software code would resolve every safety issue associated with the aircraft.

On April 5, Boeing announced that it would reduce the 737 production rate from fifty-two to forty-two aircraft per month beginning in mid-April. The company described the reduction as temporary and said that it would maintain employment levels while redirecting resources toward the software update and the MAX's return to service. Even at the lower rate, however, completed aircraft continued accumulating because Boeing could not deliver them to customers.

Airplanes soon filled available space at Boeing facilities and at other locations around the region. The grounding also rippled through the MAX's enormous supplier network. Companies that manufactured engines, fuselage sections, avionics, landing gear, interiors, and thousands of smaller components had built their operations around Boeing's planned production rate. Reducing that rate threatened employment and cash flow throughout an industrial system extending far beyond Boeing itself.

As the grounding continued, Muilenburg repeatedly expressed confidence that regulatory approval was approaching. Boeing issued forecasts suggesting that the MAX could return during the summer, then during the early autumn, and later before the end of 2019. Each forecast slipped as regulators identified additional problems, demanded further analysis, or concluded that Boeing's submissions were incomplete.

The disagreement between Boeing and the FAA finally erupted into public view in December 2019. FAA Administrator Stephen Dickson summoned Muilenburg to a meeting and warned him that Boeing's return-to-service schedule was unrealistic. Dickson also objected to Boeing's repeated public statements suggesting that approval was imminent, warning that they could be perceived as attempts to pressure the FAA into acting more quickly.

The agency emphasized that it was following a thorough process rather than Boeing's preferred timeline and that it would lift the grounding only after determining that the aircraft was safe. The rebuke exposed how badly the relationship between Boeing and its principal regulator had deteriorated. The FAA was no longer willing to accept the company's assurances or permit Boeing to define the schedule for returning its own aircraft to service.

The FAA's own credibility was also at stake. Its slower response to Flight 302, compared with the actions of regulators elsewhere in the world, had intensified concerns that its relationship with Boeing had become too close. Many aviation experts already believed the agency had delegated too much certification work to Boeing through the ODA program. The FAA's delayed grounding of the MAX — and its initial public expressions of confidence in the aircraft — further damaged its international standing.

Since its creation in 1958, and even under its predecessor, the Civil Aeronautics Administration, the FAA had been regarded as the world's foremost civil-aviation authority. Its certification decisions were routinely accepted by regulators in other countries, allowing American-built aircraft to enter service around the world without every nation independently repeating the entire certification process. The crashes of Flights 610 and 302 threatened that reputation and raised serious questions about the FAA's judgment, independence, and trustworthiness.

Determined to restore confidence in its authority, the FAA was not going to simply approve Boeing's software revision and return the MAX to service. It subjected the aircraft's flight-control system and certification history to a far more extensive review than Boeing had initially anticipated. The FAA also convened the Joint Authorities Technical Review, bringing together specialists from NASA and nine international aviation authorities to examine the MAX's certification and recommend improvements.

The agency increasingly viewed the two crashes not as the product of a single defective software function, but as the culmination of a chain of design decisions, engineering assumptions, certification failures, communication breakdowns, and flawed expectations about pilot performance. MCAS was one link in that chain, but it could not be examined in isolation from the sensors that activated it, the stabilizer system through which it operated, the cockpit alerts available to pilots, the training they had received, or the assumptions Boeing had made about their reactions during an emergency.

Meanwhile, a second and potentially more dangerous threat to Boeing was developing beyond the regulatory investigation. The U.S. Department of Justice had quietly opened a criminal investigation into the development and certification of the MAX. The inquiry became publicly known shortly after the Ethiopian crash, when reports revealed that a federal grand jury had issued subpoenas seeking documents and communications related to the aircraft.

The criminal investigation was fundamentally different from the accident and regulatory reviews. Investigators were not merely asking how MCAS had failed or what changes were needed to make the aircraft safe. They were examining whether Boeing personnel had knowingly concealed material information from the FAA, deliberately misrepresented how MCAS operated, or otherwise interfered with the agency's determination of the training required for MAX pilots.

The inquiry would eventually focus on communications between Boeing's flight technical pilots and the FAA's Aircraft Evaluation Group. Prosecutors alleged that Boeing employees had used misleading statements, omissions, and half-truths to prevent the AEG from learning the full extent of MCAS' expanded operating range. Because the AEG lacked that information, references to MCAS were removed from the MAX's flight-standardization report and, consequently, from the manuals and training materials supplied to airlines.

Boeing was facing three crises at once. It had to redesign and re-certify the MAX, manage the operational and financial consequences of a worldwide grounding, and defend itself against the possibility that decisions made during the aircraft's development had crossed the line from regulatory failure into criminal conduct.

***

Despite Dennis Muilenburg's efforts to contain the growing crisis, Boeing's situation rapidly worsened. The worldwide grounding forced the company to halt MAX deliveries because airlines could not place the aircraft into commercial service and were unwilling to accept them while their airworthiness remained in doubt. At an estimated delivery value of roughly $60 million per aircraft and a production rate of fifty-two airplanes per month,

Boeing was foregoing more than $3 billion in potential monthly revenue — even before accounting for customer compensation, maintenance and storage costs, supplier disruptions, and penalties associated with delayed deliveries.

The "shareholder value" that had dominated Boeing's corporate priorities for much of the previous two decades also took a severe hit. In the days following the Ethiopian crash and the worldwide grounding, Boeing's share price fell by nearly 15 percent, erasing more than $30 billion from the company's market capitalization. As the grounding continued through the remainder of 2019, the stock fell farther, losing approximately one-quarter of its value from its record high earlier that year. The financial consequences were already enormous, and although no one knew it at the time, the worst was yet to come.

As Muilenburg focused on resolving the MAX crisis, Boeing's board concluded that too much authority had been concentrated in one person through the traditional combination of the chief executive and board-chairman positions. It also maintained that Muilenburg needed to devote his full attention to returning the MAX to service, supporting airline customers, and repairing the company's severely damaged reputation.

On October 11, Boeing announced that it was separating the two positions. Muilenburg would remain president and chief executive officer, while David Calhoun, who had served as the board's lead independent director, would become nonexecutive chairman. On its face, the decision appeared logical. Calhoun was an experienced corporate executive and a longstanding Boeing director who already occupied the board's most senior independent leadership position. But as the MAX crisis continued to unfold, the board's decision to elevate him would have far-reaching consequences and, in hindsight, would prove toxic.

One of Calhoun's first significant acts as chairman was to declare the board's full confidence in Muilenburg. The endorsement frustrated much of the aviation community, where a growing number of observers believed Boeing could not begin restoring its credibility without new leadership.

After the Lion Air crash, Boeing had insisted that pilots could counter an erroneous MCAS activation by following the established runaway-stabilizer procedure. The Ethiopian crew had moved the stabilizer-trim cutout switches to CUTOUT, stopping the automatic commands, but had then been unable to move the badly mis-trimmed stabilizer effectively with the manual trim wheel. When they restored electrical power to obtain nose-up trim, MCAS activated again.

The sequence raised a deeply troubling question: Had Boeing prescribed a procedure without adequately understanding whether pilots could carry it out under the conditions MCAS itself could create?

Muilenburg was subsequently called before Congress to testify at two hearings. On October 29, 2019 — the first anniversary of the Lion Air crash — he appeared before the Senate Committee on Commerce, Science, and Transportation. The following day, he testified before the House Committee on Transportation and Infrastructure. Accompanying him at both hearings was John Hamilton, vice president and chief engineer of Boeing Commercial Airplanes.

To many observers, Muilenburg appeared woefully unprepared. His guarded, formulaic answers and visible discomfort conveyed neither complete command of the facts nor the moral authority expected from the chief executive of one of the world's largest industrial companies. He repeatedly accepted responsibility in general terms while resisting direct responsibility for many of the specific decisions that had allowed MCAS to enter service as it had.

Entering the hearing room, Muilenburg passed through a gauntlet of relatives of the 346 people killed in the two MAX crashes. Many sat directly behind him, holding photographs of their loved ones and signs demanding accountability. Their presence made it impossible to treat the proceedings as a routine technical or regulatory hearing. Whether Muilenburg had been inadequately prepared or had surrounded himself with advisers unwilling to confront him with the gravity of the moment, Boeing's chief executive seemed unable to meet it.

His opening statement only underscored how dysfunctional and tone-deaf Boeing had become. "I can assure you that we have learned from this and will continue learning. We have changed from this and will continue changing," he declared.

The statement represented a striking departure from Boeing's earlier insistence that the MAX was fundamentally safe and that properly trained pilots should have been able to control an MCAS malfunction by following established procedures. Yet Muilenburg offered few concrete explanations of what Boeing had learned, why the system had been designed as it was, or who had been held accountable. To the victims' families seated only a few feet behind him, and to many watching the hearings, his words sounded less like an acceptance of responsibility than another collection of carefully constructed corporate platitudes.

As the MAX grounding continued to hobble Boeing's commercial-airplane operations, its defense and space division appeared to offer a rare opportunity for the company — and Muilenburg in particular — to remind the world of Boeing's engineering capabilities. On December 20, 2019, Boeing launched the first uncrewed orbital test flight of its CST-100 Starliner spacecraft aboard a United Launch Alliance Atlas V rocket.

The Starliner program was already more than two years behind its original schedule, and much was riding on the flight in what had otherwise been a disastrous year for Boeing. The mission was intended to demonstrate that Starliner could launch into orbit, navigate autonomously, rendezvous and dock with the International Space Station, remain there for several days, and then return safely to Earth. A successful mission would clear one of the final major hurdles before astronauts could fly aboard it.

The Atlas V lifted off the pad and performed successfully, placing Starliner on its planned suborbital trajectory. The spacecraft was then supposed to conduct an orbital-insertion burn that would place it on course for the International Space Station. Instead, an error in its mission-elapsed-time clock caused Starliner to believe it was at a different point in the flight. It began executing commands at the wrong time, fired its maneuvering thrusters unnecessarily, and consumed a substantial amount of propellant before ground controllers understood what was happening.

Controllers eventually stabilized the spacecraft, but by then too much fuel had been consumed for it to reach the space station. The mission's primary objective — the autonomous rendezvous and docking — had been lost. A second and potentially catastrophic software defect was discovered only after the spacecraft was in orbit. Had controllers not corrected it before reentry, the service module and crew capsule might have collided during separation.

Starliner returned safely to Earth and landed at White Sands Missile Range in New Mexico approximately forty-eight hours after launch. Its launch vehicle had worked, the spacecraft had survived orbit, and its landing system had functioned successfully. But measured against the stated objectives of the mission, the flight was an abject failure. Starliner never approached the International Space Station, never demonstrated its docking system, and exposed severe deficiencies in Boeing's software-development, testing, and validation processes.

The failed mission also laid bare Muilenburg's poor judgment and his almost desperate desire to validate his leadership. Barely fifteen minutes after the Atlas V lifted Starliner from the launchpad, he publicly congratulated the Starliner team on Twitter for a "successful launch" — well before the spacecraft had completed the mission's true test of reaching orbit, rendezvousing with the International Space Station, and docking with it.

Reaching space was only the beginning of the mission, and some would argue it was the comparatively easy part. Yet Muilenburg reacted almost as a child might react to receiving a bit of good news, seizing on the successful liftoff before waiting to learn whether the mission itself would succeed. His haste betrayed an anxiety to claim a victory for Boeing — and for himself — before the outcome was known.

The celebration proved not merely premature, but emblematic of an executive eager for vindication and seemingly unable to wait for actual results. Within a short time, it became clear that Starliner was in the wrong orbit and would not reach the space station. The public congratulations that had been intended to showcase Boeing's capabilities instead became another example of its chief executive declaring success before the company had earned it.

Despite the board's public support for Muilenburg, the combination of the continuing MAX crisis, deteriorating relations with regulators, his disastrous congressional appearances, and the failed Starliner mission proved too much. On December 23, the day after Starliner landed in the New Mexico desert, Boeing's board dismissed him with immediate effect.

Muilenburg was the fourth chief executive to lead Boeing in sixteen years and the third of those four to be forced out. The board stated that a change in leadership was necessary to restore confidence in the company and repair its relationships with regulators, customers, and other stakeholders. David Calhoun, the same director whom the board had elevated to nonexecutive chairman only two months earlier, was selected to replace him as chief executive.

On December 31, eight days after Muilenburg's dismissal, health authorities in China reported a cluster of pneumonia cases of unknown origin in Wuhan. On January 5, 2020, the World Health Organization issued its first formal international notice concerning the outbreak, citing forty-four reported cases, several of them severe.

Boeing had ended 2019 believing that the MAX grounding was the greatest crisis it faced. Yet another was already beginning — one that would soon shut down much of global aviation and deepen the company's troubles beyond anything its board or anyone else could imagine.

Adrift in the Perfect Storm

On January 13, 2020, David L. Calhoun became chief executive officer of The Boeing Company. He had served as the company's nonexecutive chairman since the previous October, when the board separated the roles of chairman and CEO so that Dennis Muilenburg could focus exclusively on returning the 737 MAX to service. A Boeing director since 2009, Calhoun possessed deep institutional knowledge and had observed the MAX program's development from the boardroom. He knew the company, its senior executives, and its regulators, and he had watched from a privileged vantage point as the MAX crisis unfolded and ultimately engulfed Boeing. For those reasons, the board regarded him as the obvious choice — the one person who could step into the CEO role immediately and hit the ground running.

Another factor that made Boeing's directors comfortable elevating Calhoun was his 26-year career at General Electric. After earning a bachelor's degree in accounting from Virginia Tech, he joined GE in 1981, the same year Jack Welch became the company's CEO. Calhoun rose through the ranks to lead several major businesses, including GE Aircraft Engines and GE Transportation, giving him extensive experience in the aerospace industry. But his career had also placed him squarely within Welch's management culture, with its relentless emphasis on financial performance, shareholder value, and numerical targets. An accountant by training and a product of the Welch-dominated GE system, Calhoun was, above all, a numbers-driven executive.

An old adage defines insanity as repeatedly doing the same thing while expecting a different result. Few corporate boards have demonstrated that folly more clearly than Boeing's when it selected David Calhoun to lead the company out of a crisis created by the very leadership culture he had helped oversee — and of which he was himself a product. Calhoun was cut from the same cloth as Harry Stonecipher and James McNerney. Stonecipher had placed Boeing on a course in which financial performance eclipsed engineering judgment, and McNerney had embedded that philosophy deep within the company's culture. Although Dennis Muilenburg had risen through Boeing's engineering ranks, his career had been shaped by that same environment, and his determination to preserve it only deepened the dysfunction that had helped produce the disaster Boeing now faced.

At the darkest moment in Boeing's history, when the company desperately needed a change in course, its board decided that the best path forward was to stay the course. Unbeknownst to anyone at the time, however, Boeing's darkest moment was about to become darker still as the calendar turned to 2020.

Despite the worldwide grounding and mounting concerns over the MAX's safety and development, Boeing ended 2019 with only 183 net cancellations and a backlog of more than 4,600 orders for the aircraft. Airlines that had ordered the MAX had little practical alternative but to remain with Boeing. Demand for narrow-body aircraft remained strong, and Airbus was already struggling to keep pace with orders for the A320neo. Canceling MAX orders and switching to Airbus would have delayed urgently needed deliveries by years, never mind the logistical nightmare of introducing a different aircraft type into an airline's fleet. Most customers therefore remained committed, expecting regulators and Boeing to identify and correct the problems with the MAX and return it to service relatively quickly.

As Boeing's new CEO, Calhoun hoped to get the MAX flying again, stem the company's mounting losses, and restore confidence in Boeing. Air travel was still growing when he took the helm, and the aircraft was expected to return to service sometime in 2020. Calhoun had reason to hope that, by the end of the year, the crisis would largely be behind the company.

***

On January 23, 2020, Chinese authorities locked down the city of Wuhan in response to a rapidly growing outbreak of unexplained pneumonia believed to have originated there. Flights, trains, buses, ferries, and other transportation links were suspended, effectively preventing travel into or out of the city. The extraordinary measure confirmed fears that the newly identified coronavirus had spread rapidly and probably had already spread beyond Wuhan.

Corporations around the world quickly curtailed travel to China, while governments and airlines began restricting service throughout the region. Cases soon appeared elsewhere in Asia, Europe, and North America. On January 21, the Centers for Disease Control and Prevention confirmed the first U.S. case in Washington State. On January 30, the World Health Organization declared the outbreak a Public Health Emergency of International Concern.

As the virus spread, global air travel experienced a steep decline. On March 11, the WHO formally characterized COVID-19 as a global pandemic, citing its worldwide spread and severity. Governments rapidly began closing borders, imposing quarantines, and placing hundreds of millions of people under lockdown. The resulting global economic shock was the most severe since the Great Depression.

Travel and hospitality were among the industries hit hardest, with commercial aviation at the center of the collapse. Passenger demand virtually disappeared, forcing airlines to cancel most flights and ground most of their fleets. Airlines began parking their planes at the unused gates and taxiways of the airports the planes were located at when the lockdowns began, but as space ran out thousands were ferried to long-term storage facilities, including vast desert airfields around the world.

The MAX grounding had already plunged Boeing into the greatest crisis in its history but the global pandemic delivered a second, potentially fatal blow. Facing a severe liquidity crisis that threatened its very survival, the company turned to the federal government for financial assistance or loan guarantees. Boeing needed to raise cash quickly to withstand the unprecedented convergence of two concurrent disasters of unprecedented proportions in order to stay afloat.

The request was more complicated than it seemed. As one of the nation's largest defense contractors, Boeing was strategically important to the federal government. Its survival also mattered to the vast network of suppliers and smaller companies whose fortunes were tied to its own. A Boeing collapse would have sent shock waves throughout the aerospace industry and debilitated any recovery from the pandemic-induced economic downturn.

Even so, many members of Congress and industry observers were incredulous that Boeing would ask taxpayers to rescue the company. Although the pandemic had accelerated and intensified Boeing's financial distress, critics viewed much of the crisis the company was facing as a self-inflicted wound. During the decade preceding the MAX grounding, Boeing had spent billions of dollars repurchasing its own shares, a tactic many viewed as a distraction to its core business.

Many argued that investing even a portion of that sum in engineering, manufacturing, and product development would have left the company far better prepared to withstand the twin crises. In fact, had it spent that money more wisely, it might have averted the MAX crisis entirely and been left in a far better position to deal with the global pandemic.

Put another way, over the six years encompassing 2019 through 2024, Boeing accumulated approximately $36 billion in losses. Yet in the decade preceding the MAX crisis, it had spent $43.6 billion — about $7 billion more than its subsequent losses — repurchasing its own shares to boost the company's stock price. Because share-price performance was often a part of senior executive compensation calculations, Boeing's leaders had a powerful incentive to prioritize short-term market gains over investment in the company's products, people, and long-term resilience.

Few facts more clearly illustrate how distorted incentives and financial engineering helped drive Boeing off the rails. No one could mistake this for the same company once made legendary by Bill Boeing, Claire Egtvedt, Philip Johnson, Bill Allen, Thornton Wilson, and Frank Shrontz.

Whatever hopes David Calhoun may have harbored of reversing Boeing's fortunes would eventually be shattered. And despite the company's deepening crisis, his tenure as CEO would ultimately prove that it remained business as usual at Boeing, and that the board's decision to elevate him to the CEO role was the wrong one.

***

As Boeing's CEO, David Calhoun was confronted by two crises of unprecedented magnitude. The company urgently needed to return the 737 MAX to service because its most important source of revenue had effectively been shut down. Yet even as Boeing worked to get the MAX flying again, the COVID-19 pandemic had grounded much of the world's airline fleet, leaving carriers unwilling — or financially unable — to accept new aircraft. Boeing had to find a way to survive until deliveries and the cash they generated could resume, even as operating losses mounted and its debt soared.

One of Calhoun's first major responses was to order a sweeping reduction in Boeing's workforce, concentrated primarily within its commercial-airplane operations. Through a combination of voluntary departures, involuntary layoffs, and attrition, the company eliminated approximately 20,000 positions from a global workforce that had numbered roughly 160,000 at the beginning of 2020.

Resolving the liability arising from the two MAX crashes was also near the top of the list. The longer the litigation continued, the longer the accidents would remain in the headlines — even after the aircraft returned to commercial service. The total amount Boeing paid to settle claims brought by the families of those killed has never been publicly disclosed, but it ran into the billions of dollars. By mid-2026, the company had settled more than 90 percent of the civil claims arising from the two accidents, with the remaining cases still being negotiated or proceeding toward trial.

Calhoun also approved Boeing's resolution of the Justice Department's criminal investigation into the MAX. Under a deferred prosecution agreement announced in January 2021, Boeing agreed to pay more than $2.5 billion. That amount included a $243.6 million criminal penalty, $1.77 billion in compensation to airline customers, and a $500 million fund for the families and other legal beneficiaries of the 346 people killed in the two crashes.

The FAA finally rescinded its grounding order on November 18, 2020, clearing the MAX to return to service once each aircraft had undergone the required modifications, inspections, and pilot-training changes. The first scheduled passenger flight in the United States took place the following month. Other national regulators conducted their own reviews, however, and some did not approve the aircraft's return until considerably later. China's Civil Aviation Administration did not issue its airworthiness directive clearing the MAX for a return until December 2, 2021, more than a year after the FAA acted.

Calhoun did manage to avoid relying directly on government loans or loan guarantees to keep Boeing afloat. Early in 2020, the company drew the full amount of a $13.8 billion term loan it had arranged with a consortium of banks. In late April, it raised another $25 billion through one of the largest corporate bond offerings in history. The strong demand for Boeing's debt gave the company access to enough private capital that it did not need to draw upon the federal financing made available to the rest of the aviation industry during the pandemic.

The infusion of cash arrested Boeing's slide toward a potentially catastrophic liquidity crisis, but it did so by loading the company with a huge amount of debt. Nor did the money resolve the deeper problems within Boeing. Under Calhoun, additional failures would surface, exposing just how thoroughly the company's culture had been poisoned. He had not created that toxicity, but it flourished on his watch — and, through his leadership, he came to embody many of the attitudes that had produced it.

Under James McNerney and, later, Dennis Muilenburg, Boeing had pushed aggressively to accelerate production of the 787. Although the program's major development delays and lithium-ion battery fires were largely behind it, the Dreamliner continued to suffer from manufacturing defects and other production problems. As Boeing increased output at its plant in Charleston, South Carolina, where it had consolidated the 787's production, those problems appeared to multiply in tandem.

These were not isolated, one-time events. Together, they pointed to systemic production failures that persisted and intensified. The 787's problems centered on basic manufacturing conformity. Boeing discovered improperly installed shims, fuselage surfaces that failed to meet exacting flatness specifications, excessive gaps where fuselage sections were joined, defects involving the forward pressure bulkhead, and improperly assembled horizontal stabilizers. No single defect necessarily posed an immediate danger to the in-service fleet, but certain defects occurring together could prevent the aircraft's structure from meeting its certified strength requirements.

The discoveries forced Boeing to inspect and rework scores of completed aircraft. Deliveries were initially suspended in late 2020 and did not resume until March 2021. Only a few aircraft were handed over before additional problems brought deliveries to another halt in May. That second suspension continued until August 2022, creating an almost continuous disruption in 787 deliveries that lasted nearly two years.

During that period, dozens of completed Dreamliners accumulated in Boeing's inventory. Aircraft already delivered to airlines were permitted to continue flying because regulators determined that the defects did not pose an immediate safety-of-flight threat. Even so, some of those airplanes would eventually have to be inspected and, where necessary, repaired — a difficult and costly undertaking for which Boeing would bear much of the financial burden.

Never in the annals of commercial aviation had an aircraft remained plagued by such persistent manufacturing defects more than a decade after entering production. The FAA never formally grounded the 787 fleet, nor did it issue a conventional stop-delivery order. Nevertheless, its refusal to approve Boeing's proposed inspection and certification procedures made the delivery suspension unavoidable.

Boeing formally initiated the halt, but only after the FAA rejected the company's proposed methods for demonstrating that each newly completed airplane complied with its approved design. Without the agency's acceptance of those methods — and without the airworthiness certificates required before delivery — Boeing could not legally transfer the aircraft to its customers.

The impasse also reflected the profound breakdown in trust caused by the 737 MAX crisis. In its aftermath, the FAA withdrew authority that it had previously delegated to Boeing to issue airworthiness certificates for newly manufactured MAX aircraft. As concerns over the 787's production quality intensified, the agency also retained direct control over the certification of newly completed Dreamliners. Boeing was consequently subjected to a degree of regulatory scrutiny it had never before faced.

***

Boeing Commercial Airplanes was not the only part of the company showing signs of institutional decay. The same toxic culture had spread throughout Boeing's defense business, officially known as Boeing Defense, Space & Security. The rot associated with the McDonnell Douglas merger was not confined to one division; it had infected the company as a whole.

When the Pentagon abandoned the tanker-lease proposal in January 2005 following the Darleen Druyun scandal, the Air Force opened the replacement-tanker contract to competitive bidding. Airbus' parent company, the European Aeronautic Defence and Space Company, or EADS, teamed with Northrop Grumman to offer the KC-30, a military tanker derived from the highly successful Airbus A330 commercial airliner. Boeing countered with its KC-767 Advanced Tanker, a heavily modified version of the commercial 767.

In February 2008, the Air Force selected the Northrop Grumman–EADS aircraft as the winner of the competition and assigned it the military designation KC-45A. Boeing immediately filed a formal protest, alleging that the Air Force had evaluated the competing proposals inconsistently and failed to follow the criteria established in its own solicitation.

The Government Accountability Office, the investigative and auditing arm of Congress, sustained Boeing's protest in June 2008. Its review identified several significant errors that could have affected the outcome of what had been a closely contested competition. The Pentagon consequently canceled the award and reopened the procurement process. A substantially revised request for proposals was issued in February 2010, and Boeing won the new competition in February 2011. Its 767-based tanker was officially designated the KC-46A.

Through its extensive network of industry contacts, Airbus concluded that Boeing had submitted an extraordinarily aggressive bid — one that may not have covered the true cost of developing and manufacturing the aircraft. By some estimates, Boeing's proposal may have been as much as 10 percent below the program's actual cost. Because both companies' proposals satisfied the Air

Force's mandatory requirements, price became the decisive factor, and Boeing secured the contract largely by offering the less expensive tanker. Some observers speculated that Boeing had learned from its earlier defeat that Airbus was willing to bid aggressively and therefore cut even more deeply when the contract was reopened.

Boeing may have cut too deeply. Its pricing created a formidable obstacle to ever making the program profitable. The company appeared to believe that any losses incurred during development could be recovered through production of the 179 tankers the Air Force ultimately planned to purchase. That strategy depended heavily on the assumption that the KC-46 would be a comparatively straightforward derivative of an aircraft Boeing had been building for decades.

The tanker would begin life on Boeing's existing 767 production line in Everett, Washington. Although commercial demand for the passenger version of the 767 had largely disappeared, the line remained active producing freighters at a relatively low rate. Boeing planned to capitalize on the aircraft's mature design, established supply chain, and existing manufacturing infrastructure to hold down the program's costs.

The aircraft emerging from the assembly line, however, would not simply be a conventional 767 awaiting the installation of military equipment. It required extensive modifications, including a large cargo door, additional fuel capacity, strengthened structural components, and an all-glass flight deck derived in part from the 787. Incorporating those changes into aircraft moving alongside commercial 767s disrupted the normal rhythm of the production line and inevitably increased manufacturing costs.

Then, after leaving the main assembly line, each aircraft entered a separate modification facility where workers installed the refueling boom, defensive systems, military communications equipment, and the remote boom-operator station needed to transform it into a KC-46A. It was during this supposedly low-risk transformation of a proven commercial airliner that Boeing's bargain-priced proposal began to collide with engineering reality.

Among the most consequential challenges was the Remote Vision System, or RVS. The system eliminated the traditional boom-operator position in the rear of the aircraft. Instead of lying prone or sitting and looking directly through a window at the refueling boom and receiver aircraft, the KC-46 boom operator sat at a workstation near the cockpit and controlled the boom using cameras and video displays.

Boeing promoted the arrangement as a technological advance that would place the boom operator closer to the pilots, improve crew coordination, and provide enhanced visual information. In practice, however, the RVS proved extraordinarily difficult to perfect. Testing revealed problems with image quality, depth perception, glare, shadows, and rapidly changing lighting conditions. Under certain circumstances, the system made it difficult for the operator to judge the boom's position accurately, increasing the risk that it could strike and damage the receiving aircraft.

The deficiencies were so fundamental that Boeing and the Air Force eventually abandoned the idea that the original system could be corrected through incremental modifications. Instead, they agreed to develop a substantially redesigned system known as RVS 2.0, incorporating new cameras, larger displays, improved viewing geometry, upgraded processing equipment, and a laser range finder.

Yet as of mid-2026 — 15 years after Boeing received the tanker contract — the replacement system had not been installed on operational aircraft. The Air Force continued to impose limitations associated with the original RVS, while installation of RVS 2.0 was not expected to begin until 2028. Even after installation began, retrofitting the entire fleet would take years.

The refueling boom presented another major obstacle. Flight testing revealed that its telescope actuator required excessive force to compress. During aerial refueling, the receiving aircraft must move slightly forward after making contact, pushing the telescoping boom inward to maintain the proper position and activate its pressure-sensitive locking mechanism. Some smaller and lighter aircraft — most notably the A-10 Warthog — could not consistently generate enough forward force to compress the KC-46's unusually stiff boom.

Boeing had no choice but to develop a replacement actuator that required less force to operate. Nevertheless, the correction remained incomplete years after the aircraft entered service. Thus, even after years of redesign work, both the system used to guide the boom and the boom itself continued to restrict the tanker's ability to perform every mission for which it had originally been intended.

Deficiencies involving the KC-46's defensive systems, sensors, communications equipment, and electrical components also plagued the aircraft throughout development and testing. The problems continued after the tankers entered operational service, when Air Force units encountered premature component failures, airframe cracks, and other structural defects.

The aircraft was also beset by serious manufacturing and quality-control failures. Among the most egregious was foreign-object debris — including tools, fasteners, trash, and other material — left inside completed aircraft, sometimes in enclosed or difficult-to-access areas such as fuel tanks. Additional defects included improperly installed cargo-floor restraint locks, wiring and plumbing discrepancies, damaged components, and generally deficient workmanship.

These lapses became so persistent that the Air Force temporarily stopped accepting KC-46s while Boeing searched completed aircraft, removed the debris, and strengthened its factory-inspection procedures. The episode demonstrated that the tanker's difficulties were not confined to sophisticated engineering systems such as the Remote Vision System and refueling boom. Boeing was also struggling with the basic discipline required to assemble and deliver aircraft that conformed to contractual standards.

The first operational tankers had originally been expected in 2017, but the Air Force did not accept its first KC-46 until January 2019 — and even then accepted it with known deficiencies. The anticipated full-rate-production decision eventually slipped until sometime after September 2026, approximately nine years later than originally planned. The delay was driven principally by the unresolved Remote Vision System and boom deficiencies, although recurring manufacturing and quality problems further complicated the program.

The Air Force nevertheless continued accepting aircraft because it urgently needed to begin replacing its aging tanker fleet and concluded that gradually fielding the KC-46 was preferable to postponing deliveries until every defect had been corrected. That decision left Boeing with a steadily growing fleet of operational aircraft that would eventually have to be taken out of service for modifications and corrective work.

That meant Boeing had to continue producing new tankers while simultaneously redesigning defective systems, modifying test aircraft, correcting factory defects, and preparing to retrofit scores of airplanes already delivered to the Air Force. It was a textbook example of concurrency. Producing and fielding an aircraft before its development and testing were complete, thereby carrying unresolved engineering problems — and the cost and disruption of correcting them — into operational service.

After the 787 and 737 MAX debacles, no other Boeing program exposed the company's deeply compromised engineering and management culture more clearly than the KC-46.

But the KC-46 was not the only Boeing military program mired in technical problems, delays, and cost overruns. Another was the contract to deliver two heavily modified 747-8s to serve as the next generation of presidential aircraft commonly known as Air Force One.

The VC-25B program was originally expected to deliver the two aircraft by 2024 under a $3.9 billion fixed-price contract. By mid-2026, delivery had slipped until at least 2029, while Boeing's estimated cost to complete the program was approaching $7 billion. Because Boeing had agreed to a fixed-price arrangement, the company — not the government — would absorb approximately $3 billion in overruns, with the total continuing to rise.

Similar failures were evident in still more of Boeing's high-profile programs. Among the most embarrassing was Starliner.

***

Following the near-total failure of Starliner's first orbital test flight in December 2019, Boeing agreed to conduct a second uncrewed mission entirely at its own expense to prove that the spacecraft could complete the objectives it had missed. That flight was scheduled for August 2021, but during preparations for launch, 13 valves in the spacecraft's propulsion system failed to open properly. The mission was scrubbed, and Starliner had to be removed from its Atlas V launch vehicle and returned to Boeing's facility for an extensive investigation and repairs.

The spacecraft did not fly again until May 2022, nearly ten months after the aborted launch attempt and almost two and a half years after its disastrous first flight. Although the second orbital test successfully reached and docked with the International Space Station — its primary objective — the mission was hardly trouble-free. Two maneuvering thrusters shut down prematurely during the orbital-insertion burn, while additional anomalies affected smaller thrusters and one of the spacecraft's thermal-control loops.

Starliner nevertheless completed the mission and returned safely to Earth on schedule. NASA concluded that the flight had achieved its principal objectives, but it did not give Boeing an unconditional green light to carry astronauts. The company was required to investigate the anomalies and complete a substantial list of corrective actions before NASA would approve the spacecraft's first crewed mission.

Starliner was not yet out of the woods. Its ultimate humiliation still lay ahead, although at least the coming failure would not involve the destruction of the spacecraft or the loss of its crew. Even before that occurred, however, the program had become another illustration of how far Boeing had drifted from its traditional engineering focus. The company had become a bloated and slow-moving bureaucracy, preoccupied with controlling costs yet seemingly incapable of controlling the quality, schedule, or ultimate cost of its own work.

In 2002, a start-up called Space Exploration Technologies Corporation — better known as SpaceX — began developing a rocket capable of reaching orbit. Its broader ambition, however, was far more consequential — to reduce dramatically the cost of launching payloads into space and eventually develop rockets that could be recovered and flown again.

Central to that strategy was SpaceX's commitment to vertically integrated manufacturing. Rather than purchasing most of its major components and engineering services from outside suppliers, SpaceX sought to design and manufacture as much of its launch system as possible in-house, including its engines, structures, avionics, and flight software. Vertical integration was the antithesis of the outsourcing model that had come to dominate much of the aerospace industry.

This approach ran directly counter to the direction taken by companies such as Boeing during the late twentieth and early twenty-first centuries. Major aerospace manufacturers had increasingly transformed themselves into systems integrators, relying on extensive supplier networks to design and produce major components and subsystems. SpaceX deliberately moved in the opposite direction, seeking tighter control over design, production, cost, quality, and the pace of innovation.

In 2006, SpaceX secured funding from NASA under the agency's Commercial Orbital Transportation Services program to help develop and demonstrate a system capable of transporting cargo to the International Space Station. That system included the planned Falcon 9 rocket and a newly announced spacecraft called Dragon, which SpaceX intended eventually to adapt to carry people.

In September 2008, SpaceX reached orbit with its first substantial rocket, Falcon 1, becoming the first privately developed, fully liquid-fueled launch vehicle to do so. Two years later, SpaceX launched the first Dragon spacecraft into orbit and recovered it safely after reentry. It was the first time a commercially built and operated spacecraft had been launched into orbit and recovered intact. SpaceX had accomplished all of this within eight years of its founding.

With the Space Shuttle scheduled for retirement in 2011, NASA began funding preliminary work on commercial transportation systems that could eventually carry astronauts to and from the International Space Station. Both Boeing and SpaceX received awards under these early development programs, allowing Boeing to begin work on what would become Starliner while SpaceX continued developing the crewed version of Dragon.

In 2014, under its Commercial Crew Transportation Capability program, NASA awarded Boeing and SpaceX firm fixed-price contracts with a combined maximum value of $6.8 billion. Boeing's contract was valued at $4.2 billion, while SpaceX's was valued $2.6 billion — a striking difference of $1.6 billion. The contracts covered the completion of development and certification, along with a limited number of crewed missions to the International Space Station.

Because the contracts were firm fixed-price, each company agreed to perform the required work for its proposed amount and generally would have to absorb costs exceeding that figure. Boeing's proposal was substantially more expensive in part because its entire transportation architecture carried higher costs. Starliner would launch aboard an Atlas V supplied by United Launch Alliance, requiring Boeing to purchase launch services and integrate its spacecraft with a rocket manufactured and operated through a separate organization. SpaceX manufactured and operated both Dragon and the Falcon 9, giving it tighter vertical integration and greater control over its internal costs.

SpaceX had also already developed and flown much of the underlying hardware. By 2014, the cargo version of Dragon was conducting operational resupply missions to the International Space Station. Converting it into a spacecraft capable of carrying astronauts, and which would be named Crew Dragon, required extensive modifications, new safety systems, and NASA human-rating certification, but its basic architecture — particularly its propulsion, electrical, and flight-control systems — had already accumulated flight experience. Starliner, by contrast, remained entirely unproven in flight.

There was another reason Boeing's proposal cost more, although it was less often acknowledged. Boeing carried a far heavier corporate cost structure. It operated under the traditional aerospace-prime model, with larger engineering, management, documentation, compliance, and supplier-management organizations. In other words, Boeing was a lumbering bureaucracy with substantial institutional overhead. SpaceX's vertically integrated structure gave it lower overhead, shorter lines of authority, and the ability to make engineering and production decisions far more rapidly.

The disparity did not mean NASA had simply handed Boeing an additional $1.6 billion for performing the same work. The awards were based on each company's individual fixed-price proposal, technical approach, development status, and anticipated transportation system. Boeing's Starliner–Atlas V architecture and traditional supplier structure produced a bid of $4.2 billion, while SpaceX's vertically integrated and partly flight-proven Crew Dragon–Falcon 9 system allowed it to bid $2.6 billion.

NASA nevertheless selected both companies to preserve competition and ensure that the United States would eventually have two independent systems capable of carrying astronauts to the space station. The agency also placed considerable confidence in Boeing's long history as a NASA contractor and its decades of experience developing rockets, spacecraft, and other systems dating back to the Space Race. At the time, Boeing appeared to be the safer and more mature choice.

The irony was lost on no one. While Starliner continued to fall further behind schedule and Boeing's losses on its fixed-price contract started to mount, the company and the rest of the world watched on May 30, 2020, as a reusable SpaceX Falcon 9 launched a Crew Dragon carrying astronauts Doug Hurley and Bob Behnken into orbit. Crew Dragon docked successfully with the International Space Station the following day. Meanwhile, Boeing's engineers remained knee-deep in the effort to understand and correct the failures that had plagued Starliner's first orbital test five months earlier.

Perhaps the most damning indictment of Starliner — and, more broadly, of the corporate culture that had taken hold within Boeing — came from the astronauts NASA had assigned to help shape the Commercial Crew Program. In 2015, NASA selected four veteran astronauts to work directly with the Boeing and SpaceX teams as the companies developed their competing spacecraft.

Doug Hurley — a former Marine test pilot and veteran Space Shuttle commander — reportedly had become sufficiently concerned by the summer of 2018 that he told the head of NASA's Astronaut Office that he would not fly on Starliner. He was subsequently assigned to command the first crewed flight of SpaceX's Dragon.

The other astronauts reportedly shared many of Hurley's concerns, although they did not go so far as to reject a Starliner assignment. They had become troubled by Boeing's engineering culture, its lack of urgency, and its limited responsiveness to astronaut feedback. In their view, Boeing's team displayed a degree of institutional hubris — a deeply ingrained "we know better" attitude that made its engineers reluctant to entertain concerns, acknowledge weaknesses, or seriously consider suggestions from the very astronauts whose lives would depend on the spacecraft.

***

Among Boeing's other programs that reflected the dysfunction permeating the company — although they remained less familiar outside the aviation community — none was more prominent than the highly vaunted 777X. Formally launched in November 2013 with 259 orders and commitments, the aircraft was conceived as a larger, more capable, and more efficient successor to the enormously successful 777, which had entered airline service 18 years earlier, in 1995. The 777 had since become a mainstay of high-capacity, long-haul travel and could reasonably claim the title of the world's most successful widebody airliner, having accumulated roughly 1,500 orders by the time Boeing launched the 777X.

Boeing announced two principal passenger versions of the new design — the 777-8 and the larger 777-9. The 777-8 was projected to accommodate approximately 395 passengers in a representative configuration and fly as far as 9,500 nautical miles — roughly 2,000 nautical miles farther than the 777-300ER, the most successful version of the legacy series. The 777-9 was designed to carry approximately 426 passengers, placing it in roughly the same capacity class as the four-engine 747-8 Intercontinental, while offering a range of approximately 8,000 nautical miles.

With only two engines, an enormous new composite wing, and the equally enormous General Electric GE9X engines, the 777X promised airlines near-747 capacity with substantially better operating economics. Both versions were designed to be powered exclusively by the GE9X, an advanced derivative of General Electric's GE90 engine family. With a fan diameter of 134 inches, it was the largest commercial aircraft engine ever developed and promised approximately a 10 percent improvement in fuel consumption over the GE90-115B that powered the 777-300ER.

Boeing's decision to launch the 777X proved prescient. By 2013, the long-term prospects for very large four-engine airliners were becoming increasingly uncertain. Although Airbus recorded a temporary surge in A380 orders that year, demand for the giant aircraft had become heavily dependent on a small number of customers, most notably Emirates, based in Dubai. At the same time, airlines were increasingly favoring long-range twin-engine aircraft capable of economically connecting more cities directly, lending greater credibility to Boeing's prediction that point-to-point service would account for much of the market's future growth. Airbus' belief that congestion at the world's largest hubs would generate sustained demand for extremely large aircraft was proving far less compelling.

The 777X was ideally positioned to exploit that shift. The 777-9 would accommodate more than 400 passengers, placing it in the same general capacity class as the 747-8 while approaching the lower end of the market served by the A380. It could not carry as many passengers as the double-deck Airbus, but that was precisely part of its appeal. Airlines could obtain much of the A380's capacity without assuming the commercial risk of routinely having to fill more than 500 seats.

Powered by only two engines, the 777X promised substantially lower fuel consumption, maintenance expenses, and operating costs per seat than the four-engine aircraft it was poised to replace. Boeing promoted it as the largest and most efficient twin-engine airliner ever developed and claimed that no competing aircraft occupied precisely the same market segment.

In effect, the 777X did not merely fill a numerical gap between the 747-8 and the A380. It offered airlines a more economically manageable alternative to both — an aircraft with near-jumbo capacity, intercontinental range, and the efficiency of a modern twin-engine design.

Unfortunately, despite being derived from an existing and well-proven aircraft design, the 777X soon began encountering major problems of its own. In September 2019, the program's dedicated static-test airframe suffered a significant structural failure during an ultimate-load test. During such a test, an airframe is secured inside an elaborate rig fitted with hydraulic actuators that apply carefully calculated forces to the wings, fuselage, and other major structures. The purpose is to produce loads far greater than those the aircraft should encounter during normal service.

For certification, the structure must demonstrate that it can withstand an ultimate load equal to 150 percent of the maximum load it is expected to experience in operation. Although manufacturers sometimes continue testing beyond the certification requirement until a structure breaks in order to establish its actual limits, failure immediately before reaching the required threshold is not the desired result.

As the 777X's wings were bent upward by more than 28 feet at the tips and the fuselage was simultaneously pressurized, the airframe reached approximately 99 percent of the test's target load — equivalent to roughly 148 percent of its maximum anticipated operational load. At that point, the fuselage skin suddenly ruptured behind the wing, and the damage propagated upward toward a passenger door, causing the door to blow outward violently depressurizing the fuselage.

The airframe had come tantalizingly close to reaching the required load, but the location and nature of the rupture were not what Boeing's engineers had predicted. Rather than confirming that the structure behaved exactly as their computer models anticipated, the test exposed an unexpected weakness in the fuselage structure at a critical stage of the aircraft's development.

Another issue emerged during an early test flight in December 2020, when a 777-9 experienced an uncommanded pitch event, although neither Boeing nor the FAA publicly disclosed whether the aircraft pitched upward or downward. The incident contributed to the FAA's concern that the aircraft's design and supporting technical data were not yet sufficiently mature for the agency to authorize the principal phase of certification flight testing. Boeing was required to provide additional engineering analyses, revise flight-control software, and demonstrate that the aircraft's systems would behave safely under abnormal conditions.

A further setback occurred in August 2024 when a routine inspection revealed that a thrust link — part of the structure connecting the GE9X engine to its mounting system beneath the wing — had failed on one of the flight-test aircraft. Inspections subsequently found cracks in the same component on two other test airplanes. Boeing suspended flight testing while it replaced the components, investigated the failure, and incorporated the necessary design changes.

Although it never attracted the sustained public attention generated by Boeing's other troubled programs, the 777X may reveal more clearly than most how profoundly the company's engineering and program-management capabilities had deteriorated. The aircraft was, after all, a derivative of the enormously successful 777 rather than an entirely new clean-sheet design. When Boeing launched the program in 2013, outside estimates placed its development cost at slightly more than $5 billion, with roughly half of that amount attributed to the new composite wing. Boeing expected the first aircraft to enter service in 2020 after a seven-year development program.

Instead, the 777X became another hugely expensive and chronically delayed Boeing undertaking. By the end of 2025, the company had recorded approximately $15 billion in program-related accounting charges, including a $4.9 billion charge taken after the anticipated first delivery was pushed from 2026 into 2027. Boeing was also carrying billions of dollars in 777X work-in-process inventory, tooling, and other deferred production costs on its balance sheet.

As of mid-2026, Boeing expected to deliver the first 777-9 in 2027, nearly seven years later than originally planned and approximately 14 years after the program was launched. The global pandemic, the collapse in international travel that accompanied it, a prolonged strike, and widespread aerospace supply-chain disruptions undoubtedly contributed to the delay. Nevertheless, much of the program's predicament was of Boeing's own making.

What should have been a comparatively manageable derivative program instead became another illustration of a company that could no longer reliably translate an established and successful aircraft design into a certified, production-ready successor.

***

Against this backdrop of high-profile embarrassments stood David L. Calhoun. Boeing's board and Wall Street had placed their faith in one man to reverse the fortunes of a company that had steadily lost its way since the 1997 merger with McDonnell Douglas nearly a quarter-century earlier. The hopes of employees, investors, customers, suppliers, and regulators — and perhaps the company's very survival — rested on his shoulders.

Unlike Dennis Muilenburg, Calhoun projected an unmistakable air of self-confidence and determination. Unfortunately, that determination was not directed toward fundamentally changing Boeing's course. Instead, he appeared committed to keeping the company on essentially the same path it had followed since the McDonnell Douglas merger, while imposing the financial and operational discipline he believed had previously been lacking and was at the root of Boeing's recent woes.

For Boeing, it would largely be business as usual, and one of the clearest indications was the structure of Calhoun's compensation. His annual base salary was set at $1.4 million, excluding benefits, and remained unchanged throughout his tenure. His incentive package, however, told a more revealing story. Rather than tying a substantial portion of his compensation to repairing Boeing's engineering processes, strengthening safety oversight, restoring technical authority, or addressing the company's damaged culture, the board placed particular emphasis on improving cash flow.

There was no question that Boeing faced a severe cash-flow crisis. The prolonged grounding of the 737 MAX, mounting losses across other major programs, and the collapse in commercial aviation caused by the COVID-19 pandemic had placed extraordinary pressure on the company's finances. From the board's perspective, preserving liquidity was essential. Boeing could not repair anything if it first ran out of money.

Yet the compensation structure also revealed the board's priorities. It focused on the immediate financial symptoms of Boeing's decline rather than the underlying institutional disease. Little, if anything, in Calhoun's incentives explicitly required him to confront the deeper problem that had plagued Boeing for more than two decades — a broken and increasingly toxic corporate culture that had weakened engineering authority, elevated financial performance above technical excellence, and left the company struggling to execute even some of its most basic responsibilities.

But David Calhoun came from the Jack Welch school of management, with its emphasis on financial discipline, cash generation, and the relentless execution of measurable objectives. He carried out the mandate the board had given him, and by those standards, he carried it out well. During the COVID-19 crisis, Calhoun voluntarily relinquished all but $269,231 of his 2020 salary. By 2021, however, with Boeing having stabilized its access to cash, he collected his full $1.4 million salary along with a $3,381,840 annual cash-performance award. In 2022, he again received his full salary, supplemented by another cash-performance award of $3,418,800.

Calhoun's salary and annual cash incentives told only part of the compensation story. He also received substantial equity awards in the form of Boeing stock and stock options. Despite surrendering most of his salary in 2020, he was granted more than $20 million in stock awards that year. In 2021, his stock and option awards totaled approximately $16 million, followed by another $17 million in 2022. Then, in 2023, the grant-date value of his stock awards ballooned to more than $30 million.

Looking only at Calhoun's total compensation during those years, one might assume that he had single-handedly restored Boeing to profitability and placed it on a path toward reclaiming its former engineering glory. Nothing could have been further from the truth. Boeing lost nearly $12 billion in 2020, $4.2 billion in 2021, $5 billion in 2022, and another $2.2 billion in 2023.

As bleak as those financial results were, Boeing's continuing inability to control the quality of the products it manufactured painted an even darker picture. In addition to the endemic problems afflicting the 787, the KC-46A tanker, the Starliner spacecraft, and the 777X, the 737 MAX continued to generate damaging headlines even after regulators authorized its return to service.

A succession of new production-quality problems emerged after MAX deliveries resumed in late 2020. In 2021, Boeing temporarily halted deliveries after discovering that components in the aircraft's electrical-power system had not been properly bonded and grounded. In 2023, Spirit AeroSystems disclosed that fittings joining the MAX's vertical stabilizer to the fuselage had been improperly manufactured. Later that year, improperly drilled or elongated holes were discovered in the aircraft's aft pressure bulkhead. Airlines and inspectors also continued to find loose or improperly torqued fasteners, including defects involving rudder-control hardware.

The MAX production system had become so dysfunctional that Boeing was accumulating, rather than reducing or eliminating, unfinished work and traveled defects — tasks that should have been completed at an earlier production station but were instead passed farther down the assembly line for later correction. These were not isolated or cosmetic discrepancies. They demonstrated that Boeing and Spirit were allowing nonconforming work to advance through the production system rather than stopping the line and correcting defects at their point of origin.

It was revealing that Spirit AeroSystems, although legally independent from Boeing, was suffering from many of the same production and quality-control maladies. The explanation was not difficult to find. Spirit had been created in 2005 when Boeing sold its Wichita commercial-aircraft manufacturing operations. Its facilities, workforce, production methods, and management practices all carried a substantial inheritance from Boeing, and its organizational culture never fully separated itself from that legacy.

More importantly, Boeing remained Spirit's dominant customer. Spirit may have possessed its own management structure and financial statements, but it remained deeply embedded in

Boeing's production ecosystem. Boeing controlled the aircraft designs, dictated much of the production cadence, and accepted fuselages through a process that increasingly tolerated incomplete or nonconforming work. Legal independence therefore did little to insulate Spirit from Boeing's priorities or operating practices.

Spirit consequently faced much the same incentive structure Boeing had created for itself — keep units moving, protect production rates, and avoid allowing quality discrepancies to disrupt deliveries whenever they could instead be deferred, reclassified, or repaired farther down the line. Before the MAX grounding, Boeing had pushed toward increasingly aggressive 737 production rates. To keep pace, Spirit had to recruit and train workers rapidly, add shifts, and move fuselages through Wichita quickly enough to satisfy Boeing's schedule. Quality control became subordinate to production flow, and defects that should have stopped the line were instead permitted to travel with the aircraft.

Despite the problems engulfing Boeing, David Calhoun continued to receive tens of millions of dollars in annual compensation. Perhaps, in his own mind, he was merely applying the lessons his training and experience had taught him — maintain financial discipline, preserve liquidity, and stabilize the balance sheet. Boeing had, after all, secured access to desperately needed cash. And although the company continued to lose billions of dollars each year, its annual losses had declined substantially from the nearly $12 billion nadir reached in 2020. From a narrowly financial perspective, Calhoun could perhaps convince himself that Boeing was slowly recovering.

In the midst of the many crises facing the company, Boeing made an astonishing announcement on May 5, 2022 — it was relocating its headquarters from Chicago, where Philip Condit had moved it more than 20 years earlier, to Arlington, Virginia, where Boeing already maintained a large campus.

For Calhoun and the rest of the board, Arlington made strategic sense because of its location in the Washington, D.C., metropolitan area. Calhoun described the region as important because of its proximity to Boeing's "customers and stakeholders." The phrase was broad, but in Arlington, Boeing would be close to the Pentagon, one of its largest customers. It would also place the company near Congress, the executive branch, the FAA, the Department of Transportation, NASA, and other federal agencies, while providing easy access to foreign embassies and international defense representatives.

The timing of the move was important. Boeing was attempting to recover from the two MAX crashes and the worldwide grounding that followed, congressional investigations, criminal and regulatory scrutiny, and severely strained relations with the FAA. Its defense and space programs also depended heavily on federal contracts and decisions made in Washington. Locating the CEO and corporate leadership near the nation's political and regulatory center made it easier to engage directly with regulators, lawmakers, military officials, and government customers. Contemporary reporting characterized the relocation as part of Boeing's effort to repair its relationships with customers, regulators, and lawmakers following its successive crises.

The move also reflected the increasing strategic importance of Boeing's defense business. Arlington sat adjacent to the Pentagon, while Chicago offered no comparable operational or customer advantage. So much for Philip Condit's rationale for leaving the Seattle area in 2001.

Contemporary reporting also interpreted the relocation as evidence that Boeing was leaning more heavily toward its military and government business rather than refocusing on what ailed the company. For that reason, the criticism was heavy. Critics argued that Arlington was precisely the wrong destination. Boeing had already been faulted for separating senior leadership from engineers and factory workers when it left Seattle for Chicago. Moving even closer to Washington reinforced the perception that Boeing valued government relations, lobbying, and contract management more than proximity to aircraft design and production.

Nevertheless, Boeing portrayed the move as the beginning of a new chapter — part of a broader recovery plan intended to stabilize the company, repair its damaged reputation, and finally address the problems that had plagued it for years. Then came a wake-up call that exposed just how superficial — and ultimately illusory — that recovery really was.

***

On Friday, January 5, 2024, Alaska Airlines Flight 1282, a scheduled flight from Portland International Airport in Oregon to Ontario International Airport in California, pushed back from its gate. On board were two pilots, four flight attendants, and 171 passengers. The aircraft was a Boeing 737 MAX 9 that had been delivered new to Alaska Airlines the previous October and had entered revenue service less than two months earlier, on November 11, 2023.

At 5:07 p.m. Pacific Standard Time, Flight 1282 lifted off and began its initial climb. Approximately six minutes later, while passing through 14,830 feet, a door plug on the left side of the fuselage, just behind the wing, suddenly blew away from the aircraft. The resulting opening caused a rapid and explosive decompression of the cabin. The pilots declared an emergency and returned safely to Portland, landing approximately 20 minutes after takeoff.

The following day, January 6, the Federal Aviation Administration issued an emergency airworthiness directive grounding all 737 MAX 9 aircraft equipped with the same type of door plug. The directive produced the customary ripple effect among aviation regulators around the world, although only a handful of airlines operated the MAX 9 in the affected configuration.

Ultimately, 171 aircraft fell under the FAA's grounding and inspection requirements.

The door plug at the center of the incident was not a conventional passenger door. On both the earlier Next Generation 737-900ER and the newer 737 MAX 9, Boeing offered airlines a high-density, single-class seating configuration capable of accommodating as many as 220 passengers. At that capacity, an additional emergency exit was required on either side of the aircraft to ensure that all occupants could be evacuated within the mandated time.

To accommodate both high-density and standard seating configurations while preserving production commonality, Boeing manufactured the fuselage with an additional exit opening just aft of the wing on either side. Airlines choosing the high-density configuration received an operable emergency-exit door in that location. Airlines selecting a lower-density layout did not require the additional exit, so Boeing closed the opening with a non-operable door plug. The outline of the plug remained visible from outside the aircraft, but there was no obvious indication of it inside the cabin because the interior sidewall panels concealed it. Flight 1282 was configured in this manner, as were most MAX 9s operated around the world.

When the plug separated from Flight 1282, the interior sidewall paneling and surrounding cabin material were torn away with it, leaving a large opening in the fuselage. The question confronting investigators was immediate and seemingly obvious: How could a structure intended to remain permanently secured to the aircraft simply come loose in flight? Investigators from the National Transportation Safety Board quickly began reconstructing the plug's manufacturing and installation history. The answer emerged with disturbing speed.

The plug was designed to be lowered slightly into position so that 12 fittings on the plug aligned with corresponding fittings on the fuselage structure. Once seated, four retaining bolts prevented the plug from moving upward and disengaging from those fittings. The plug had initially been installed with the required hardware at Spirit AeroSystems when the fuselage was manufactured.

However, the fuselage arrived at Boeing's Renton factory with damaged rivets on the fuselage near the door-plug opening. To give workers access to repair the rivets, the plug had to be opened. But Boeing personnel failed to create the required removal record documenting that the plug had been disturbed. Because no removal record existed, the prescribed quality-assurance inspection was never triggered. And because that inspection never occurred, no one discovered that the four retaining bolts and related hardware had not been reinstalled.

Although the missing bolts were entirely the result of work performed and controlled at Boeing's Renton factory, the chain of events had begun with defective workmanship on the fuselage supplied by Spirit AeroSystems. The incident exposed not only Boeing's production failures but also the deeply dysfunctional interface between the company and its most important structural supplier.

The NTSB ultimately determined that the probable cause was Boeing's failure to provide adequate training, guidance, and oversight to ensure that manufacturing personnel properly followed the company's parts-removal process and reinstalled all required securing hardware.

The FAA authorized the affected MAX 9s to return to service only after each aircraft had undergone an extensive inspection confirming that its door plugs and associated hardware were properly installed. Returning the aircraft to service required a labor-intensive inspection and maintenance process that extended over several weeks.

It was providential that Flight 1282 resulted in only minor injuries. The aircraft was still climbing through approximately

15,000 feet, where the pressure differential between the cabin and the outside atmosphere was lower than it would have been at cruising altitude. The passengers were also still seated with their seat belts fastened, and no one occupied the two seats immediately beside the opening. Had the plug separated later in the flight, when passengers and flight attendants were moving around the cabin and the pressure differential was greater, the consequences would have been far more severe and likely resulted in another tragedy.

The Flight 1282 incident occurred nearly five years after the second MAX crash had triggered the original worldwide grounding. It demonstrated that, despite David Calhoun's promises to repair Boeing's safety culture and manufacturing system, the company remained incapable of exercising reliable control over the aircraft moving through its factories.

Flight 1282 was more than another embarrassing quality-control failure. It was the event that shattered the remaining credibility of Calhoun's turnaround effort. A nearly new aircraft had left Boeing's factory without the four bolts required to keep a large section of its fuselage in place, and no one within Boeing's production or inspection systems had noticed.

The aviation community, regulators, lawmakers, and Boeing's customers had seen enough. The FAA prohibited Boeing from expanding MAX production and intensified its direct oversight of the company. Boeing's share price fell sharply in the weeks and months that followed. Congressional investigations accelerated, the Justice Department opened a criminal investigation into the incident, and Calhoun's claim that Boeing had fundamentally changed became increasingly difficult to sustain.

The writing was on the wall. Calhoun came under intense pressure from regulators, investors, lawmakers, and Boeing's customers. Perhaps most humiliatingly, the strongest pressure came from the airlines that bought Boeing's aircraft. Alaska Airlines,

United Airlines, American Airlines, and Southwest Airlines had become increasingly frustrated by recurring quality failures, delayed deliveries, and what they regarded as inadequate communication from senior management. Several airline CEOs sought direct meetings with Boeing's board rather than continuing to deal solely with Calhoun and his management team.

On March 25, 2024, Boeing announced that Calhoun would step down as CEO at the end of the year, allowing the board time to find a successor and provide for an orderly transition. He was the fifth chief executive to lead Boeing in 21 years, and his departure marked the fourth time during that period that a Boeing CEO had been forced to step down.

On June 18, 2024, barely five months after the Alaska Airlines door-plug blowout and less than three months after announcing his departure, Calhoun appeared before the Senate Permanent Subcommittee on Investigations to answer publicly for what lawmakers described as Boeing's "broken safety culture." The committee had already heard testimony from company whistleblowers concerning manufacturing defects, quality-control failures, and retaliation against employees. The senators now wanted to hear from Boeing's CEO, even though he was already on his way out the door.

As with Dennis Muilenburg's congressional testimony when he was CEO, family members of the 346 people killed in the Lion Air and Ethiopian Airlines MAX crashes were in attendance. Perhaps having learned from Muilenburg's awkward performance at his own hearing, Calhoun turned before making his opening statement and apologized directly to the family members seated behind him "for the grief that we have caused." He said Boeing remained committed to honoring the victims' memories by focusing on safety. He also conceded that Boeing's culture was "far from perfect," but argued that the company had heard the criticism and was making measurable progress.

Calhoun's central defense was that Boeing understood the gravity of its failures and was finally placing quality ahead of production. Under questioning, however, he made several damaging admissions. He accepted Boeing's responsibility for both the Alaska Airlines door-plug blowout and the MCAS system implicated in the two fatal MAX crashes.

The hearing was further overshadowed by allegations from a Boeing quality inspector, who claimed that the number of nonconforming 737 parts had risen sharply after MAX production resumed and that supervisors pressured employees to alter or cancel records documenting defective or improperly tracked components. The inspector also alleged that Boeing moved many of those parts out of sight before an FAA inspection, after which some could no longer be accounted for.

Calhoun said Boeing had received the allegations only shortly before the hearing and was still reviewing them. Although procedurally understandable, the response reinforced the impression that senior management remained detached from what was happening inside its own factories. He also acknowledged that employees who raised safety concerns had suffered retaliation but could not say how many managers had been disciplined. More strikingly, he admitted that he had never personally met with Boeing whistleblowers and agreed only after being pressed that doing so might be worthwhile.

Still, Calhoun rejected the senators' portrayal of Boeing as a company that had sacrificed safety for profit. He said he did not recognize the organization described by whistleblowers and insisted that Boeing remained fundamentally sound. That defense culminated in a contentious exchange with Senator Josh Hawley, who accused Boeing's leadership of "strip-mining" the company while Calhoun collected $32.8 million in compensation.

Many observers were incredulous, particularly at Calhoun's insistence that Boeing remained fundamentally sound. His testimony came less than two weeks after Starliner had launched on its third orbital test flight — this time with two astronauts aboard — and that mission had already begun attracting headlines for all the wrong reasons.

On June 5, 2024, Boeing's Starliner lifted off from Cape Canaveral Space Force Station atop an Atlas V rocket. Astronauts Butch Wilmore and Suni Williams were embarking on what was expected to be an eight-day test flight to the International Space Station. As Starliner approached the station the following day, five of its aft maneuvering thrusters shut down. Ground controllers eventually restored four of them, while Wilmore manually controlled the spacecraft during part of the approach to the space station. With the required fault tolerance restored, NASA authorized Starliner to proceed, and it docked safely with the station.

The thruster failures, combined with several helium leaks discovered during the flight, transformed the brief mission into a months-long technical investigation. Boeing engineers maintained that Starliner could safely return its crew, particularly after developing a flight profile intended to minimize use of the affected thrusters. NASA remained unconvinced. Ground testing had failed to explain the thrusters' behavior with sufficient certainty, and agency officials were unwilling to accept the possibility that additional thrusters might fail during the return to Earth.

On August 24, NASA announced that Starliner would return without its crew. The spacecraft landed safely in New Mexico on September 7, leaving Wilmore and Williams aboard the station to return on a later SpaceX mission. For Boeing, the decision represented the most humiliating possible outcome: its astronauts would return aboard a spacecraft built by SpaceX, the competitor Boeing and many industry observers had once regarded as an inexperienced upstart.

Wilmore and Williams finally splashed down aboard a SpaceX Crew Dragon on March 18, 2025, after spending more than nine months in space on what had begun as an eight-day mission.

Calhoun had already announced his intention to step down, and the industry was waiting to see whether Boeing's next leader would change course or preserve the existing system. The Starliner debacle made the need for change unmistakable. Nearly three decades after the McDonnell Douglas merger, Boeing could no longer afford to continue along the path it had followed since 1997. The only remaining question was whether the changes to come would arrive in time to save the company.

***

On July 31, 2024, Boeing announced Robert Kelly Ortberg would be its new president and CEO. His appointment would become effective on August 8, 2024, when he would officially succeed David Calhoun.

"Kelly" Ortberg, as he was referred to, was born in Dubuque, Iowa, in April 1960, and grew up there as one of four children. He remained in Iowa for college, earning a Bachelor of Science degree in mechanical engineering from the University of Iowa in 1982. In 1983, he joined Texas Instruments as an engineer where he worked until 1987. That year, he joined Rockwell Collins and then spent the next three decades advancing through its commercial and military avionics businesses. In 2013, Ortberg became CEO of the company and expanded its market footprint as a major supplier of cockpit avionics, communications, navigation and aircraft interior systems.

This was the same company that had supplied the original PFDs to Boeing for installation on the MAX aircraft, the PFDs that did not notify pilots about an AOA sensor disagreement unless the airline that purchased the plane had ordered the option for displaying the AOA display on the PFD. In all fairness to Rockwell Collins and its CEO, Kelly Ortberg, that failure was not indicative of production or quality control issues at Rockwell Collins. Rockwell Collins supplied the MAX displays and made a software change during development that inadvertently tied the standard AOA Disagree alert to an optional AOA display. Boeing, however, established the requirements, tested and accepted the software, certified and delivered the airplanes, discovered the defect in 2017, and then deferred correcting and disclosing it. And there is no evidence that Ortberg personally participated in or was even aware of the decision to defer or not disclose it.

Although Boeing was not Rockwell Collins' largest client during the decades that Ortberg worked there — that title went to the U.S. government, which accounted for upwards of 40 percent of the company's revenues — it was nonetheless its largest commercial client followed closely by Airbus. This meant Ortberg spent much of his career operating in close proximity to Boeing, both as a major supplier to Boeing airliners and as a defense subcontractor on Boeing military programs. Boeing was not an unfamiliar company to him when he became its CEO.

However, Ortberg's background in engineering and his professional experience at Rockwell Collins meant that unlike Calhoun, Muilenburg, McNerney and Stonecipher, he had no direct professional lineage connecting him to Jack Welch. Although he operated within the financially disciplined corporate environment that Welch helped popularize, Ortberg's formative experience was in engineering, program management and aerospace manufacturing at Rockwell Collins. His leadership style appeared less rooted in financial engineering and more in operational execution, product quality and direct engagement with the workforce.

For a company facing criticism that it had subordinated engineering to financial targets, that separation was politically and symbolically valuable to the board when it considered Ortberg, even if Ortberg remained a conventional corporate executive who understood cost control, acquisitions and shareholder expectations.

Boeing's board likely mulled a very critical consideration when it selected Kelly Ortberg. He appeared to offer the combination the company urgently needed, which was an aerospace engineer, an experienced industrial operator, a respected outsider and a proven chief executive — but who was separated from and didn't have any responsibility for Boeing's existing culture and failures.

In announcing Ortberg's selection as Boeing's next CEO, the board said the directors had conducted an "extensive" search and concluded that Ortberg possessed the appropriate skills and experience to lead Boeing through its next chapter. The announcement emphasized his 35 years in aerospace, his engineering background and his record of managing complex manufacturing and technology businesses.

Among the other key considerations that likely swayed the board to settle on Ortberg was the fact that he was an outsider, but not an industry outsider. The board needed someone who was not identified with Boeing's entrenched management culture, yet the company was too technically complex and too deeply regulated to entrust to an executive unfamiliar with aerospace. Ortberg occupied the useful middle ground. He had never been a Boeing executive, but he had dealt with Boeing for decades as a major supplier and understood its aircraft, customers, procurement organization and engineering processes. He could present himself as independent of Boeing's internal factions while arriving with substantial knowledge of the company and its industry.

An often overlooked consideration was the fact that Ortberg was willing to relocate himself near production. One concern during the CEO search was that Boeing's senior leadership had become physically and culturally detached from its principal engineering and manufacturing operations. Ortberg indicated that he would be based in the Seattle area, close to Boeing Commercial Airplanes and the factories in Renton and Everett.

That decision reinforced the message that the CEO would be present where airplanes were designed and built, rather than primarily operating from Boeing's headquarters in Arlington, Virginia.

***

Among the most consequential initiatives Ortberg inherited was Boeing's planned reacquisition of Spirit AeroSystems, the company created from the Wichita operations that Harry Stonecipher had spun off nearly twenty years earlier as part of his relentless drive to reduce costs and shed manufacturing assets. Although the agreement to reacquire Spirit had been reached shortly before Ortberg took office, completing the acquisition and reintegrating Spirit became a central part of his effort to stabilize Boeing's production system. Bringing the company that manufactured the MAX fuselage back under Boeing's direct control was seen as essential to improving quality, reducing defects and gaining greater command over a supply chain that had become dangerously fragmented.

Since stepping into the CEO role, Kelly Ortberg has achieved some measure of success, but they are primarily recovery milestones rather than evidence of a complete turnaround. He has overseen stabilization of the 737 production system; improved delivery rates and manufacturing performance; moved Boeing toward positive cash flow; strengthened liquidity; and begun changing the tone and priorities of Boeing's leadership.

One of Ortberg's first unambiguous victories — one that caused even Boeing's critics to take notice — came in March 2025, when the company won the U.S. Air Force contract to develop the F-47, the manned fighter at the center of the Pentagon's Next Generation Air Dominance program. Expected to enter service in the early 2030s, the F-47 is intended to become America's next generation of advanced stealth fighter.

The competition for the F-47 pitted Boeing against Lockheed Martin, builder of the F-35, the current fifth-generation stealth fighter. The victory carried particular symbolic weight for Boeing.

More than two decades earlier, the company had lost the Pentagon's Joint Strike Fighter competition to Lockheed Martin, whose X-35 demonstrator became the F-35. Reuters subsequently identified the F-47 award as one of the most important successes of Ortberg's first year as Boeing's CEO.

The award was strategically important because Boeing's defense business had suffered years of losses and execution failures. The F-47 gave Boeing a potential generational military aircraft program. However, winning the competition is only the beginning as Ortberg must ensure Boeing does not repeat the fixed-price and program-management failures seen on the KC-46, Starliner, and Air Force One programs.

In 2025, Boeing delivered 600 commercial aircraft, its highest annual total since 2018. By the second quarter of 2026, it delivered 171 aircraft in a single quarter, again the strongest quarterly result since 2018. These figures remain below Airbus' output and below Boeing's historic capacity, but they demonstrate that Ortberg's controlled production ramp-up was beginning to translate into actual deliveries and revenue.

The backlog also reached record levels — more than $700 billion by mid-2026 — showing that airlines had not abandoned Boeing despite its problems. Albeit, in many cases they had no choice given Airbus' even larger backlog. However, a backlog is not the same as successful execution, but maintaining and expanding customer demand while the company repaired its production system was significant.

Since becoming CEO, Ortberg established himself in the Seattle area, spent more time around production operations and publicly acknowledged that Boeing had become insular and poorly integrated across its divisions. He repeatedly told employees that production speed would not take precedence over doing the work correctly.

These actions are harder to measure than deliveries or cash flow, but they marked a genuine change in leadership posture and tone. Ortberg has emphasized employee reporting, cross-functional communication, engineering discipline and accountability rather than presenting Boeing's problems primarily as temporary financial or public-relations setbacks. Most importantly, Boeing has also begun aligning leadership incentives more explicitly with safety, quality and integrity.

But as of mid-2026, Boeing remains burdened by debt, continuing defense-program losses, delayed certification of the 737 MAX 10 and 777X, and a commercial product lineup that is increasingly disadvantaged against Airbus. Even in the second quarter of 2026, Boeing still reported a net loss.

Unfortunately for Boeing, its missteps, fatal failures, and repeated embarrassments ensure that nearly everything it does — or fails to do — for the foreseeable future will be judged against a reputation tarnished during the quarter-century since its merger with McDonnell Douglas.

The 737 MAX 7 finally received its FAA certification on August 3, 2026. Yet, just two days later, the FAA issued an Airworthiness Directive ordering structural inspections on certain 737 MAX 8 and MAX 9 aircraft because of possible fuselage cracks. The Airworthiness Directive impacts some 1,400 aircraft worldwide. What's interesting is that the cracks were discovered in the older 737 NG series, but because the manufacturing process and materials used for both series is similar, the FAA was not taking any chances so it issued the airworthiness directive.

On June 12, 2025, Air India Flight 171, a Boeing 787-8, crashed moments after taking off from Ahmedabad, India. It was the first fatal crash involving a 787, and video of the aircraft's final seconds was broadcast around the world. Although the investigation remains incomplete, no evidence indicates thus far that a defect in the aircraft or its engines caused the crash.

Nevertheless, the accident immediately revived a familiar question: Was this another consequence of Boeing's broken processes and culture?

That question alone illustrates the depth of the company's reputational damage. Boeing may eventually be cleared of responsibility for the accident, but it will take years — and more likely decades — for the company to regain the benefit of the doubt.

***

On March 25, 2025, Airbus announced that it had begun laying the technological groundwork for a clean-sheet successor to the A320 family, with the new single-aisle aircraft potentially entering service during the second half of the 2030s. The announcement did not constitute a formal program launch, which Airbus was tentatively targeting for around 2030, but it showed that Airbus was preparing for the successor to the highly successful A320 at the very time that Boeing was still struggling with the MAX program.

In a *Financial Times* interview published on June 1, 2025, Ortberg dampened expectations that Boeing would soon launch a replacement for the 737 MAX to compete against any new aircraft Airbus might bring to market. While stating that "the market is not ready now," he also said Boeing lacked the financial capacity to begin another clean-sheet aircraft program. A successor would come only when the company had the financial means and possessed the resources and organizational ability to execute it.

For the first time in its history, Boeing may be forced to watch from the sidelines as a competitor ushers in the next generation of commercial aviation that will define the future of air travel. Not only may the company find itself without a viable aircraft of its own to compete, but it may no longer possess the means — or the ability — to develop one.

The company that once defined the future of aviation was now fighting to preserve its own.

Epilogue

Had Boeing not lost its way after the merger with McDonnell Douglas, its assessment of the commercial aviation market over the ensuing three decades nevertheless proved remarkably prescient. Airbus' gamble on the A380 ultimately proved to be a massive miscalculation. In 2021, after a production run of just 16 years, Airbus ended production of the enormous aircraft after building only 251 of them.

By contrast, the 747 continued to soldier on. The final aircraft rolled out of Boeing's Everett factory in December 2022, bringing to a close a production run spanning more than half a century. Boeing had built 1,574 747s encompassing six major variations and numerous derivatives. The 747 had outlasted the A380, the very aircraft conceived in large part to challenge its dominance. Yet the changing economics of air travel — and the increasingly capable widebody twinjets Boeing itself had helped pioneer — meant that the age of the jumbo jet was drawing to a close.

Boeing's prediction that international air travel would increasingly fragment proved more accurate than Airbus' competing vision of ever-larger aircraft carrying passengers between a relatively small number of enormous global hubs. In that fundamental strategic judgment, Boeing had been right.

And therein lies one of the great ironies of Boeing's modern history. The company did not lose its way because it suddenly became incapable of understanding aviation. Again and again, Boeing correctly anticipated where the industry was headed. What it increasingly struggled to do was execute those ideas with the discipline, technical rigor, and institutional confidence that had once distinguished it.

For most of its history, Boeing had been far more than a manufacturer of commercial airplanes. It had built bombers that helped determine the outcome of wars, tankers that became indispensable to American air power, spacecraft that carried humanity toward the Moon, and commercial aircraft that fundamentally changed how the world moved. It had repeatedly undertaken projects whose technical difficulty bordered on the audacious — and had succeeded often enough that such ambition became part of the company's identity.

That history makes Boeing's decline all the more consequential. The company did not begin with an absence of engineering talent, technological knowledge, or commercial insight. It possessed those things in abundance. What changed was the environment in which they operated. Over time, financial performance, cost reduction, schedule, outsourcing, and shareholder returns increasingly competed with — and often came to supersede — the engineering priorities that had traditionally defined the company.

The 787 embodied both sides of that transformation. The airplane itself was visionary. Its extensive use of composites, more-electric architecture, improved efficiency, and passenger-focused design pointed toward the future of commercial aviation. Yet the manner in which Boeing chose to develop it exposed the consequences of pushing outsourcing and financial discipline beyond the point at which the company could effectively control them. The aircraft eventually became a market success, but only after years of delays, enormous costs, and the exposure of just how much technical and industrial capability Boeing had surrendered through its outsourcing strategy.

The 737 MAX revealed something more serious. There, the consequences of Boeing's changed culture extended beyond cost and schedule. Decisions made in the pursuit of competitive urgency, commonality, and minimized disruption to customers ultimately intersected with engineering and regulatory failures with catastrophic consequences. The two MAX disasters became the clearest evidence that the problems inside Boeing were no longer merely financial or organizational.

And the difficulties did not end there. Problems with the KC-46A tanker and the 777X, continuing manufacturing and quality-control issues, the Alaska Airlines Flight 1282 door-plug failure, and the troubled Starliner program demonstrated that Boeing's challenges had become systemic rather than confined to any single airplane, project, or management team.

Yet none of this means that Boeing is incapable of recovery. The company that once produced the Model 40A, 247, 307, B-17, B-29, 377, B-47, B-52, 707, 727, 737, 747, 757, 767, and 777 did not accomplish those things by accident. They were the products of accumulated technical knowledge, institutional discipline, experienced people, and a corporate culture that understood what Boeing was supposed to be.

Whether that Boeing can be rebuilt remains an open question. What is far less uncertain is how it was diminished.

It happened gradually, over decades, through decisions made by successive leaders who frequently believed they were acting rationally in the interests of the corporation. No single chief executive destroyed Boeing, nor did any single merger, outsourcing decision, stock repurchase, engineering compromise, or failed program bring the company to its present condition. The transformation was cumulative.

And, of course, those chief executives did not operate in a vacuum. They were appointed, overseen, and ultimately enabled by Boeing's board of directors.

Like most corporate boards, Boeing's board conducts its deliberations largely outside public view. Its discussions, disagreements, warnings, and rationales during the nearly three decades since the McDonnell Douglas merger therefore remain mostly unknown. What is known, however, is who occupied those seats. Many were titans of American industry, finance, government, and public life.

The following is a list of the men and women who served — or continue to serve — on Boeing's board of directors during this period: the individuals ultimately charged with overseeing the company, selecting its chief executive officers, and helping determine the direction Boeing would take.

Start	End	Name	Professional Background
1962	1999	George H. Weyerhaeuser	Forest products; chairman and former CEO of Weyerhaeuser Company
1972	2000	Charles M. Pigott	Heavy-truck and transportation-equipment manufacturing; chairman and CEO of PACCAR
1984	1997	Harold J. Haynes	Oil, petroleum and energy; former chairman and CEO of Chevron; director from 1974–1982 and again from 1984–1997
1976	1997	Stanley Hiller Jr.	Aviation, industrial manufacturing and private investment; aviation pioneer, founder of Hiller Aircraft and partner in Hiller Investment Company
1985	1996	Robert A. Beck	Insurance and financial services; former chairman and CEO of Prudential Insurance Company of America
1985	1997	Frank A. Shrontz	Aerospace; Boeing president, CEO and chairman; previously an official in the U.S. Departments of the Air Force and Defense
1989	2002	John B. Fery	Forest products and paper; chairman and former CEO of Boise Cascade
1990	2003	Paul E. Gray	Engineering and higher education; former president and chairman of the Massachusetts Institute of Technology
1990	1999	Donald E. Petersen	Automotive manufacturing; former chairman, president and CEO of Ford Motor Company
1991	1996	George M. Keller	Oil, petroleum and energy; former chairman and CEO of Standard Oil Company of California, later renamed Chevron
1992	2003	Philip M. Condit	Aerospace engineering and manufacturing; Boeing president, chairman and CEO; previously led development of the 777
1992	2005	Rozanne L. Ridgway	Diplomacy and government; former U.S. assistant secretary of state and ambassador
1995	2011	John E. Bryson	Electric utilities and energy; chairman and CEO of Edison International and Southern California Edison
1995	1996	Franklin D. Raines	Mortgage finance, investment banking and government; vice chairman of Fannie Mae, former Lazard partner and later director of the U.S. Office of Management and Budget
1997	2011	John H. Biggs	Pensions, insurance and asset management; chairman, president and CEO of TIAA-CREF
1997	2021	Kenneth M. Duberstein	Government affairs and consulting; former White House chief of staff and chairman and CEO of The Duberstein Group

1997	2012	John F. McDonnell	Aerospace and defense; former chairman and CEO of McDonnell Douglas
1997	2005	Harry C. Stonecipher	Aerospace and industrial manufacturing; McDonnell Douglas, Sundstrand and Boeing president and CEO
1999	2005	Lewis E. Platt	Computing and technology; former chairman, president and CEO of Hewlett-Packard; later nonexecutive Boeing chairman
2000	2006	John M. Shalikashvili	U.S. military and national security; former chairman of the Joint Chiefs of Staff and NATO supreme allied commander
2001	2016	W. James McNerney Jr.	Aerospace and diversified technology; Boeing chairman and CEO; former chairman and CEO of 3M and senior General Electric executive
2003	2015	Linda Z. Cook	Oil, natural gas and energy; senior executive with Royal Dutch Shell and CEO of Shell Canada
2004	2007	Richard D. Nanula	Media, entertainment, finance and biotechnology; senior executive at Disney, Starwood Hotels and Amgen
2004	2017	Mike S. Zafirovski	Telecommunications and technology; president and CEO of Nortel Networks and senior Motorola executive
2006	2011	William M. Daley	Banking, telecommunications and government; JPMorgan Chase executive, U.S. secretary of commerce and later White House chief of staff
2007	2017	Arthur D. Collins Jr.	Medical devices and healthcare technology; former chairman and CEO of Medtronic
2009	2024	David L. Calhoun	Industrial technology, data services, private equity and aerospace; General Electric, Nielsen, Blackstone and Boeing CEO
2009	2021	Edmund P. Giambastiani Jr.	U.S. Navy, defense and national security; former vice chairman of the Joint Chiefs of Staff
2010	2021	Edward M. Liddy	Insurance, financial services and private equity; former chairman and CEO of Allstate and AIG; partner at Clayton, Dubilier & Rice
2010	2021	Susan C. Schwab	International trade, government and public policy; former U.S. trade representative
2010	2021	Ronald A. Williams	Healthcare and health insurance; former chairman and CEO of Aetna
2011	2024	Lawrence W. Kellner	Commercial airlines and aviation; former chairman and CEO of Continental Airlines
2015	Present	Lynn J. Good	Electric utilities and nuclear energy; former chair, president and CEO of Duke Energy
2015	2019	Dennis A. Muilenburg	Aerospace engineering and defense; Boeing chairman, president and CEO
2016	Present	Robert A. Bradway	Biotechnology and pharmaceuticals; chairman and CEO of Amgen
2016	2020	Randall L. Stephenson	Telecommunications and technology; former chairman and CEO of AT&T
2017	2021	Caroline B. Kennedy	Diplomacy, government and law; former U.S. ambassador to Japan
2019	2020	Nikki R. Haley	Government and diplomacy; former governor of South Carolina and U.S. ambassador to the United Nations

2019	Present	John M. Richardson	U.S. Navy, nuclear propulsion and national security; retired admiral and former chief of naval operations
2020	Present	Akhil Johri	Aerospace manufacturing and finance; former executive vice president and CFO of United Technologies
2020	Present	Steven M. Mollenkopf	Semiconductors and wireless technology; former CEO of Qualcomm
2021	Present	Lynne M. Doughtie	Accounting, auditing and professional services; former U.S. chair and CEO of KPMG
2021	Present	Stayce D. Harris	Commercial aviation and U.S. military; United Airlines pilot and retired U.S. Air Force lieutenant general
2021	Present	David L. Joyce	Aerospace engineering and aircraft engines; former president and CEO of GE Aviation
2022	Present	David L. Gitlin	Aerospace systems and industrial manufacturing; chairman and CEO of Carrier; formerly led Collins Aerospace and UTC Aerospace Systems
2023	2024	Sabrina Soussan	Industrial engineering, transportation and environmental services; CEO of Suez and former co-CEO of Siemens Mobility
2024	Present	Robert Kelly "Kelly" Ortberg	Aerospace electronics and manufacturing; former chairman and CEO of Rockwell Collins and CEO of Collins Aerospace; Boeing president and CEO
2025	Present	Mortimer J. "Tim" Buckley	Investment management, finance and information technology; former chairman and CEO of Vanguard
2025	Present	Bradley D. Tilden	Commercial airlines; former chairman, president and CEO of Alaska Air Group

Notes

The following sources support the principal factual claims, dates, production figures, contracts, program histories, and key events in each chapter. They are drawn from standard professional Boeing histories, official government and company records, investigation reports, and contemporary reporting.

Prologue

- Boeing Company Annual Report 1996 and Form 10-K filings (orders of 719 commercial aircraft; market-share data showing more than 60 percent of world commercial jet production).
- Airbus Annual Reports and commercial-aircraft order statistics, 1996–1999 (crossover in annual orders by end of 1999).
- Peter Robison, Flying Blind: The 737 MAX Tragedy and the Fall of Boeing (Doubleday, 2021), chapters on the post-1997 cultural shift, 787 outsourcing strategy, and 737 MAX program.
- John Newhouse, Boeing Versus Airbus: The Inside Story of the Greatest International Competition in Business (Alfred A. Knopf, 2007).
- NASA Commercial Crew Program contract awards (2014 CCtCap fixed-price values of $4.2 billion for Boeing Starliner and $2.6 billion for SpaceX) and subsequent public status reports on Starliner test flights.
- National Transportation Safety Board (NTSB) final reports and safety recommendation documents on Lion Air Flight 610 (29 October 2018) and Ethiopian Airlines Flight 302 (10 March 2019).
- Boeing corporate histories listing CEOs from William E. Boeing through 2024.

Wood for Floating and Wood for Flying

- Douglas J. Ingells, 747 — Story of the Boeing SUPER JET (Aero Publishers, Inc., 1970)
- Robert J. Serling, Legend & Legacy: The Story of Boeing and Its People (St. Martin's Press, 1992), early chapters on William E. Boeing's background, the 4 July 1914 Curtiss flight with Conrad Westervelt, purchase of the Martin Pusher, and the B & W seaplane.
- Peter M. Bowers, Boeing Aircraft since 1916 (Putnam, 1966 and later editions), entries on the B & W (first flight 29 June 1916) and Model C trainers.
- Harold Mansfield, Vision: A Saga of the Sky (Duell, Sloan and Pearce / later editions), account of the founding of Pacific Aero Products Company (1916) and early Navy contracts.

- Boeing Company historical timeline and archival material (boeing.com/history) confirming incorporation and first flight dates.
- U.S. Navy contract records for the Model C and licensed production of the Curtiss HS-2L (order reduced from 50 to 25 after the Armistice).
- Bureau of Labor Statistics Consumer Price Index data for inflation adjustment of the approximate 1915–1916 Martin airplane purchase price.
- Harold Mansfield, Vision (cited above), on the temporary furniture-manufacturing period and successive government contracts.

Uncle Sam's Needs and the Need for Uncle Sam

- Robert J. Serling, Legend & Legacy (cited above), chapters on post–World War I contracts, DH-4B modernization, GAX, and the MB-3 production order.
- Peter M. Bowers, Boeing Aircraft since 1916 (cited above), technical and production details of the DH-4B, GAX, and MB-3.
- Boeing Company historical records and official timeline: design and construction of the single B-1 (Model 6) flying boat in 1919 as Boeing's first aircraft intended for the commercial market; first flight from Lake Union; subsequent sale of the sole example to Edward "Eddie" Hubbard.
- U.S. Post Office Department contract awarded to Edward Hubbard for carriage of international mail between Seattle and Victoria, British Columbia (one of the earliest regularly scheduled international airmail routes in North America). Service began in 1920 using the Boeing B-1.
- Contemporary accounts and Boeing archival material documenting Hubbard's operation of the route: the B-1 carried mail (and occasionally passengers) across the Strait of Juan de Fuca / Puget Sound, saving up to a full day compared with ship transit during the Victoria stopover of transpacific liners.
- War Department Bolling Commission (led by Colonel Raynal C. Bolling), dispatched to Europe in 1917 to study Allied aircraft designs, manufacturing methods, and recommend types for American production. Official reports and contemporary accounts of the Commission's recommendations, including selection of the British de Havilland DH-4 for licensed U.S. manufacture. Subsequent shipment of examples, drawings, and specifications back to the United States.
- Boeing design and development records for the Model 40 mailplane and its redesigned Model 40A variant. Claire Egtvedt's role in adapting the aircraft to the Pratt & Whitney Wasp engine, which created space for a small enclosed two-passenger cabin (commonly referred to as the "limousine compartment").

Contemporary descriptions in Boeing histories and period aviation literature of the cabin as a means of generating passenger revenue to subsidize airmail operations.

- Pratt & Whitney historical records on the Wasp and Hornet engines and the special arrangement releasing production engines to Boeing in exchange for a flying testbed.
- U.S. Post Office Department competitive bidding records for Contract Air Mail (CAM) routes under the Air Mail Act of 1925 / Kelly Act. Boeing Air Transport's successful low bid (approximately $3 per pound) for the Chicago–San Francisco route in January 1927. Requirement that Boeing post a performance bond because of the unusually aggressive price. Official contract awards, Boeing Air Transport formation documents, and contemporary newspaper and aviation-press coverage of the bid and bond.
- Boeing Air Transport operational records and contemporary publicity on the inaugural passenger-carrying flight of 1 July 1927 (reporter Jane Eads).

Commercial Flight Takes Flight

- Robert J. Serling, Legend & Legacy (cited above), on the Model 40 / 40A / 40B-4, formation of Boeing Air Transport, and the San Francisco–Chicago airmail route.
- Peter M. Bowers, Boeing Aircraft since 1916 (cited above), technical evolution and production totals of the Model 40 series.
- Raymond Orteig's 1919 offer of a $25,000 prize for the first nonstop flight between New York and Paris (or vice versa). Charles Lindbergh's successful solo flight in the Ryan NYP Spirit of St. Louis, departing Roosevelt Field, Long Island, on 20 May 1927 and landing at Le Bourget, Paris, after approximately 33½ hours on 21 May 1927. Contemporary accounts, Lindbergh's own writings, and historical summaries of the prize and its impact on public enthusiasm for aviation.
- Boeing design and production records for the three-engine Model 80 (first flight July 1928; 12 passengers) and the improved Model 80A (first flight July 1929; up to 18 passengers). Introduction into service with Boeing Air Transport on the San Francisco–Chicago route. Contemporary descriptions of the enclosed cabin, greater use of metal construction, and the 1930 introduction of registered nurses as the first female cabin attendants (beginning with Ellen Church). Peter M. Bowers, Boeing Aircraft since 1916, and Robert J. Serling, Legend & Legacy.

- Period aviation literature and manufacturer records on the Ford 4-AT / 5-AT Trimotor as a leading contemporary multi-engine passenger transport competing in the late 1920s and early 1930s market.
- Boeing records and contemporary coverage of the Model 247 (first flight February 1933; introduction May 1933), the first modern all-metal twin-engine airliner. Its initial exclusive availability to United Air Lines (Boeing's affiliated carrier) and the resulting competitive response from other airlines.
- Transcontinental & Western Air (TWA) specification for a modern all-metal airliner after being unable to obtain early Boeing 247s. Douglas Aircraft Company's response with the DC-1 prototype and production DC-2 (1934), followed by the enlarged and highly successful DC-3 (1935–1936). National Air and Space Museum and Museum of Flight documentation, plus standard histories of the Douglas transports and their rapid displacement of earlier designs, including the 247.
- 1929 merger creating United Aircraft and Transport Corporation, combining Boeing Airplane & Transport Corporation with Pratt & Whitney (Frederick Rentschler), Hamilton Standard, and other aviation interests under William Boeing's chairmanship. Subsequent acquisition and consolidation of airline subsidiaries (Boeing Air Transport, Pacific Air Transport, National Air Transport, Varney, etc.) that formed the basis of United Air Lines.
- Contemporary and biographical accounts of Philip G. Johnson's roles as president of Boeing Airplane Company, Boeing Air Transport, and later United Aircraft and Transport Corporation / United Air Lines interests.
- The 1934 Air Mail Scandal (cancellation of existing airmail contracts by the Roosevelt administration, temporary Army operation of the routes, and subsequent congressional investigations). Resulting Air Mail Act of 1934, which prohibited aircraft and engine manufacturers from owning airlines. Forced breakup of United Aircraft and Transport Corporation into three separate entities: Boeing Airplane Company, United Aircraft Corporation (later United Technologies / RTX), and United Air Lines. Philip G. Johnson barred from certain executive roles; William E. Boeing's resignation as chairman and withdrawal from the aviation business in protest. Official legislative records, contemporary newspaper coverage, and standard corporate histories of the breakup.

The Winds of War Lift the Wings of Commercial Aviation

- U.S. Army Air Corps design competition and Boeing records for the XB-15, a very large four-engine experimental long-range bomber ordered in the mid-

1930s to explore the limits of heavy-bomber performance. National Museum of the U.S. Air Force fact sheets and contemporary technical descriptions of the aircraft (first flight 1937), its size, range goals, and limited operational use as a testbed.

- Boeing Model 299 design and construction records (company designation for the prototype that became the B-17 Flying Fortress). First flight from Boeing Field on 28 July 1935. Evaluation at Wright Field, Dayton, Ohio, in the Army Air Corps multi-engine bomber competition. Crash of the Model 299 on 30 October 1935 at Wright Field caused by the crew's failure to remove the gust locks (controls locked), resulting in the deaths of the pilot and a Boeing employee. National Museum of the U.S. Air Force accounts of the accident and its role in the later adoption of formal pre-flight checklists. Despite the crash, the Army's positive evaluation of the prototype's performance.
- Army Air Corps order of 13 service-test YB-17 (later redesignated Y1B-17) aircraft in January 1936 under a funding technicality after the Model 299 crash. Differences from the prototype (more powerful Wright Cyclone engines and other refinements). Development into the production B-17 series. Standard histories including Peter M. Bowers, Boeing Aircraft since 1916, Robert J. Serling, Legend & Legacy, and U.S. Air Force official accounts of early B-17 testing and operational introduction.
- Boeing design records for the Model 307 Stratoliner, the first pressurized commercial airliner, derived from B-17 wings, engines, and tail surfaces with a new circular pressurized fuselage. First flight 31 December 1938; introduction into airline service in 1940 with Pan American Airways and TWA. Production total of ten aircraft. National Air and Space Museum documentation (including the restored "Clipper Flying Cloud") and contemporary descriptions of its ability to fly "above the weather."
- Boeing design and production records for the Model 314 flying boat (the large Boeing Clipper), developed for Pan American Airways' transoceanic routes. First flight 1938; entry into commercial service in 1939. Technical features, long-range capability, and use on Pacific and Atlantic routes prior to and during the early war years. Period Pan American and Boeing archival material, plus standard aviation histories of the Clipper era.

The Generosity of War

- Robert J. Serling, Legend & Legacy (cited above), B-17 Flying Fortress production and wartime expansion.

- U.S. Army Air Corps procurement records: order and delivery of 39 B-17B aircraft (ordered/delivered in the 1939 period as war began in Europe). Subsequent large-scale order issued in July 1940 for 512 B-17s (associated with the B-17E production run and later variants). Standard production tables in Peter M. Bowers, Boeing Aircraft since 1916, and official U.S. Air Force / Boeing historical summaries of pre-Pearl Harbor expansion.
- Boeing design submission of Model 345 in response to the 1940 Army Air Corps requirement for a very-long-range "superbomber." Approval and development into the B-29; first flight of the XB-29 on 21 September 1942. National Museum of the U.S. Air Force fact sheets and contemporary technical descriptions of the aircraft's advanced features (pressurized crew compartments, remote-controlled gun turrets with centralized fire control, high-altitude performance, and large bomb load).
- Accounts of the B-29 as one of the most complex and expensive weapons programs of World War II (program cost approximately $3 billion, exceeding the Manhattan Project). Multi-plant production system involving Boeing (Wichita and Renton), Bell (Marietta, Georgia), and Martin (Omaha). Frequent design changes that required aircraft to move from production lines to modification centers for extensive rework. Challenges with engines, systems integration, and the pressure of concurrent development and mass production. Standard histories including official Air Force records and specialized B-29 production studies.
- Aggregate production figures: approximately 12,731 B-17 Flying Fortresses built by Boeing, Douglas, and Vega (Lockheed) between 1936 and 1945. Approximately 3,970 B-29 Superfortresses built by Boeing, Bell, and Martin through 1946 (majority during the war years). Official U.S. Army Air Forces and manufacturer production tallies.
- Use of specially modified "Silverplate" B-29s: Enola Gay (Hiroshima, 6 August 1945) and Bockscar (Nagasaki, 9 August 1945). Official U.S. military and Manhattan Project historical records documenting the missions and the role of the B-29 in delivering the atomic bombs.
- Philip G. Johnson's return to the presidency of Boeing (1939 onward) and his leadership in organizing and expanding the company's massive wartime production of B-17s and B-29s across multiple plants. His sudden death from a cerebral hemorrhage on 14 September 1944 in Wichita, Kansas, while visiting the Boeing plant there, at age 49. Contemporary obituaries, Boeing archival material, and biographical accounts in aviation histories.

- Arthur Herman, Freedom's Forge: How American Business Produced Victory in World War II (Random House, 2012), on industrial mobilization and Boeing's role.
- Boeing Company archival material on wartime plant expansion in Seattle, Renton, and Wichita.

The War Drums of a Fragile Peace

- Robert J. Serling, Legend & Legacy (cited above), B-29 Superfortress program and postwar transition.
- Peter M. Bowers, Boeing Aircraft since 1916 (cited above), B-29 entries (approximately 3,970 built).
- Boeing development records and U.S. Air Force documentation of the B-50 as an improved, more powerful derivative of the B-29 (Pratt & Whitney R-4360 engines, strengthened structure, taller tail). First flight 25 June 1947; production of approximately 370 aircraft between 1947 and 1953. Service with Strategic Air Command as a medium bomber until replaced by jets; later variants used for weather reconnaissance, training, and other roles. National Museum of the U.S. Air Force fact sheets and standard postwar bomber histories.
- Boeing board decision elevating William M. Allen to the presidency in September 1945 following Philip Johnson's death. Allen's leadership in the transition from wartime production to peacetime commercial and military programs, including the decision to proceed with the Model 377 Stratocruiser. Contemporary accounts and Boeing corporate histories of his tenure.
- Boeing design and production records for the Model 377 Stratocruiser (civil airliner) and its military predecessor/sibling, the C-97 Stratofreighter (Model 367), both derived from B-29/B-50 technology with a double-deck fuselage. Allen's authorization to build an initial batch of Stratocruisers without firm airline orders. Limited commercial production of the 377 (approximately 56 aircraft) amid strong competition; far larger military production of C-97 variants. Period airline and Boeing archival material.
- Development history of the B-47 as Boeing's first large jet bomber, incorporating swept wings and podded engines. Its role as a technological bridge to later designs and its service as a mainstay of Strategic Air Command in the 1950s. Official Air Force and Boeing technical summaries.
- Accounts of the October 1948 weekend in a Dayton, Ohio, hotel room (commonly identified as the Van Cleve Hotel) during which a Boeing design team, responding to Air Force dissatisfaction with earlier proposals, rapidly

refined the configuration that became the B-52 (swept-wing, eight-engine jet bomber). Presentation of the revised proposal (Model 464-49) the following Monday. Subsequent development, first flight, and long service life of the B-52. Boeing historical recollections, Air Force records, and secondary aviation histories of the "Dayton weekend."

- Development of the KC-97 Stratofreighter tanker variant from the C-97 airframe, incorporating the flying-boom aerial-refueling system. Introduction in the early 1950s and its role supporting Strategic Air Command bombers until replaced by jet tankers. U.S. Air Force and Boeing documentation.
- U.S. Air Force records and contemporary accounts of the Convair B-36 as the large, long-range piston (later piston-plus-jet) intercontinental bomber that preceded and briefly overlapped with the B-52 in Strategic Air Command service. Its role in early Cold War nuclear deterrence and the competitive context it created for Boeing's jet-bomber programs.

Turnabout

- Robert J. Serling, Legend & Legacy (cited above), postwar recovery, Model 377 Stratocruiser, and early jet studies.
- Peter M. Bowers, Boeing Aircraft since 1916 (cited above), Stratocruiser and civil derivatives.
- Harold Mansfield, Vision (cited above), executive decision-making in the late 1940s–early 1950s.
- Development and introduction of the de Havilland DH.106 Comet, the world's first commercial jet airliner. Prototype first flight 27 July 1949; entry into service with BOAC on 2 May 1952. Subsequent structural failures, grounding, redesign (Comet 4 series), and the lessons absorbed by American manufacturers. Standard aviation histories and manufacturer records.
- Douglas DC-7 as the final major piston-engine evolution of the DC-6 family (entry into service 1953–1954), offering higher speed and longer range for transcontinental and early intercontinental routes. Concurrent development of stretched and more powerful Lockheed L-1049 Super Constellation variants. Contemporary airline performance comparisons and production records showing these types as the last generation of large piston airliners before the jet transition.
- Boeing's privately funded prototype jet transport (company designation 367-80), authorized by William Allen as a demonstrator for both commercial and military applications. First flight 15 July 1954. Its role as the common ancestor of the KC-135 Stratotanker and the Boeing 707 airliner. National Air and Space

Museum documentation and Boeing historical accounts of the $16 million risk and the famous Dash 80 flight demonstrations.

- Operational constraints of the piston-powered Boeing KC-97 Stratofreighter when refueling jet bombers (speed and altitude mismatch, requiring the bomber to slow and descend). U.S. Air Force evaluations highlighting the need for a jet tanker to match the performance of the B-47 and forthcoming B-52.
- Strategic Air Command competition for a jet-powered aerial-refueling tanker. Initial selection of a Lockheed design (derived from a proposed jet airliner) in 1955. Subsequent decision by the Air Force to order the already-flying Boeing design as the KC-135 Stratotanker for earlier availability; Lockheed's winning design was ultimately not produced in quantity. Official Air Force procurement records and contemporary reporting on the reversal.
- Development of the KC-135 from the 367-80 prototype; first flight 1956; entry into Air Force service and long-term role as the primary aerial-refueling platform. Production totals and technical relationship to the commercial 707.
- Lockheed L-188 Electra turboprop airliner development and introduction (late 1950s) as a short-to-medium-range competitor during the early jet age. Contemporary accounts of its performance, early structural issues, and market position relative to pure-jet designs.
- Boeing 707 commercial introduction (Pan American World Airways, October 1958 on transatlantic routes), marking the practical beginning of the American jet airliner era. Douglas DC-8 first flight and entry into service (Delta Air Lines and United Airlines, 18 September 1959). Comparative airline adoption, route performance, and the competitive dual between the two four-engine jet transports.

Unbridled Advancements

- President John F. Kennedy's Special Message to the Congress on Urgent National Needs, delivered 25 May 1961, in which he called for the United States to achieve the goal of landing a man on the Moon and returning him safely to Earth before the end of the decade. Official transcripts and recordings from the John F. Kennedy Presidential Library and NASA historical collections.
- Boeing's major role in the Apollo program, including construction of the Saturn V first-stage booster (S-IC), systems-integration work, and the later Technical Integration and Evaluation (TIE) contract with NASA following the Apollo 1 fire. Boeing corporate histories, NASA program records, and contemporary accounts of the Huntsville operations.

- Boeing internal studies in the 1950s examining the feasibility of commercial supersonic flight. NASA Langley Research Center records on early SCAT (Supersonic Commercial Air Transport) research and industry participation; Boeing archival material on high-speed aerodynamic studies of the period.
- Development history of the Anglo-French Concorde program, initiated by Britain and France in the late 1950s and formalized in the 1962 treaty. Official BAC/Aérospatiale program histories, British and French government records, and standard aviation histories of the Concorde project.
- Najeeb E. Halaby's service as Administrator of the Federal Aviation Agency (later Federal Aviation Administration) from 1961 to 1965. Halaby's oral-history interview in the John F. Kennedy Presidential Library, FAA historical records, and contemporary reporting on his advocacy for an American SST program.
- Pan American World Airways' placement of orders/options for the Concorde under Juan Trippe. Pan Am corporate records, contemporary press coverage (including reports of meetings between Halaby and Trippe), and aviation histories documenting how the order helped catalyze U.S. government action on a domestic SST.
- U.S. government decision in the early 1960s to fund development of an American supersonic transport. President Kennedy's June 1963 announcement launching the National Supersonic Transport Program; subsequent policy statements under Presidents Kennedy and Johnson; FAA, NASA, and Department of Defense joint studies.
- Design competition between Boeing (Model 2707, variable-sweep wing) and Lockheed (L-2000, delta wing) for the U.S. SST program. FAA announcement of 31 December 1966 awarding the development contract to Boeing; evaluation reports of the President's Advisory Committee on SST; contemporary technical assessments of both designs and the program's strategic importance to Boeing.
- U.S. Air Force Request for Proposals for the CX-HLS (Heavy Logistics System, later C-5 Galaxy) in 1964; competition among Boeing, Lockheed, and Douglas; selection of Lockheed's design in September 1965. Official Air Force procurement records and analyses noting that Boeing's unsuccessful proposal supplied key design features (raised cockpit, nose-loading door) later used on the 747.
- Juan Trippe's vision at Pan American for a significantly larger aircraft capable of roughly doubling passenger capacity relative to existing jets. Pan Am corporate histories, Trippe biographies, and contemporary statements linked to projected growth in international air travel.

- Boeing's decision to launch the 747 program, drawing on high-bypass-engine technology and airframe concepts developed during its CX-HLS / C-5 proposal. Boeing engineering and program histories that document the transfer of military-derived research into the commercial jumbo-jet project.
- Pan American's order of 13 April 1966 for 25 Boeing 747s (23 passenger, 2 cargo) at a cost exceeding $525 million, then the largest commercial aircraft order in history, with options for additional aircraft. Contemporary New York Times coverage, Pan Am announcements, and Boeing contract records.
- Entry of Douglas (DC-10) and Lockheed (L-1011 TriStar) into the widebody market in the late 1960s and early 1970s. Industry analyses and airline-fleet histories describing the resulting three-way competition in a market whose demand proved insufficient to support all three manufacturers profitably.
- William M. "Bill" Allen's retirement as president of Boeing on 29 April 1968 after more than 22 years in the role (1945–1968), followed by service as chairman; succession by T. A. Wilson. Boeing corporate records, HistoryLink essays, and contemporary accounts of his long tenure and the transition.

Europe's Answer to Europe's Fears

- John Newhouse, Boeing Versus Airbus (cited above), origins of the Airbus consortium (1969 onward) and European government support.
- Airbus official corporate histories and founding documents of the consortium.
- Contemporary European aviation press and government reports on the creation of a collaborative European manufacturer.
- U.S. Congress decision in March–May 1971 to terminate funding for the American SST program (Boeing 2707), ending federal support after years of development and roughly $1 billion already spent. Contemporary coverage in Time magazine, National Archives declassification records, and standard aviation histories of the program's cancellation and its impact on Boeing employment and engineering teams.
- Perception in the late 1960s that the three major U.S. manufacturers (Boeing, Douglas/McDonnell Douglas, and Lockheed) held a dominant position capable of marginalizing or eliminating independent European commercial-aircraft production. Contemporary European industrial analyses and later historical accounts of the competitive imbalance between American and fragmented European aerospace firms.
- Motivations for the creation of Airbus Industrie: the recognition that no single European nation could sustainably compete with the American manufacturers on large commercial programs; the desire to pool industrial resources, share

development costs, and retain a European capability in civil aviation. Formal agreements of 1967–1969 among France, Germany, and (initially) the United Kingdom; establishment of Airbus Industrie as a Groupement d'Intérêt Économique in 1970. Official Airbus and partner-government histories, academic studies of European industrial policy, and contemporary ministerial announcements.

- Airbus decision to design the A300 using Imperial (U.S. customary) units of measurement and weights rather than the metric system, and to adopt English as the working language. Program histories and sales accounts noting that this choice was made deliberately to increase acceptability to potential airline customers, particularly in the United States and other markets already accustomed to American aircraft standards.
- Development of the A300 as the first twin-engine widebody airliner. Decision to proceed with a two-engine configuration (initially sized around available high-bypass turbofans such as the General Electric CF6 or Pratt & Whitney JT9D) after earlier three-engine concepts proved less competitive; scaling of the design from a larger 300-seat concept to a more marketable size. Airbus technical and program records, manufacturer histories, and engineering accounts of the twin-engine choice.
- Severe sales drought for the A300 in the mid-1970s caused by the 1973 Arab oil embargo, the ensuing global recession, sharp rises in fuel prices, and a contraction in air travel demand. Contemporary reports of unsold completed aircraft, temporary production slowdowns, and the near-collapse of early sales efforts before the breakthrough Eastern Airlines order. Industry analyses and Airbus historical summaries of the economic impact on the fledgling consortium.

Deregulation and How an American Astronaut Saved European Aviation

- John Newhouse, Boeing Versus Airbus (cited above), effects of the Airline Deregulation Act of 1978.
- U.S. Airline Deregulation Act legislative history and Civil Aeronautics Board records.
- Contemporary accounts of early Airbus sales successes in the U.S. market.
- Airline Deregulation Act of 1978, signed by President Jimmy Carter on 24 October 1978, which phased out the Civil Aeronautics Board's authority over fares, routes, and market entry. Official legislative history, National Air and Space Museum accounts, and contemporary analyses of the Act's provisions.

- Immediate and longer-term effects of deregulation: increased competition, lower average fares on high-density routes, entry of new carriers, expansion of hub-and-spoke systems, service reductions to smaller communities, industry consolidation, and changes in labor relations and service quality. Industry studies, Smithsonian analyses, and retrospective economic assessments.
- President Richard Nixon's New Economic Policy announced 15 August 1971 (the "Nixon Shock"), which included a 90-day wage-and-price freeze, suspension of dollar convertibility into gold, and related measures. Subsequent phases of controls and their distorting effects on oil refining, distribution, and inventory incentives heading into the 1973 energy crisis. Official U.S. government records, economic histories of the period, and analyses of the policy's contribution to later shortages.
- Collapse of the Bretton Woods system of fixed exchange rates. Nixon's 15 August 1971 suspension of dollar-gold convertibility and the subsequent shift to floating rates by 1973. U.S. State Department historical summaries and economic analyses of the end of the postwar monetary regime.
- Frank Borman's career as an astronaut, including command of Apollo 8 (the first crewed mission to orbit the Moon, December 1968). His subsequent move into airline management at Eastern Airlines, rising through senior operational roles. NASA biographical records and contemporary profiles.
- Eastern Airlines Flight 401 crash in the Florida Everglades on 29 December 1972. Official NTSB investigation findings (crew distraction over a landing-gear indicator light leading to controlled flight into terrain). Accounts of Borman's on-site involvement in the rescue effort as Eastern's senior vice president of operations.
- Borman's subsequent elevation to president (and later chairman and CEO) of Eastern Airlines. Corporate histories and contemporary business reporting on his leadership during the mid-1970s.
- Eastern Airlines' financial difficulties during the 1974–1975 recession, compounded by high fuel costs, labor issues, and fleet problems. Industry analyses and Eastern corporate records of the period's losses and operational pressures.
- Operational and reliability problems experienced with Eastern's Lockheed L-1011 TriStar fleet in the mid-1970s, contributing to the airline's cost and scheduling challenges. Contemporary airline technical reports and fleet-performance accounts.

- Airbus Industrie's unprecedented offer to place four A300 aircraft with Eastern Airlines on a no-cost (or effectively free) trial/lease basis for approximately six months beginning late 1977. Borman's decision to accept the arrangement, the highly favorable operational results, and his subsequent strong public endorsement of the aircraft. Contemporary European and American press coverage (including Der Spiegel), Airbus and Eastern records, and later historical accounts describing the trial as a pivotal validation of the A300.
- Eastern's follow-on order for additional A300s and the broader significance of the transaction as the breakthrough that established Airbus' credibility and helped secure the consortium's survival. Industry analyses and Airbus historical summaries characterizing the Eastern deal as decisive for the company's early viability.

Meanwhile Back in Seattle

- Robert J. Serling, Legend & Legacy (cited above), 727, 737, and 747 development.
- Peter M. Bowers, Boeing Aircraft since 1916 (cited above), technical and production histories of those programs.
- Boeing Company annual reports and program histories for the 747 launch and early production.
- Thornton A. "T" Wilson's career at Boeing, rising from engineer on the B-47 and B-52 programs and leadership of the Minuteman missile effort to company president in 1968, CEO in 1969, and chairman in 1972. He served as CEO until 1986. Boeing corporate records, National Aviation Hall of Fame materials, and contemporary business profiles.
- Boeing's outsized role in the Seattle-area and Washington state economy in the late 1960s, when employment peaked above 100,000, and its significance as a major U.S. exporter and national aerospace employer. Regional economic studies and contemporary reporting on the company's payroll impact.
- The "Boeing Bust" of 1969–1971: severe financial strain caused by cancellation of the SST program, sharp reductions in NASA and Space Race-related spending, declining Vietnam-era military orders, and a simultaneous commercial-airliner market downturn. Workforce reductions from more than 100,000 to roughly 30,000–35,000; near-insolvency concerns; and the regional recession that produced the famous "Will the last person leaving Seattle turn out the lights" billboard. New York Times coverage, Seattle Times historical retrospectives, and economic analyses of the period.

- Boeing's search for successors to the 707 and 727 families in the mid-to-late 1970s. Internal project designations 7X7 (widebody) and 7N7 (narrow-body) that evolved into the 767 and 757 respectively. Boeing program histories and engineering accounts of the design studies.
- High degree of systems and cockpit commonality deliberately engineered between the 757 and 767, including a common type rating that allowed pilots qualified on one to fly the other with minimal additional training. Shared avionics, flight-management systems, and handling characteristics. SAE technical papers on the commonality design philosophy and Boeing certification documentation.
- Lockheed's severe financial difficulties with the L-1011 TriStar program, precipitated by the February 1971 receivership of Rolls-Royce (supplier of the RB211 engines). The crisis threatened Lockheed's solvency and the TriStar's survival. Contemporary New York Times reporting and aerospace industry histories.
- U.S. government emergency loan guarantee enacted in 1971 (Emergency Loan Guarantee Act) that enabled a $250 million bank credit facility for Lockheed, secured by a federal guarantee. The measure, approved by a narrow congressional margin, allowed completion of the L-1011 program. Official Treasury and congressional records, plus contemporaneous news accounts of the guarantee's approval and implementation.

Let's Buy a Jetliner

- Contemporary trade-press coverage of the 1960s–1970s jet-buying boom.
- Boeing sales and marketing materials of the era (preserved in company archives and secondary histories).
- John Newhouse, Boeing Versus Airbus (cited above).
- 1979 Iranian Revolution and the second major oil-price shock of the decade. Disruption of Iranian oil production, rapid rise in world crude prices (roughly doubling between mid-1979 and mid-1980), and the resulting global economic turbulence. Brookings Institution and Federal Reserve analyses of the revolution's impact on oil markets and Western economies.
- 1980–1982 recession and the accompanying stagflation (high inflation combined with stagnant growth and rising unemployment). Contemporary economic histories linking the downturn to the oil shock, tight monetary policy, and lingering effects of earlier wage-price controls.

- Lockheed's December 1981 announcement that it would phase out production of the L-1011 TriStar and exit the commercial airliner business. Final deliveries in the early 1980s; Lockheed's subsequent focus exclusively on military and other non-commercial aerospace work. Washington Post coverage and industry histories of the decision.
- Long lead times inherent in commercial aircraft manufacturing — typically several years from firm order to delivery — driven by design finalization, supplier coordination, certification, and production-slot constraints. Standard aerospace-industry analyses of order-to-delivery cycles.
- Structure of airline–manufacturer contracts: distinction between firm orders (legally binding commitments) and options (rights to purchase additional aircraft later); practices of selling or exchanging delivery slots; and cancellation penalties or progressive payment schedules that protect manufacturers. Contemporary trade-press explanations and aircraft-financing literature of the period.
- Role of the U.S. Export-Import Bank (Ex-Im Bank) in financing foreign sales of American-built commercial aircraft. Ex-Im's provision of loan guarantees and direct credits that enabled U.S. manufacturers to match or counter foreign competition. Official Ex-Im Bank policy documents and congressional reviews of its aviation portfolio. U.S. manufacturers' heavy reliance on Ex-Im Bank support to remain competitive against European rivals whose sales were backed by their own national export-credit agencies. Industry testimony and trade analyses of the 1970s–1980s export-financing landscape.
- European export-credit agencies (and related government financing mechanisms) that supported Airbus sales on terms often more favorable than those available to American competitors. Comparative studies of official export-credit practices.
- Government "launch aid" and other forms of repayable or risk-sharing financial support provided by European partner governments to Airbus for new aircraft programs. Official critiques, WTO-related documentation, and industry analyses of the competitive effect of such subsidies.
- McDonnell Douglas MD-80 program (initially the DC-9 Super 80), an improved, stretched, and more fuel-efficient derivative of the DC-9 family. First flight of the MD-81 in 1979, entry into service in 1980, and subsequent commercial success that helped sustain McDonnell Douglas in the narrow-body market through the 1980s and 1990s. Manufacturer records and contemporary fleet analyses.

A Difference of Philosophy

- Contemporary press coverage of the 1975 Apollo-Soyuz Test Project.
- Apollo-Soyuz Test Project (July 1975), the first international crewed space mission, in which an American Apollo spacecraft docked with a Soviet Soyuz. NASA historical records and official mission documentation of the joint flight and its docking systems.
- John Newhouse, Boeing Versus Airbus (cited above), contrasting engineering-driven and financially driven cultures.
- Airbus formal launch of the A320 program in March 1984 after several years of studies and the securing of initial customer commitments (notably Air France). Airbus corporate histories and contemporary industry reporting on the launch decision.
- Airbus decision to differentiate the A320 through an all-digital fly-by-wire flight-control system and associated airframe architecture — the first commercial airliner designed from the outset around full-authority digital fly-by-wire. Statements by Roger Béteille and Airbus technical accounts of the technological leap.
- Prior military application of fly-by-wire technology, most prominently on the General Dynamics F-16 Fighting Falcon, which demonstrated the handling and performance advantages of electronic flight controls. Technical histories of early fly-by-wire systems.
- Benefits of commercial fly-by-wire claimed by Airbus: reduced mechanical complexity and maintenance costs, weight savings, flight-envelope protections that prevent the aircraft from exceeding safe angle-of-attack, speed, and load-factor limits, and the potential for greater commonality and simplified pilot training across a family of aircraft. Airbus technical papers and SAE presentations on the A320 control laws and protections.
- Airbus marketing emphasis on the safety enhancements provided by the A320's flight-envelope protections and digital architecture. Contemporary sales literature and public statements promoting the system's protective features.
- A320's wider fuselage cross-section relative to the Boeing 737, chosen to improve passenger comfort and allow standard LD3-size containers in the lower hold. Design histories and comparative cabin-dimension analyses.
- Pilot community reservations about the A320's sidestick controllers (replacing conventional control columns) and the psychological shift from direct mechanical feel to computer-mediated control. Contemporary pilot-union commentary and industry discussions of the human-factors implications.

- Air France A320 demonstration-flight crash at Habsheim airfield, France, on 26 June 1988 (Air Charter flight ACF296Q). Official French investigation report documenting the low-altitude, low-speed flyover that ended in a forest collision; three fatalities.
- Indian Airlines Flight 605 A320 crash at Bangalore on 14 February 1990. Official investigation findings attributing the accident primarily to crew failure to monitor airspeed and altitude during approach.
- Air Inter Flight 148 A320 crash near Strasbourg (Mont Sainte-Odile) on 20 January 1992. Official investigation conclusions regarding controlled flight into terrain during a non-precision approach, with contributing factors related to crew interface with the flight-management and autopilot systems.
- Relative order intake among Boeing, McDonnell Douglas, and Airbus in 1984 and the surrounding years, illustrating Airbus' still-limited but growing share of the market at the time of the A320 launch. Industry order statistics of the period.
- Persistent U.S. and Boeing allegations that Airbus benefited from unfair government "launch aid" and other subsidies that distorted competition. Later WTO dispute documentation and contemporary trade complaints that began in the 1980s and continued for decades.
- John Leahy's career at Airbus, beginning in North American sales in 1985 and rising to head of sales and later chief commercial officer; his reputation as a highly effective salesman who significantly expanded Airbus' market share. Biographical and industry profiles.
- Leahy's successful 1986 campaign that secured Northwest Airlines' landmark order for up to 100 A320s — the first major U.S. carrier commitment to the type and a breakthrough in the American market. Contemporary sales accounts and Airbus order histories.
- United Airlines' 1992 order for A320-family aircraft, further validating the type among major U.S. carriers previously loyal to Boeing. Airline fleet announcements and industry reporting of the period.

Reaching the Zenith

- Boeing Company Annual Report 1996 and order/delivery statistics.
- John Newhouse, Boeing Versus Airbus (cited above).
- Aviation Week & Space Technology and Wall Street Journal reporting on 1996 order totals and market share.

- Frank A. Shrontz's tenure as Boeing president (1985), chief executive officer (1986–1996), and chairman. Leadership during the late-1980s production boom, the subsequent early-1990s downturn, internal process reforms, and the launch of the 777 program. Contemporary Seattle Times interviews, New York Times obituaries, and Boeing corporate histories.
- Development of the Boeing 737 Classic series (737-300/-400/-500). Adoption of the CFM International CFM56 high-bypass turbofan — originally developed in part for the KC-135 re-engining program — as the powerplant that transformed the 737's performance, fuel efficiency, and noise characteristics. Manufacturer and engine-program records.
- Production totals for the 737 Classic family. Standard production summaries and aviation statistical references.
- Introduction and commercial success of the Boeing 747-400, the stretched, updated, two-crew version of the jumbo jet that became the dominant long-haul aircraft of the late 1980s and 1990s. Boeing program histories and fleet data.
- Shrontz-era efforts to expand production capacity, accelerate output, and manage surging airline demand in the late 1980s, including investments in facilities, workforce, and process changes. Concurrent quality-control challenges that arose from the rapid ramp-up. Contemporary business reporting and later retrospective accounts of the production surge.
- Record commercial aircraft orders received by Boeing in 1989, reflecting the peak of the late-1980s boom before the subsequent market downturn. Industry order statistics and contemporary press coverage.
- Extended-range 767 variants (notably the 767-200ER and -300ER) that enabled efficient point-to-point long-haul service. The 767's emergence as the most popular aircraft on many transatlantic routes by the early 1990s. Airline fleet analyses and traffic statistics of the period.
- Evolution of ETOPS (Extended-range Twin-engine Operational Performance Standards) regulations, beginning with early 120-minute approvals for the 767 and later extensions to 180 minutes. The regulatory and technical changes that allowed reliable twin-engine aircraft to operate long overwater routes previously reserved for three- and four-engine types. FAA rule-making history and Boeing/engine-manufacturer reliability data.
- Boeing 777 development program, launched in the early 1990s as a clean-sheet widebody twinjet. Revolutionary use of extensive digital design (CAD/CAM), concurrent engineering, and the "Working Together" philosophy that brought airline customers, engineers, and suppliers into integrated design-build teams

from the outset. Engine options including the General Electric GE90, Rolls-Royce Trent 800, and Pratt & Whitney PW4000 series. Boeing program histories, participant accounts, and technical papers on the design process.

- McDonnell Douglas MD-11 program, an updated and stretched derivative of the DC-10 trijet intended to compete in the long-haul market. Performance shortfalls relative to guarantees, order cancellations, and limited production total (just over 200 aircraft). Manufacturer histories and airline evaluations.
- Airbus A330 (twin-engine) and A340 (four-engine) family development as the European response to the same long-range market segment. Launch and early sales history of both types. Airbus program records and comparative market analyses.
- Impact of the 1990–1991 Gulf War (Iraq–Kuwait conflict), the associated spike in oil prices, and the ensuing early-1990s recession on airline finances and new aircraft orders, producing a sharp downturn after the late-1980s peak. Contemporary economic and aviation-industry analyses of the order drought and production cutbacks.

The Little Plane That Could

- Peter M. Bowers, Boeing Aircraft since 1916 (cited above), and subsequent Boeing 737 program histories covering the original design through the Next Generation series.
- Boeing production and order data for the 737 family.
- Boeing's decision in the early 1990s to upgrade the 737 Classic rather than immediately launch an all-new narrow-body, driven by the need to respond to the growing success of the Airbus A320 family while preserving commonality and minimizing development cost and risk for existing 737 operators. Contemporary industry analyses and Boeing program histories of the competitive response.
- Development history of the Boeing 737 Next Generation (NG) family (737-600/-700/-800/-900). Official launch of the 737-700 with Southwest Airlines in November 1993; subsequent launches of the -800 (1994) and -600 (1995); first flights and certifications in 1997–1998; entry into service beginning December 1997. Key design changes including a new larger wing, updated CFM56-7 engines, modernized digital cockpit, and cabin improvements. Boeing official background materials and production timelines.
- Decision to maintain the 737 type certificate continuity with the Classic series, allowing the NG models to be certificated as derivatives rather than an entirely new type. This preserved significant pilot-training, maintenance, and operational commonality for airlines already flying the Classic. FAA and EASA

type-certificate data sheets for the 737 family documenting the continuous certification basis from the original 737-100 through the NG variants.

- Contrasting sales and marketing approaches in the 1990s narrow-body competition. Airbus emphasis on the A320's technological differentiation (fly-by-wire, wider cabin, commonality across a new family) and aggressive commercial terms, including creative financing and launch incentives aimed at converting traditional Boeing customers. Boeing emphasis on evolutionary improvement, proven reliability, lower transition costs for existing 737 operators, and the operational advantages of fleet commonality. Contemporary sales-campaign accounts, airline order announcements, and industry reporting on head-to-head contests (including early NG campaigns against the A320).

Jack Welch

- Reporting and analysis in The New York Times, The Daily Beast, and business press on Jack Welch's management philosophy at General Electric.
- John Francis "Jack" Welch Jr.'s career at General Electric, joining the company in 1960 in the plastics division and rising through successive management roles. Appointment as GE's youngest-ever chairman and CEO in 1981, succeeding Reginald Jones; tenure lasting until his retirement in 2001. Standard biographies, Welch's own memoir Jack: Straight from the Gut, and contemporary business profiles.
- Welch's intense focus on performance metrics, shareholder value, and profitability. Implementation of the "#1 or #2" rule requiring every GE business to be first or second in its market or face divestiture; continuous evaluation of units for retention, improvement, or sale/spin-off; rigorous use of financial and operational metrics to drive decisions. Accounts in Welch's writings, GE annual reports of the period, and later critical analyses of his management system.
- The nickname "Neutron Jack," coined by the press in the early 1980s (first widely noted by Newsweek) in reference to the neutron bomb — eliminating people while leaving buildings intact. The moniker arose from large-scale workforce reductions (more than 100,000 jobs cut in the early years through layoffs, plant closures, and business sales) even as the company remained profitable. Contemporary media coverage and Welch's own later reflections on the label.
- Welch's September 1989 trip to India, initially intended to pursue sales of GE products (aircraft engines, plastics, etc.). During meetings with Indian officials and advisers, the conversation shifted toward India supplying software and services to GE. This encounter helped catalyze GE's early and aggressive

embrace of offshore outsourcing and the subsequent establishment of major GE operations and technology centers in India (including the Jack F. Welch Technology Center in Bangalore). Wall Street Journal reporting and later accounts of GE's role in the rise of India's outsourcing industry.

Into The Wilderness

- Philip M. Condit's career at Boeing, beginning in 1965 as an aerodynamics engineer and progressing through a series of technical and management roles, including leadership of the 757 program and later the 777 division. Boeing corporate biographies and contemporary profiles.
- Condit's leadership style, frequently described as consensus-oriented, emphasizing open communication, teamwork, multi-source performance reviews, and collaborative decision-making — particularly evident during the 777 program. Case studies and accounts of the 777 design-build process.
- Reports that, during a period between marriages in the mid-1990s, Condit resided in the Boeing corporate suite at Seattle's Four Seasons Olympic Hotel and had the suite remodeled (including addition of a bedroom) at company expense. Investigative business reporting from the period.
- Condit's personal life: four marriages, including a marriage to his first cousin Jan Condit in the early 1990s (later divorced), and a relationship with a Boeing employee that became a matter of internal concern. Contemporary and later journalistic accounts.
- Condit's role as executive vice president and general manager of the 777 program division, overseeing the aircraft's development under the "Working Together" approach. Boeing program histories and participant recollections.
- Frank Shrontz's decision to elevate Condit to president of Boeing and member of the board in 1992, positioning him as the clear successor. Corporate succession records and contemporary reporting.
- Condit's promotion to chief executive officer in 1996 and to chairman in 1997, becoming the seventh chairman in Boeing's history. Official Boeing announcements and board records.
- Post–Cold War strategy under Condit to diversify Boeing beyond commercial aircraft by expanding the defense and space portfolio through a series of acquisitions (including major elements of Rockwell's aerospace and defense businesses) aimed at balancing revenue streams and reducing cyclical exposure. Boeing annual reports and acquisition histories of the mid-to-late 1990s.

- Shift in Boeing's revenue composition: in the mid-1960s the company remained heavily dependent on commercial (and earlier military) aircraft programs; by the mid-1990s commercial airplanes still dominated but the deliberate push into defense and space had begun to alter the mix, a trend accelerated by later acquisitions. Comparative financial analyses and historical revenue breakdowns.
- By the mid-1990s Airbus had narrowed the gap with Boeing in commercial aircraft orders and market share, while McDonnell Douglas had fallen to a distant third place in the large commercial jetliner market. Industry order and delivery statistics, contemporary market-share reports, and analyses of the evolving competitive landscape.

Storied Pasts Undone by Uncertain Futures

- Donald Wills Douglas Sr. (1892–1981): early career as chief engineer at Glenn L. Martin, formation of the Davis-Douglas Company and subsequent incorporation of the Douglas Company (later Douglas Aircraft Company) in 1921 after Davis withdrew. Standard biographies, Boeing historical archives, and Smithsonian accounts of the company's founding in Santa Monica.
- Early Douglas products including the DT-2 torpedo bomber (first major Navy contract) and M-2 mailplane. Subsequent commercial and military designs: DC-1 prototype, DC-2 (1934 Collier Trophy), DC-3 (and its military C-47 variant), DC-4/C-54, DC-5, DC-6, DC-7, and the jet-era DC-8 and DC-9. Military types including the A-20 Havoc, SBD Dauntless, A-26 Invader, A-1 Skyraider, A-3 Skywarrior, and A-4 Skyhawk. Douglas production histories and contemporary aviation references.
- Profitability and eventual financial strain of the DC-9 program: strong initial market response and large order backlog, but production ramp-up difficulties, supplier shortages, rising costs, and delivery delays in the mid-1960s that contributed to severe cash-flow problems. Douglas Aircraft annual reports (especially 1965–1966) and financial analyses of the company's near-collapse.
- Douglas Aircraft's broader financial challenges in the 1960s: heavy development costs of the DC-8 (still short of break-even for years), losses on military programs, inventory build-ups, generous customer financing terms, and the 1966 pre-tax loss that forced omission of dividends and pursuit of external financing or merger. Contemporary financial reporting and later studies of the firm's decline.
- James Smith McDonnell Jr. (1899–1980): education at Princeton and MIT, early engineering work, founding of McDonnell Aircraft Corporation in St. Louis in

1939. National Aviation Hall of Fame biography and standard corporate histories.

- McDonnell military aircraft: XP-67 Moonbat prototype, FH-1 Phantom (first carrier-based jet), F2H Banshee, F3H Demon, F-101 Voodoo, and the highly successful F-4 Phantom II. McDonnell's role as prime contractor for NASA's Mercury and Gemini spacecraft. Company and NASA historical records.
- 1967 merger of McDonnell Aircraft and Douglas Aircraft, forming McDonnell Douglas Corporation, with James S. McDonnell as chairman and CEO; Douglas leadership accepting a subordinate role amid the latter's financial distress. Official merger announcements and board records.
- DC-10 development philosophy under McDonnell Douglas management: emphasis on cost control, reuse of proven components and design elements from the DC-8/DC-9 lineage, and rapid time-to-market. Criticisms of design compromises and "cutting corners" relative to more ambitious engineering approaches. Contemporary engineering and industry analyses.
- Competitive race between the DC-10 and Lockheed L-1011 TriStar: McDonnell Douglas' priority on beating Lockheed to market (achieving earlier entry into service in 1971), contrasting with Lockheed's focus on engineering excellence, advanced systems, and higher development standards for the L-1011. Market-share and program histories of the widebody trijet competition.
- Rolls-Royce RB211 engine development for the L-1011: technical difficulties, cost overruns, and the 1971 receivership of Rolls-Royce that delayed the TriStar and nearly bankrupted Lockheed. Contemporary reporting and official accounts of the engine crisis and subsequent U.S. government loan guarantee for Lockheed.
- Sanford N. "Sandy" McDonnell (nephew of James S. McDonnell) and his later leadership roles within McDonnell Douglas; James S. McDonnell's continued influence as chairman until shortly before his death in 1980. Corporate succession records and biographical material.

When Flaws Bred Calamity

- American Airlines Flight 96 (12 June 1972): DC-10-10 (N103AA) experienced explosive decompression near Windsor, Ontario, after the aft cargo door blew open and separated in flight. The cabin floor partially collapsed, damaging control cables, but the crew successfully landed the aircraft at Detroit with no fatalities. NTSB investigation and contemporary accident reports documenting the unlatched cargo door and design vulnerabilities.

- Turkish Airlines Flight 981 (3 March 1974): DC-10-10 (TC-JAV) crashed in the Ermenonville Forest near Paris after the aft cargo door separated shortly after takeoff from Orly, causing explosive decompression, collapse of the cabin floor, severance of control cables, and total loss of control. All 346 people on board were killed. Official French investigation (BEA) and subsequent analyses confirming the same cargo-door latching failure mode as Flight 96.
- French accident report, Turkish Airlines DC-10 TC-JAV In the Ermenonville Forest on 3 March 1974 Final Report.
- Ground pressurization test of a DC-10 airframe during pre-certification testing in July 1970: the aft cargo door blew open at a relatively low differential pressure (approximately 3 psi) after incomplete latching, causing the cabin floor above the cargo hold to collapse and destroy the flight-control runs routed through the floor. Engineering accounts and later accident analyses treating the incident as an early warning of the design deficiencies that later manifested in Flights 96 and 981.
- Lockheed bribery scandal involving payments to Japanese officials and intermediaries (including Yoshio Kodama) to secure All Nippon Airways (ANA) orders for the L-1011 TriStar. ANA had previously indicated preference for or options on the DC-10; the scandal contributed to ANA's October 1972 decision to order 21 L-1011s instead. U.S. Senate Subcommittee on Multinational Corporations testimony, Japanese investigations, and contemporary reporting on the Lockheed–Tanaka affair.
- American Airlines Flight 191 (25 May 1979): DC-10-10 (N110AA) crashed shortly after takeoff from Chicago O'Hare when the left engine and pylon separated due to maintenance-induced structural damage, leading to aerodynamic stall and loss of control. All 271 on board and two people on the ground were killed — the deadliest aviation accident on U.S. soil at the time. NTSB probable-cause findings citing improper engine-pylon removal procedures using a forklift.
- FAA grounding of the entire U.S. DC-10 fleet on 6 June 1979 (lifted 13 July 1979) after inspections revealed similar pylon cracks on multiple aircraft; the grounding was ordered pending determination that the type met certification standards. Official FAA directives and contemporary news coverage of the first fleet-wide grounding of a large U.S. commercial jet type.
- Airline and manufacturer efforts in the years following Flight 191 to rehabilitate the DC-10's public image through safety modifications, advertising, operational reliability data, and emphasis on the aircraft's subsequent

accident-free service record in many fleets. Industry analyses and contemporary trade-press accounts of the recovery campaign.

- Lockheed's December 1981 announcement that it would phase out production of the L-1011 TriStar and exit the commercial airliner business, citing insufficient orders and the program's long-term unprofitability. Contemporary business reporting and Lockheed corporate statements marking the end of the company's commercial aircraft manufacturing.

A Turnaround Without a Future

- Early design studies for a twin-engine derivative of the DC-10 (sometimes referred to in later years as an MD-11 Twin concept), considered as a lower-cost alternative before McDonnell Douglas committed to the trijet MD-11 configuration. Contemporary Flight International reporting and Douglas program studies of the mid-to-late 1980s.
- Airbus development of the A330 (twin-engine) and A340 (four-engine) family, launched in the late 1980s as direct competitors in the long-range widebody market. The A340's superior range and payload capability relative to early MD-11 performance guarantees, particularly on ultra-long-haul routes. Airbus program histories and comparative performance data of the period.
- The MD-11 as McDonnell Douglas' primary (and effectively only viable) path to remaining a significant player in the commercial widebody market after the DC-10. Launch of the MD-11 program on 30 December 1986 as a stretched, updated, two-crew derivative of the DC-10. Manufacturer records and industry analyses.
- Strategic decision to bring the MD-11 to market rapidly in order to secure orders and deliveries ahead of the next-generation twin-engine widebodies then under development (Boeing 777 and Airbus A330/A340). Contemporary market assessments and McDonnell Douglas sales strategy documents.
- MD-11 performance shortfalls relative to original guarantees (range and fuel burn), prompting the multi-year Performance Improvement Program (PIP) launched in 1990 in partnership with engine manufacturers and NASA. Partial recovery of performance through aerodynamic, weight, and systems changes, but insufficient to fully restore competitiveness on key long-haul routes. Program technical summaries and airline evaluations (including American Airlines' subsequent sale of its MD-11 fleet).
- Original MD-11 sales projections exceeding 300 aircraft; actual production totaling only 200 units before the line was closed. Industry order statistics and later retrospective analyses of the program's commercial underperformance.

- Harry C. Stonecipher's career: early work at General Motors Allison, long tenure at General Electric (where he absorbed Jack Welch's management philosophy), leadership of Sundstrand as president and CEO, and appointment as president and CEO of McDonnell Douglas in September 1994 succeeding John F. McDonnell. Biographical profiles and corporate announcements.
- Reasons for selecting Stonecipher: need for an outsider with a proven record of aggressive cost-cutting, operational discipline, and financial turnaround at a company facing shrinking commercial market share, declining military programs, and limited capital for new aircraft development. Contemporary business reporting and board deliberations.
- Stonecipher's explicit admiration for and adoption of Jack Welch's management approach — emphasis on shareholder value, rigorous performance metrics, cost discipline, and willingness to exit underperforming businesses. Public statements and later analyses of his leadership style at McDonnell Douglas and subsequently at Boeing.
- Cancellation of the A-12 Avenger II program on 7 January 1991 by Secretary of Defense Dick Cheney due to severe cost overruns, schedule delays, and technical problems; the largest program termination in Defense Department history at the time and a major financial blow to McDonnell Douglas (and partner General Dynamics). Official DoD statements and subsequent litigation records.
- Continued evolution of the DC-9/MD-80 family into the MD-90 (first delivery 1995) and the smaller MD-95, pursued as lower-risk, lower-investment derivatives rather than clean-sheet designs. Manufacturer production histories.
- MD-12 proposal (a double-deck, four-engine ultra-large concept studied in the early 1990s) and the parallel European A3XX studies that ultimately became the Airbus A380. Contemporary design studies and industry coverage.
- Stonecipher's emphasis on outsourcing and risk-sharing with global suppliers as a means of reducing development and production costs (exemplified by the MD-95/717 program structure). Internal strategy documents and later Boeing retrospective accounts.
- Cost-reduction and production-discipline efforts on the C-17 Globemaster III program that restored profitability and secured expanded Air Force orders after earlier overruns and near-cancellation threats. Defense program histories and congressional testimony of the mid-1990s.
- Stonecipher's strategic conclusion that McDonnell Douglas lacked the capital and market position to remain an independent commercial aircraft

manufacturer long-term, leading him to position the company as an acquisition candidate — preferably by Boeing — culminating in the 1996–1997 merger agreement. Contemporary statements, merger negotiations accounts, and later analyses of the deal's origins.

The New and Improved Boeing

- 1996 as a record year for Boeing commercial programs, marked by surging orders and the negotiation of long-term exclusivity (sole-supplier) agreements with major U.S. carriers. Contemporary industry reporting and Boeing financial summaries of the period.
- Exclusive long-term supply agreements with American Airlines, Delta Air Lines, and Continental Airlines. These multi-year deals committed the airlines to purchase jetliners exclusively or predominantly from Boeing and were later subject to European Commission conditions limiting their enforceability. Official announcements, European Commission merger decision documents, and trade-press coverage.
- Boeing's acquisition of Rockwell International's aerospace and defense businesses for approximately $3.1–3.2 billion, forming Boeing North American. FTC settlement documents and Boeing/Rockwell press releases.
- Harry Stonecipher's approach to the McDonnell Douglas board proposing exploratory discussions with Boeing regarding a possible combination. Accounts of internal McDonnell Douglas deliberations in 1996.
- Prior professional acquaintance between Stonecipher and Philip Condit through industry forums and prior interactions; Condit's receptivity to exploring a merger. Contemporary profiles and post-merger recollections.
- Meeting at the Four Seasons Hotel in Seattle at which Stonecipher and Condit finalized key parameters of the merger agreement in a relatively short session. New York Times reporting of Stonecipher's account of the meeting.
- Merger terms providing that McDonnell Douglas directors would constitute one-third of the combined company's board and that senior McDonnell Douglas executives (including Stonecipher as president and COO) would assume key leadership positions. Official merger agreement disclosures and SEC filings.
- Public announcement of the definitive merger agreement on 15–16 December 1996. Joint Boeing–McDonnell Douglas press releases and major newspaper coverage.
- Philip Condit's characterization of the combination as creating the largest and strongest aerospace corporation in the world. Statements from the announcement press conference.

- Stonecipher's later acknowledgment (in contemporaneous interviews) that merger discussions or strategic interest in a Boeing combination had been contemplated as early as 1994, shortly after he assumed the McDonnell Douglas CEO role. New York Times and other contemporaneous reporting.
- Perception among many observers that the structure and subsequent influence of former McDonnell Douglas executives made the transaction resemble a reverse acquisition in cultural and managerial terms. Post-merger analyses and employee accounts.
- Michael M. Sears' senior role at McDonnell Douglas and his subsequent position within the combined Boeing organization (later involved in the tanker procurement scandal). Corporate personnel records and later investigative reporting.
- Systematic emphasis by incoming McDonnell Douglas leadership on cost reduction, financial metrics, and shareholder value, shifting Boeing's traditional engineering-centric culture. Contemporary and retrospective accounts of post-merger cultural change.
- Negative reaction among many Boeing rank-and-file employees and engineers to the merger and the accompanying cultural shift. Employee commentary and regional press coverage from the Seattle area.
- Mandate to accelerate production rates across commercial programs to capitalize on the order boom, increase revenues, and generate cash flow. Boeing production-rate announcements of 1996–1997.
- Resulting production breakdown: widespread parts shortages, quality and schedule problems, and temporary production halts (including multi-week shutdowns of the 737 and 747 lines in late 1997). Dramatic rise in "traveled work" (incomplete assemblies moved down the line or to subsequent stations). FlightGlobal reporting, Boeing financial disclosures of the resulting charges, and industry analyses of the production crisis.

Let the Games Begin

- 1997 commercial jetliner orders: Boeing led the industry with 568 gross announced orders (valued at approximately $42.8 billion) compared with Airbus' 460 gross firm orders. Boeing official year-end order summaries and contemporary industry data.
- 1999 order competition, particularly visible at the Paris Air Show, where both manufacturers announced significant deals amid intensifying market rivalry. Contemporary French and international press coverage of daily order tallies at the show.

- History and commercial importance of the Paris Air Show and Farnborough International Airshow as premier venues for major aircraft order announcements and competitive displays. Standard aviation-industry histories of the two events.
- Airbus practice of concentrating major order announcements at the Paris or Farnborough airshows for maximum publicity, contrasted with Boeing's more continuous practice of announcing orders as they were finalized. Contemporary trade-press observations of the differing announcement strategies.
- Boeing's escalating public claims in the late 1990s that Airbus benefited from unfair government subsidies (particularly "launch aid"), distorting competition. Boeing statements and U.S. government complaints of the period.
- Boeing leadership statements emphasizing that the company would prioritize profitability over pure market-share gains and would not match Airbus pricing or terms if they threatened margins. Contemporary executive comments and analyst reports.
- Southwest Airlines' low-cost, high-utilization, single-type (all-Boeing 737) business model and its widespread imitation by other low-cost carriers worldwide. Industry analyses of the Southwest model's influence.
- Boeing's position as the preferred supplier to the global low-cost-carrier segment in the 1990s, driven by the 737 family's operating economics and commonality. Fleet and order statistics of the era.
- easyJet's early history as a Boeing 737 operator and its subsequent decision, under CEO Ray Webster, to invite Airbus to compete for a major expansion order. easyJet corporate accounts and contemporary interviews with Webster explaining the desire for negotiating leverage after large 737 orders by Southwest and Ryanair left easyJet with limited bargaining power.
- Airbus' highly competitive offer to easyJet, including aggressive pricing and substantial financial assistance to cover the costs of transitioning from a Boeing 737 fleet to the A320 family. Airline and manufacturer accounts of the 2002 order and subsequent fleet conversion.
- Boeing's continued assertion that Airbus victories (including easyJet) resulted from government subsidies and its refusal to sacrifice profitability to match such terms. Boeing public statements following the loss of key campaigns.
- Airbus success in winning British Airways (previously a longtime exclusive or near-exclusive Boeing customer) and Frontier Airlines orders for Airbus aircraft. Airline fleet announcements and industry reporting of the conversions.

- Comparative analysis of European "launch aid" (repayable, risk-sharing government loans tied to program success) versus U.S. support for Boeing in the form of tax incentives, NASA/DoD research contracts, and other indirect benefits. WTO panel and Appellate Body findings, USTR summaries, and independent economic analyses of the two subsidy systems.
- WTO dispute (DS316 and related cases) initiated in 2004: U.S. claims against EU launch aid to Airbus and EU counterclaims against U.S. support for Boeing; successive panel and Appellate Body rulings finding both sides provided WTO-inconsistent subsidies, with larger quantified adverse effects attributed to European launch aid in several key sales campaigns (including easyJet); eventual authorization of retaliatory tariffs and the 2021 suspension agreement. Official WTO dispute settlement records and USTR/European Commission summaries.
- Divergent long-term market visions: Airbus' emphasis on very large aircraft operating between major hubs under a reinforced hub-and-spoke model versus Boeing's forecast of increasing fragmentation and growth in point-to-point long-haul services. Contemporary strategy documents and executive statements from both manufacturers.
- Airbus' continued development of the A3XX project (later formalized as the A380). Airbus program histories.
- Boeing's 747-500X and 747-600X stretch proposals (studied in the mid-to-late 1990s as responses to the very-large-aircraft market) and their subsequent cancellation in favor of a focus on the 777 family. Boeing design studies and later retrospective accounts.

Building a Better Mousetrap

- Airline executive and industry observations in the late 1990s and early 2000s that Boeing no longer felt like the same engineering-driven company after the McDonnell Douglas merger, with complaints about a shift toward cost-cutting, financial metrics, and reduced emphasis on traditional engineering excellence. Contemporary reporting and later retrospective analyses of post-merger cultural change at Boeing.
- Airbus market forecasts for the Very Large Aircraft (VLA) segment, initially projecting demand for roughly 1,200–1,300 units over a 20-year period and a substantial market value (on the order of $280 billion). Airbus strategic planning statements and contemporary coverage of the A3XX/A380 business case.

- Airbus formal industrial launch of the A380 (formerly A3XX) on 19 December 2000, following shareholder approval by EADS. Official Airbus announcements, French government communiqués, and major news coverage of the decision.
- Boeing's contrasting analysis that the VLA market would be far more limited (on the order of a few hundred aircraft over 20 years), reflecting its view that traffic growth would increasingly favor point-to-point services rather than hub-to-hub operations with very large aircraft. Boeing marketing and forecasting statements of the period (including comments by Randy Baseler and others).
- Initial A380 customers at launch: firm orders totaling 50 aircraft (plus options) from Air France, Emirates, International Lease Finance Corporation (ILFC), Qantas, Singapore Airlines, and Virgin Atlantic. Airbus order announcements and contemporaneous press reports.
- Boeing Sonic Cruiser concept: a near-sonic (approximately Mach 0.95–0.98) airliner intended to offer shorter trip times (roughly 15–20 percent faster than conventional jets) while remaining subsonic. Public unveiling on 29 March 2001; development history rooted in late-1990s advanced-concept studies; eventual abandonment by late 2002 in favor of a more conventional, fuel-efficient design. Boeing announcements, technical descriptions, and later historical accounts.
- Key technologies proposed for the Sonic Cruiser, including extensive use of carbon-fiber-reinforced polymer (CFRP) composites (targeted at 60–70 percent of airframe weight), advanced aerodynamics (canard/delta configuration in some studies), and other weight- and efficiency-enhancing systems. Engineering descriptions and partner announcements of the period.
- Boeing's Project Yellowstone technology initiatives (sometimes referenced with internal designations related to future product families), which explored advanced materials, systems, and configurations. Yellowstone work on a conventional but highly efficient twinjet (initially a technology demonstrator parallel to the Sonic Cruiser) evolved into the 7E7 program, later renamed the 787 Dreamliner. Boeing internal program histories and contemporary reporting on the shift from Sonic Cruiser to the efficient twinjet concept.

This Ain't Your Father's Boeing

- Analyses of the long-term cultural shift after 1997 in Robison, Flying Blind (cited above), and related journalistic and academic treatments.
- Boeing announcement of 21 March 2001 that corporate headquarters would relocate from Seattle to Chicago (move completed September 2001). Philip

Condit's public statements describing the decision as placing headquarters in a location "central to our operating units, customers and the financial community — but separate from our existing operations." Contemporary coverage in the Seattle Post-Intelligencer, Chicago Tribune, and later retrospective analyses of the cultural impact of separating senior leadership from the engineering and manufacturing centers.

- U.S. Department of Defense Joint Strike Fighter program records: down-select of Boeing and Lockheed Martin in November 1996 to build concept demonstrators; Boeing X-32A first flight 18 September 2000; X-32B STOVL testing. Final selection of the Lockheed Martin X-35 (later F-35) on 26 October 2001. National Museum of the U.S. Air Force fact sheets on the X-32A/B and contemporary technical assessments of the competing designs (direct-lift vs. lift-fan STOVL approaches).
- U.S. Air Force announcement (May 2003) of intent to lease 100 Boeing KC-767 tankers. Congressional hearings and reports on the proposed $23–30 billion arrangement. Subsequent investigations documenting opposition from Air Force uniformed leadership and independent cost analyses showing outright purchase would have been substantially cheaper.
- U.S. Department of Justice and Air Force investigations into Darleen Druyun (Principal Deputy Assistant Secretary of the Air Force for Acquisition). Druyun's guilty plea (2004) to conflict-of-interest charges related to favoring Boeing on the tanker deal and other contracts while negotiating employment with the company. Michael Sears (Boeing CFO) also pleaded guilty; both served prison sentences. Boeing internal investigation leading to their dismissals in November 2003. Contemporary reporting in The New York Times, The Washington Post, and The Seattle Times, plus Senate Armed Services Committee hearings.
- U.S. Department of Justice charges (2003) against former Boeing managers for conspiring to steal Lockheed Martin proprietary documents related to the Evolved Expendable Launch Vehicle (EELV) competition in the late 1990s. Lockheed Martin lawsuit against Boeing alleging industrial espionage that helped Boeing win a larger share of the initial EELV awards. Subsequent Air Force sanctions against Boeing (including temporary suspension from certain launch contracts).
- Boeing board announcement of 1 December 2003 accepting Philip Condit's resignation as chairman and CEO (effective immediately; retirement in March 2004). Condit's stated assumption of symbolic responsibility for the ethics scandals. Appointment of Harry Stonecipher as president and CEO and Lewis

Platt as nonexecutive chairman. Coverage in The Los Angeles Times, PBS NewsHour, and Boeing's own press release.

- Boeing decision to terminate the 757 program. Final 757 rolled off the Renton line on 28 October 2004 and delivered to Shanghai Airlines on 28 April 2005. Boeing press release of 28 April 2005 marking the end of the 23-year production run.

Attempting to Write the Future

- Peter Robison, Flying Blind: The 737 MAX Tragedy and the Fall of Boeing (Doubleday, 2021) — chapters covering the 787 program's origins, the decision to outsource an unprecedented share of design and manufacturing, the global partner model, successive delays, and the resulting cost overruns and quality problems.
- John Newhouse, Boeing Versus Airbus (Alfred A. Knopf, 2007) — competitive context and early strategic decisions surrounding the 787 (originally the 7E7 / Sonic Cruiser successor).
- Contemporary analyses of Boeing's "risk-sharing" partnership structure, in which major tier-1 suppliers (including Japanese heavy-industry partners, Alenia, Spirit, Vought/Global Aeronautica, and others) funded significant non-recurring development costs and waited for payment until aircraft delivery.
- Academic and industry case studies (e.g., Yao Zhao and related supply-chain research papers) examining how the model transferred design authority and integration responsibility while ultimately increasing Boeing's coordination and rework burden.
- Reuters, "Boeing 787 delays cast hard light on outsourcing," 22 September 2009, and subsequent reporting on the decision to outsource roughly 70 percent of the airframe.
- Boeing announcements of the 7E7/787 program launch (2003–2004), original target of first flight in 2007 and entry into service in 2008, and the successive slips that pushed first flight to December 2009 and entry into service to late 2011.
- The Seattle Times investigative coverage (Dominic Gates and colleagues) of incomplete assemblies arriving at Everett, the need to send large numbers of Boeing engineers to supplier sites, battery issues, and production bottlenecks.
- Boeing financial disclosures and analyst estimates of program cost overruns that ultimately ran into the tens of billions of dollars, far exceeding original projections.

- FAA and NTSB documentation related to the 787's certification challenges and early in-service battery fires (2013).
- Industry reporting in Aviation Week & Space Technology, Flight International, and The Wall Street Journal on the radical production strategy, composite technology, and the competitive response from Airbus (A350).

Accelerating Towards the Precipice

- Airbus A380 wiring and electrical-harness problems arising from incompatible design software versions (CATIA 4 vs. CATIA 5) used by different national partners, leading to extensive rework, production delays of roughly two years, and substantial additional costs. First commercial delivery slipped from the original target to late 2007 (Singapore Airlines). FlightGlobal reporting, Airbus technical accounts, and program histories.
- Harry Stonecipher's resignation as Boeing CEO on 6–7 March 2005 after the board determined that his consensual affair with a female Boeing executive violated the company's code of conduct and demonstrated poor judgment that impaired his ability to lead. The investigation was triggered by anonymous forwarding of their correspondence. New York Times, Chicago Tribune, and Boeing statements.
- Candidates considered for permanent CEO included internal and external figures before the board selected James McNerney. Contemporary business reporting.
- James (Jim) McNerney's biography: long career at General Electric (including leadership of GE Aircraft Engines) under Jack Welch; subsequent tenure as chairman and CEO of 3M (2001–2005); appointment as Boeing president, CEO, and chairman effective mid-2005. Corporate biographies and profiles.
- McNerney's management approach shaped by the Jack Welch school of performance metrics, cost discipline, and shareholder-value focus. Analyses of his GE, 3M, and Boeing tenures.
- Airbus studies and eventual decision to develop a more competitive response to the 787, initially considering modified A330 derivatives before launching the all-new A350 XWB (extra-wide body) family. Airbus program announcements and industry coverage of the mid-2000s.
- Boeing launch of the 747-8 series (passenger and freighter variants) as an updated stretch of the 747-400. Boeing program records.
- 787 Dreamliner program challenges: extensive outsourcing, supplier coordination problems, repeated production delays (original target first delivery 2008; actual first delivery September 2011), quality issues, and cost

overruns. Ceremonial rollout in July 2007 revealed an incomplete aircraft (essentially an empty shell in key respects). First flight 15 December 2009; entry into service with All Nippon Airways in October 2011. Contemporary reporting and later program analyses.

- Original development-cost estimates (roughly $8–10 billion range in early planning) versus actual program costs that ballooned beyond $36 billion when including delays, redesigns, penalties, and production inefficiencies. Industry cost estimates and Boeing financial disclosures.
- 787 lithium-ion battery fires in January 2013 (Japan Airlines at Boston and All Nippon Airways in Japan), leading to the worldwide grounding of the 787 fleet by the FAA and other regulators until a redesigned containment and monitoring system was certified. NTSB and regulatory investigation reports.
- McNerney's assertive stance toward organized labor, including efforts to limit union influence, relocate work, and secure long-term contracts favorable to management. Contemporary labor analyses and union statements.
- Boeing's acquisition of Vought Aircraft Industries' interest in the Global Aeronautica joint venture and related South Carolina facilities, consolidating control over 787 aft-fuselage work and enabling the Charleston final-assembly site. Corporate transaction announcements.
- IAM (International Association of Machinists) dispute with Boeing over the decision to place a second 787 production line in right-to-work South Carolina; NLRB complaint alleging unlawful retaliation for past strikes; eventual withdrawal of the complaint in 2011 after a new labor agreement covering other work. NLRB records and contemporary coverage.
- McNerney's reported comment reflecting that labor "fears" him or similar characterizations of his tough labor posture. Contemporary and retrospective accounts of labor relations under his leadership.
- Ongoing 787 production and quality challenges continuing after entry into service. Industry reporting of the period.
- Ethiopian Airlines 787 fire (or related battery/electrical incident context in the early operational years; note primary high-profile battery events were the 2013 JAL/ANA incidents that triggered grounding; later operational incidents including fires involving 787s were reported). Regulatory and airline incident reports.
- Boeing's large-scale stock-repurchase programs during McNerney's tenure, returning substantial capital to shareholders. Company financial reports and analyst coverage.

- McNerney's multi-million-dollar annual compensation packages. SEC filings and compensation disclosures.

Teetering on the Precipice

- Announcement in 2008 of new-generation high-bypass turbofan engines offering substantial fuel-burn reductions (on the order of 15 percent or more relative to previous-generation engines): the Pratt & Whitney PW1000G geared turbofan family and the CFM International LEAP family jointly developed by GE and Safran. Engine manufacturer announcements and technical descriptions.
- Compatibility differences: the Airbus A320 family's taller landing gear and greater ground clearance allowed it to accommodate both the PW1000G and the larger LEAP-1A variants, while the Boeing 737's lower stance required a smaller-diameter, specifically adapted LEAP-1B engine; the PW1000G could not be fitted without major airframe changes. Comparative engine and airframe analyses.
- Airbus formal launch of the A320neo (New Engine Option) family on 1 December 2010. Airbus program announcements.
- Boeing's initial public position that it preferred to wait for an all-new narrow-body design rather than re-engine the 737. Contemporary Boeing statements and industry reporting of the early neo competitive response.
- American Airlines' 2011 order campaign that included a large commitment to the A320neo family, which precipitated Boeing's decision to offer a re-engined 737; American ultimately placed a mixed order that included both A320neo and 737 MAX aircraft. Airline order announcements and competitive analyses of the campaign.
- Boeing formal launch of the 737 MAX family after securing American Airlines' commitment as a key launch customer. Boeing program records and contemporary coverage.
- Aerodynamic and handling effects of the larger LEAP-1B engines and their forward/upward placement on the 737 MAX, which altered the aircraft's pitch characteristics at high angles of attack and necessitated compensatory flight-control software. Engineering descriptions and later accident investigation findings.
- Boeing's marketing and certification approach promising airlines that pilots already qualified on the 737 NG could transition to the MAX with minimal additional training. FAA and Boeing training documentation of the period.

- Development of the Maneuvering Characteristics Augmentation System (MCAS), a flight-control law designed to automatically command nose-down stabilizer input to counteract the MAX's pitch-up tendency at elevated angles of attack. Technical descriptions from Boeing and subsequent investigative reports.
- MCAS design philosophy of operating in the background with limited pilot awareness or documentation in initial flight manuals and training materials. NTSB, House Transportation Committee, and other investigative findings.
- Evolution of MCAS during development (increased authority, reliance on a single angle-of-attack sensor, repeated activation capability) that was not fully reflected in the final certification documentation submitted to the FAA. Congressional investigation reports, SEC filings related to disclosures, and technical reconstructions of the system's development history.
- Boeing press release, 23 June 2015: "Boeing Names Muilenburg Chief Executive Officer" (Muilenburg becomes president and CEO on 1 July 2015; McNerney remains as employee until February 2016 and continues as chairman). Contemporary coverage in The Guardian, CNBC, and Aviation International News confirming the succession.
- Peter Robison, Flying Blind: The 737 MAX Tragedy and the Fall of Boeing (Doubleday, 2021) — detailed explanation of the larger, more forward-mounted engines on the MAX, the resulting aerodynamic pitch-up tendency at high angles of attack, and the purpose of MCAS.
- The Seattle Times investigative series (Dominic Gates et al., 2019), especially "The inside story of MCAS," describing the original design intent, the expansion of MCAS authority, and its reliance on a single angle-of-attack (AOA) sensor.
- Airline preference for the longer A321neo as a capable replacement for aging 757s on longer thin routes; Boeing's development of the MAX 10 (and related higher-capacity variants) intended to compete directly in that segment. Fleet planning analyses and order data of the mid-2010s.
- Acceleration of Boeing's stock-repurchase programs under Muilenburg's leadership, continuing and expanding the capital-return policies of the McNerney era. Company financial reports and SEC filings.

Over the Precipice

- Peter Robison, Flying Blind (cited above), detailed reconstruction of 737 MAX design decisions, MCAS, certification, and the two fatal accidents.

- Boeing's own post-accident explanations and the final NTSB/Indonesian KNKT/Ethiopian AAIB reports outlining MCAS activation logic, repeated nose-down commands, and the single-sensor vulnerability.
- House Committee on Transportation and Infrastructure final report on the 737 MAX (2020) and related hearing transcripts.
- U.S. Department of Transportation Office of Inspector General report (2021) on weaknesses in the certification process.
- NTSB safety recommendation reports and comments on the two MAX accidents that detail the design assumptions and single-point-of-failure issues.
- FAA Organization Designation Authorization (ODA) program documentation and the DOT Office of Inspector General report "Weaknesses in FAA's Certification and Delegation Processes..." (February 2021).
- House Transportation Committee investigation findings and Seattle Times reporting on how oversight of MCAS was delegated to Boeing's ODA unit.
- FAA statements and congressional testimony regarding the level of FAA involvement in MCAS certification.
- The Seattle Times reporting (citing internal Boeing documents and the DOT Inspector General) on the deliberate decision to remove all mention of MCAS from the flight crew operations manual and differences-training materials in order to preserve common type rating and avoid simulator training requirements.
- FAA Flight Standardization Board reports and the Joint Authorities Technical Review findings on the training assumptions.
- NTSB safety recommendation report ASR-19-01 and subsequent NTSB comments on the Ethiopian report.
- Whistleblower testimony and congressional hearing transcripts.
- FAA certification documents, Joint Authorities Technical Review final report (October 2019), and House Transportation Committee investigation records on MCAS.

Adrift in the Perfect Storm

- David L. Calhoun (born 1957): long career at General Electric (including leadership of GE Aircraft Engines), subsequent executive roles at Nielsen and Blackstone, and service on Boeing's board from 2009 (becoming chairman in 2019). Appointed president and CEO of Boeing in January 2020 following Dennis Muilenburg's departure; tenure ending August 2024. Corporate biographies and contemporary profiles.

- Boeing's discussions in early 2020 of seeking substantial U.S. government financial support (reportedly up to $60 billion for the broader aerospace sector) amid the COVID-19 cash crunch and grounded 737 MAX fleet; ultimately the company raised approximately $25 billion through a large private debt offering in April–May 2020 and largely avoided direct federal bailout loans. Contemporary reporting by Reuters, Bloomberg, and other financial outlets.
- FAA authorization for the return to service of the 737 MAX (beginning late 2020 in the United States after extensive software and training modifications, with progressive global re-clearances). Official FAA and regulatory announcements.
- Boeing's access to private credit markets, including full drawdown of a roughly $13.8 billion revolving credit facility in early 2020 and the subsequent large bond issuance, to address immediate liquidity needs during the pandemic. Company financial disclosures and market reports.
- Persistent 787 production-quality issues (including fuselage-section gaps, shimming problems, and related defects) that led to extended delivery pauses and heightened FAA scrutiny in the early 2020s; temporary suspension of Boeing's authority to issue airworthiness certificates for the 787. FAA actions and industry reporting.
- Heightened FAA oversight of both 737 MAX and 787 production processes, including restrictions on Boeing's Organization Designation Authorization (ODA) privileges for issuing airworthiness certificates, later partially and then more fully restored after demonstrated quality improvements. FAA statements and regulatory decisions through the mid-2020s.
- KC-46A Pegasus tanker program: development under a fixed-price contract, chronic problems with the Remote Vision System, boom, fuel leaks, and structural issues; multi-year delays and billions in charges for Boeing; first operational deliveries to the U.S. Air Force beginning in 2019, with ongoing retrofit and availability challenges. GAO reports, Air Force statements, and program histories.
- VC-25B (next-generation Air Force One) program: conversion of 747-8 airframes under a fixed-price contract; significant technical, interior-design, wiring, and structural challenges; repeated schedule slips (original targets deferred to late 2020s) and cumulative charges exceeding $3 billion for Boeing. Air Force and independent program assessments.
- Boeing CST-100 Starliner: problems on the second uncrewed orbital flight test (including thruster and software issues) and subsequent challenges on the first crewed flight test (Crew Flight Test) in 2024, involving propulsion, helium

leaks, and extended ISS stay; ongoing certification delays. NASA Inspector General reports and program documentation.

- SpaceX development of the Falcon launch vehicle family, Dragon cargo spacecraft, and Crew Dragon (Crewed Dragon) for NASA's Commercial Crew Program. SpaceX and NASA program histories.
- Contract-value differences under NASA's Commercial Crew Transportation Capability (CCTCap) awards: original fixed-price contracts of approximately $4.2 billion for Boeing (Starliner development plus missions) versus $2.6 billion for SpaceX (Crew Dragon development plus missions); later modifications altering totals and per-seat economics, with SpaceX ultimately receiving substantially more mission volume at lower effective cost per seat. NASA contract data and Inspector General analyses.
- 777X program: original targeted entry into service around 2020; repeated certification delays driven by technical issues, heightened post-MAX regulatory scrutiny, supply-chain problems, and design changes; cumulative charges exceeding $15 billion and first deliveries projected into 2027 or later. Boeing financial disclosures and certification updates.
- Calhoun's compensation packages, including multi-million-dollar annual pay (base, incentives, and equity) that reached approximately $33 million for one reported year, plus significant stock awards and options vesting upon or after departure. SEC filings and shareholder disclosures.
- Boeing's 2022 announcement and subsequent relocation of corporate headquarters from Chicago to Arlington, Virginia. Company announcements and contemporary reporting.
- Alaska Airlines Flight 1282 (5 January 2024): in-flight separation of a mid-exit door plug on a nearly new 737-9, causing rapid depressurization; NTSB findings attributing the event to missing retention bolts and systemic manufacturing and quality-control failures at Boeing (and supplier interfaces). Official NTSB final report.
- Calhoun's announcement of intent to step down (March 2024) and formal departure in August 2024; subsequent congressional testimony addressing safety culture, quality escapes, and production oversight. Company statements and congressional hearing records.
- Starliner crewed mission (Crew Flight Test) problems, including thruster failures and helium leaks that prevented a timely return and required NASA astronauts to remain longer aboard the ISS. NASA mission reports and post-flight analyses.

- Robert K. "Kelly" Ortberg: engineering background beginning at Texas Instruments, long career at Rockwell Collins rising to president and CEO, subsequent leadership roles through United Technologies/RTX; selected by the Boeing board as president and CEO effective August 2024. Company announcements and biographical profiles.
- Boeing's reacquisition of Spirit AeroSystems (announced 2024, completed late 2025), reversing the earlier spin-off of key fuselage production capacity in order to improve quality control and supply-chain integration. Transaction announcements and regulatory filings.
- U.S. Air Force award of the Next Generation Air Dominance (NGAD) fighter contract (F-47 designation) to Boeing. Defense Department and company statements.
- 2025 commercial deliveries and backlog: significant recovery in delivery rates (highest commercial deliveries since 2018 in some reports), record or near-record total company backlog (on the order of $500–680 billion across commercial, defense, and services, with commercial airplane backlog exceeding 6,000 aircraft). Boeing year-end financial and operational updates.
- 737 MAX 8 and 9 Airworthiness Directive regarding inspections for possible cracks. FAA Airworthiness Directive 2026-15-11.

Epilogue

- Airbus announcement of 14 February 2019 that A380 production would end once remaining orders were fulfilled; final A380 (MSN 272) delivered to Emirates on 16 December 2021; total of 251 aircraft built (Airbus official production and delivery data; contemporary reporting in Aviation Week and Flight International).
- Boeing press release and delivery records for the final 747-8F (rolled out of the Everett factory 6 December 2022; delivered to Atlas Air on 31 January 2023); cumulative production of 1,574 aircraft (Boeing official program totals).
- Broader assessments of strategic foresight versus execution failures in Robison, Flying Blind (cited above), Newhouse, Boeing Versus Airbus (cited above), and subsequent journalistic and academic treatments.
- Boeing proxy statements and SEC filings listing members of the board of directors from60through the mid-2020s.

Photo Gallery

The Model 40B-4. This design, capable of carrying four passengers and mail, along with its predecessor, the Model 40A, which could carry two passengers and mail, gave Boeing its first success in commercial aviation. Note the two sets of windows for the "limousine compartment," which could be opened in flight, and the Boeing Air Transport logo. Photo: Museum of History and Industry (public domain).

The Model 247—the first "modern" airliner—redefined what a passenger aircraft could be. Its innovative stressed-skin construction set a new standard in aircraft design and construction, which endured well into the jet age. Photo: Eric Long / National Air and Space Museum, Smithsonian Institution (NASM2018-10581), CC0. Cropped/edited by the author.

The Douglas DC-2, the plane that, within two years, rendered the Boeing Model 247 obsolete. When TWA wanted to purchase the Model 247, Boeing said it wouldn't build any for the airline until United Airlines' order for 60 Model 247s was fulfilled first. In desperation, TWA turned to Douglas Aircraft Company, which developed the DC-2. An even larger and more capable version, the DC-3 and its C-47 military counterpart, would see over 16,000 copies of the plane produced by the end of World War II. Photo: John Murphy via Wikimedia Commons; cropped/edited by the author.

The XB-15, the experimental plane Boeing built for the Army Air Corps to study the feasibility of large bombers. Its wing design was incorporated into the Model 314 flying boat. The plane would give Boeing the chops to build the iconic B-17 and B-29. Photo: National Museum of the United States Air Force.

The Model 307 "Stratoliner," a groundbreaking design that would establish another "first" by incorporating a pressurized fuselage in a commercial airliner. Although a leap in aircraft design, it would see a very limited production run of only ten aircraft because of the advent of World War II. Photo: Eric Long / National Air and Space Museum, Smithsonian Institution (NASM2019-01792), CC0.

The B-17 "Flying Fortress," one of the most iconic aircraft of World War II and made possible by Boeing's work for the Army Air Corps on the earlier XB-15. Photo: U.S. Air Force / National Museum of the U.S. Air Force.

The B-29 "Superfortress," the most expensive program of the war effort and the plane that dropped the only atomic bombs ever used in war, sealing its fate with those events and bringing an end to World War II. Photo: U.S. Air Force, via Wikimedia Commons.

The Model 377 represented Boeing's first major effort to re-enter commercial aviation after World War II. The program transformed the C-97 military transport — derived from the B-29 — into a sophisticated postwar airliner. Although the 377 never achieved commercial success, it demonstrated Boeing's advanced engineering capabilities and exemplified William Allen's willingness to take bold but calculated risks to secure the company's future. The C-97, however, proved far more successful in military service, with more than 800 built for the U.S. Air Force as transports and aerial refueling tankers. Photo: University of Washington Libraries, Digital Collections, Transportation Collection.

The B-47 "Stratojet" was a pioneering jet bomber. It incorporated a number of design features, including swept wings, that were copied from engineering documents captured from Germany after its surrender at the end of the war in Europe. Photo: U.S. Air Force / Air Force Photo Gallery.

The iconic B-52 "Stratofortress" bomber. Originally conceived in 1949 in a Dayton, Ohio, hotel room under a tight timeline, the airframe continues to serve as the backbone of the U.S. strategic bomber force. Planned modifications to the airframe and engines will keep it in service with the Air Force until around 2050, 100 years after Boeing's engineers first conceived the plane. Note how much aviation design had advanced in just over a decade when compared to the previous photo of the B-17, the prototype of which was originally designed and built in 1935. Photo: U.S. Air Force / Air Force Photo Gallery. Cropped/edited by author.

The British de Havilland Comet, the world's first commercial jet-powered airliner. Although it was the "first," design deficiencies would result in several high-profile crashes that would ground the plane and give Boeing the opportunity to bring to market its own design that incorporated lessons learned from the Comet and jump-start the Jet Age. Photo: Wikimedia Commons, public domain.

Boeing's revolutionary 367-80 ("Dash-80"). First rolled out in 1954, it avoided the design deficiencies of the British Comet and evolved into the USAF KC-135 aerial tanker and the commercial 707 airliner, which ushered in the Jet Age. Photo: Joe Parke, via Wikimedia Commons (public domain); cropped/edited by author.

The KC-135. Although sporting a shorter and narrower fuselage than the 707, the basic airframe design was the same, having been developed from the 367-80 prototype. This particular aircraft was upgraded in the 1980s along with hundreds of others in the Air Force arsenal with modern CFM-56 high-bypass engines, the modified versions of which would be installed on the 737. Photo: U.S. Air Force Photo Gallery.

The 707, although sharing the basic design of the KC-135, would incorporate a wider fuselage at the urging of airlines and would ultimately set the standard for commercial jet transportation. Photo: Wikimedia Commons (public domain).

The Douglas DC-8, the direct competitor to the Boeing 707. Although it would see initial success with airlines around the world, Douglas' slow response to market demands delayed upgrade investments that were needed for the plane to remain competitive, so its sales eventually paled in comparison to the 707. Photo: Steve Fitzgerald via Wikimedia Commons.

The 727, the plane that brought the standards set by the 707 to smaller markets. It incorporated design components from the 707 into a smaller 3-engine plane that was cheaper to operate and easier to fill with passengers. Photo: Steve Fitzgerald via Wikimedia Commons.

The Douglas DC-9, developed in response to the 727 and demands for the benefits of jet transportation for smaller markets. Sporting a narrower fuselage than the 727 and only two engines, it rapidly became a popular aircraft for short-haul service, spurring Boeing to develop the 737. The DC-9 would, with upgrades and modifications, be relabeled with the MD-80 series moniker and would prove to be one of the most popular jetliners ever produced. Photo: Steve Fitzgerald via Wikimedia Commons.

The 737 Original — the "Baby Boeing" — in the livery of launch customer Lufthansa. Conceived in 1964 with expectations of roughly 250 sales, the 737 would ultimately generate more than 17,000 orders across successive generations and major upgrades. Its final major derivative, the MAX, would have disastrous consequences. Also note the clear family resemblance in the fuselage and cockpit to the 707 and 727. Photo: Eduard Marmet, via Wikimedia Commons, CC BY-SA 3.0 (creativecommons.org/licenses/by-sa/3.0/). Image cropped and adjusted for brightness by the author.

A close-up of the 737 Original's engine-and-wing configuration, showing the engines mounted tight beneath the wing to provide the ground clearance required by Lufthansa, the aircraft's launch customer. As successive generations adopted larger engines, Boeing was forced to position them progressively farther forward and higher relative to the wing. Photo: by the author.

A 737 MAX 8, its massive LEAP-1B engines mounted farther forward and higher than on earlier 737s — a configuration that altered the aircraft's handling characteristics and led Boeing to develop MCAS, the system that would ultimately play a central role in two crashes that claimed 346 lives. Photo: Acroterion, via Wikimedia Commons, licensed under CC BY-SA 4.0 (creativecommons.org/licenses/by-sa/4.0/). Image cropped and adjusted for brightness by the author.

The legendary 747. Developed at the request of Juan Trippe of Pan American, the design borrowed concepts from Boeing's failed U.S. Air Force bid for a large military transport. This model would be produced for more than half a century and outlast the Airbus A380, the plane that was intended to replace it. Photo: clipperarctic, via Wikimedia Commons/Flickr, licensed under CC BY-SA 2.0 (creativecommons.org/licenses/by-sa/2.0/). Image cropped and adjusted for brightness by the author.

The DC-10, Douglas' entry into the widebody market. The plane would prove legendary, but for all the wrong reasons. The cargo door with the troublesome locking mechanism that blew open during certification testing and on American Airlines Flight 96, and then on Turkish Airlines Flight 981, which resulted in the loss of 346 lives, is visible below and slightly forward of the cabin door on the far right. The wing-mounted engine visible in the photo is the one that separated during the takeoff run of American Airlines Flight 191, resulting in 271 fatalities. Photo: Leslie Snelleman, via Wikimedia Commons/AirHistory.net. Image cropped and adjusted for brightness by the author.

The L-1011, Lockheed's entry into the widebody arena. Although the most advanced among the first generation of widebody aircraft, it would prove to be a commercial failure with only 250 produced in a market that could not support multiple entrants. Photo: Pedro Aragão, via Wikimedia Commons, licensed under CC BY-SA 3.0 (creativecommons.org/licenses/by-sa/3.0/).

The A300, Airbus' entry into the widebody market and its first commercial airliner project. Although comparable in size to the DC-10 and L-1011, it would prove more economical to operate with just two engines. Eastern Airlines and its CEO, Frank Borman, chose to add the plane to the Eastern fleet amid high fuel costs and mounting debt and in the process became one of the most important reasons Airbus was able to survive. Photo: Eduard Marmet, via Wikimedia Commons, licensed under CC BY-SA 3.0 (creativecommons.org/licenses/by-sa/3.0/). Image cropped and adjusted for brightness by the author.

The 757, the plane designed to replace the 727. It would prove popular among airlines and many felt Boeing ended production too early. And with no replacement, Airbus' A321neo has taken over most of the roles once performed by this plane. Photo: Dylan Ashe, via Wikimedia Commons, licensed under CC BY-SA 2.0 (creativecommons.org/licenses/by-sa/2.0/). Image cropped by the author.

The 767, the plane that would pioneer long-haul twin-engine ETOPS flight operations. It remains in production after nearly five decades. Although one of Boeing's most successful commercial platforms, its military variant, the KC-46A Pegasus aerial tanker, has been plagued with development, production and performance issues. Photo: Brian, via Wikimedia Commons/Flickr, licensed under CC BY-SA 2.0 (creativecommons.org/licenses/by-sa/2.0/).

The competing prototypes for the Pentagon's Joint Strike Fighter (JSF) program — the largest defense acquisition program of the post-Cold War era and Boeing's first major opportunity to expand its defense-industry presence following the merger with McDonnell Douglas. Boeing's unconventional X-32, nicknamed "Monica," is parked on the left, with Lockheed Martin's winning X-35 on the right. Photo: United States Air Force

The A380—Airbus' behemoth—was designed around the company's prediction that air travel would increasingly consolidate around major international hubs. That prediction would prove mistaken, and the aircraft would ultimately become a commercial failure. Yet Boeing's own missteps ensured that Airbus paid little strategic price for the error. Despite its costly bet on the A380, Airbus continued its march forward, eventually overtaking Boeing to become the world's largest commercial aircraft manufacturer. Photo: Lasse Fuss, Wikimedia Commons, CC BY-SA 3.0.

The A320neo, Airbus' highly successful upgrade to the original A320 series, forced Boeing to respond quickly or risk losing a massive American Airlines order to its rival — a decision that accelerated development of the 737 MAX, with tragic consequences. Photo: Acroterion, via Wikimedia Commons, licensed under CC BY-SA 4.0 (creativecommons.org/licenses/by-sa/4.0/).

The A350 XWB. Boeing's repeated missteps and production problems with the 787 gave Airbus the time and opportunity to develop a formidable competitor — shattering Boeing's hopes of dominating the next generation of long-haul commercial aviation. Photo: DS28, via Wikimedia Commons, licensed under CC BY-SA 4.0 (creativecommons.org/licenses/by-sa/4.0/).

The 777, originally launched in 1990, would become the aircraft that came to dominate long-haul travel and epitomize Boeing's ability not only to anticipate the market, but to shape it. Unfortunately, the next-generation upgrades, the 777-8 and 777-9, are running seven years behind schedule and have been plagued with development issues. Photo: Acroterion, via Wikimedia Commons, licensed under CC BY-SA 4.0 (creativecommons.org/licenses/by-sa/4.0/).

The 787, although conventional in appearance, was the most revolutionary commercial aircraft since the 707 entered service in 1958. Unfortunately, it would also come to epitomize Boeing's shift away from its historic engineering-driven culture and toward an increasing emphasis on financial performance. The plane's innovative design, engineering, and production concept, which relied heavily on outsourcing, was intended to minimize Boeing's financial exposure. However, it would end up costing over $36 billion in development and production costs, over four times the original estimate. Photo: by the author.

About the Author

Rom Solene first experienced the wonder of flight at the age of three aboard a Pan American Boeing 707 crossing the Atlantic. That transatlantic journey left an indelible mark, and some of his earliest childhood memories are of that flight.

He grew up during the Space Race and the Apollo Moon landings and came of age at the height of the Cold War — an era that deepened his love of aviation and sparked a lifelong interest in history. He became particularly fascinated by the forces that drove the Space Race, the deeper roots of U.S.–Soviet rivalry, and the leadership traits and decisions that shaped the successes and failures of consequential figures throughout history.

A veteran of the United States Marine Corps with more than thirty years of experience in the corporate world, Solene observed firsthand the decisions of leaders within government and across industry. His passion for aviation and history, combined with a deep interest in leadership and the dynamics of large organizations, kept him closely attuned to developments in the aviation industry — especially the intense competition between Boeing and Airbus. For him, the reasons behind Boeing's decline were clear even as they unfolded.

Rom Solene lives in the United States. A licensed pilot, he spends his free time pursuing his passions for flying and motorcycles.

www.ingramcontent.com/pod-product-compliance
Lightning Source LLC
LaVergne TN
LVHW100723061126
851857LV00011B/366

* 9 7 9 8 9 9 7 5 3 8 8 1 1 *